Chemistry of Insecticides and Fungicides

◄ 3rd Revised Edition ►

Related Books

Titles	*Authors*
— Agricultural Insect Pests And Their Control *2nd Ed.*	*V.B. Awasthi*
— Agriculture Chemical Users Manual	*Geoff Cowles*
— बागवानी फसलों में समेकित रोग–कीट प्रबंधन	एस.के. सिंह
— Beekeeping: A Compressive Guide to Bees and Beekeeping	*D.P. Abrol*
— Biofertilizer Technology	*S. Kannaiyan*
— Biointensive Integrated Pest Management In Horticultural Ecosystems	*P.P. Reddy*
— Crop Protection Strategies under climate change scenarios	*P.P. Reddy*
— Cytochalasin: Incidence and Biological Activities	*S. Kiran*
— Chemistry of The Soil, *2nd Edition*	*Firman E. Bear*
— Ecological Entomology: Insect Life In Odd Environments	*T.N. Ananthakrishnan*
— Emerging Crop Pest Problems: Redefining Management Strategies	*P.P. Reddy*
— Entomology In The Doon Valley	*B.K. Tyagi*
— Fundamentals of Agriculture *2nd Ed. (Vol. 1-2) (Set)*	*R.L. Arya*
— Gall-Inducing Arthropods On Forest Trees	*John Prasanth Jacob*
— Handbook of Practical Nematology	*H.K. Bajaj*
— Insect Biodiversity: Functional Dynamics and Ecological Perspectives	*T.N. Ananthakrishnan*
— Insect Mite and Vertebrate Pests and Their Management In Horticultural Crops	*P.P. Reddy*
— Insect Pathology Text Book and Practical Manual	*N.K. Maru*
— Insect Physiology	*V.B. Wigglesworth*
— Insect Physiology (21St Century Biology And Agriculture: Textbook Series)	*K.P. Sanjayan*
— Integrated Pest Management	*A.K. Dhawan*
— Introduction to General and Applied Entomology *3rd Ed.*	*V.B. Awasthi*
— Introductory Plant Nematology	*P.P. Reddy*
— Manual of Diseases and Insect-Pests of Important Vegetables	*Satish K. Gupta*
— Medical Entomology *2nd Edition*	*B.K. Tyagi*
— Pest and Vector Management In The Tropics	*A. Youdeowei*
— Pesticide Application Equipment for use in Agriculture *(Vol. 1)*	*G.A. Matthews*
— Pests and Pollinators of Vegetable and Oilseed Crops	*D.P. Abrol*
— Plant Protection In Horticulture *(Vol. 1-3) (Set)*	*P.P. Reddy*
— Practicals In Plant Physiology and Biochemistry	*Manju Bala*
— Physiology and Biochemistry Basis of Seeds	*K. Vanangamudi*
— Technological Innovations In Integrated Pest Management Biorational andEcological Perspective	*D.P. Abrol*
— Theory and Practice of Integrated Pest Management	*A.K. Dhawan*
— The Chemistry Of Pesticides Their Metabolism, Mode of Action and Uses In Crop Protection	*K.A. Hassall*
— Vertebrate Pests In Agriculture	*S. Sridhara*

Chemistry of Insecticides and Fungicides

◂ 3rd Revised Edition ▸

SREE RAMULU
Department of Soil Science and Agricultural Chemistry
Agricultural College and Research Institure
(Tamilnadu Agriculture University)
Madurai

SCIENTIFIC®
PUBLISHERS

Published by:

SCIENTIFIC PUBLISHERS (INDIA)
5A, New Pali Road, P.O. Box 91
Jodhpur 342 001, India
E-mail: info@scientificpub.com
Website: www.scientificpubonline.com

Print 2025

ISBN: 978-93-89832-02-0
eISBN: 978-93-89832-04-4

Printed in India

To
My Professors
Dr. Parkar F. Pratt
and
Dr. Nathaniel T. Coleman
of
University of California, Riverside
California, U.S.A.

Foreword

Thanks to the introduction of fertiliser-responsive, photo-insensitive, high yielding variety of cereals and other commercial crops, "Management" in agriculture has assumed new dimensions in India. With the provision of assured irrigation facilities, intensity of land use has also been stepped up. To reduce the "risk" element in this high cost and high intensity crop management system, farmers resort to a rigid and intense plant protection schedule. Use of plant protection chemicals has increased tenfold within the past 15 years.

Plant protection chemicals, like any other chemical must be properly chosen and handled to get the maximum benefit and minimum undesirable side effects. Indiscriminate use of the chemicals, has happened in the case of cotton, will lead to a stage when it will be very difficult to control effectively the pests and diseases with the available pesticides. To pick and use the right chemicals, a detailed knowledge on the chemicals now available in the market, will be highly useful. I am happy to record that Dr. U. S. Sree Ramulu, Associate Professor of Soil Science, Tamil Nadu Agricultural University, Coimbatore, has brought out in a concise manner the different properties of pesticides now in use. I am sure this book will be treasured by all those associated with Agricultural Development and Crop Management.

Tamil Nadu Agricultural University
Coimbatore

A. Venkataraman
Vice-Chancellor

Preface to the Thrid Edition

This revision of *Chemistry of Insecticides and Fungicides* follows the same general pattern as the original edition. However in this edition, information on production, utilisation of pesticides in India and also list of important references have been uptodated. Additional information on integrated pest control, synthetic pyrethroids, and first aid measures in the event of pesticides poisoning have been added in this edition. Also certain omissions in the earlier edition have been rectified.

In addition to those mentioned in the preface to the first edition, I must mention that I am greatly indebted to my students for their constructive suggestions made for the improvement of this book. Also, my ever sincere gratitude to my Professor Dr. P. F. Pratt and my family for their continued encouragement in this task, and also to Rockefeller Foundation for training the author under specialists in this field in the U.S.A.

Madurai **Sree Ramulu**

Preface

Development of pesticides has made tremendous strides in the last three decades and the two world wars as well as the population explosion have also contributed their mite to the development of this field. Though, it is impossible for any author in this field to keep pace with these developments which have been taking place with great momentum, still, attempts have to be made to keep the gaps narrower. It is with this humble view of fulfilling a need that this book have been written. An attempt has been made to present the subject matter in a way that both beginners and expert teachers could use it to their advantage.

The present cover in detail, the Chemistry of Insecticides and Fungicides. Information on their nomenclature, glossary of terms and such other related topics that could help a beginner and an extension worker to comprehend the subject without difficulty have also been added. A list of reference books have been included at the end for further studies in the field.

As the book has been written with a focus on chemistry of insecticides and fungicides, only a limited amount of information on their metabolism has been provided. I am aware of the possibility of a few omissions relating to some recent pesticides, but that could not be avoided due to the unavoidable time lag between the preparation of the manuscript and printing of the book. I wish to be excused for the same.

Many hands and minds have helped me in selecting, organising and presenting the material with clarity and precision. All such help is sincerely and gratefully acknowledged. The contribution of my students of over a decade who have offered many useful comments, is gratefully acknowledged. Dr. G. Rangaswami (former Vice-Chancellor) and Mr. A. Venkataraman, I.A.S. (present Vice-Chancellor, Tamil Nadu Agricultural University and former Director of Agriculture, Tamil Nadu) deserve a special mention for providing me with the opportunities to work in this field, and also for encouraging me in this effort. Sincere appreciation is also due to the help and advice given by Dr. K.K. Krishnamoorthi, (Dean, Agricultural College and Research Institute, Madurai), Dr. N.G. Perur (Director of Instruction UAS, Bangalore) and Dr. A.L. Mookerjee (Cyanamid India Ltd.)

My grateful thanks are due to Mr. A. Venkataraman, I.A.S., Vice-Chancellor, Tamil Nadu Agricultural University who has given the valuable foreword to this book.

I owe a deep dept of gratitude on behalf of all the readers and myself to National Book Trust, Delhi for subsidising the cost of book and bringing down the price within the purchasing power of an ordinary student.

Finally, 1 wish to record my appreciation of the patience (which can be understood and appreciated fully only by authors and their families), help and encouragement given by my wife Mrs. Rajamani and children Vasu and Shoba over the years of writing this book.

Before closing may I request you, readers, to let me know your suggestions for further improvement and I would assure you that the same will be accepted with gratitude.

Sree Ramulu

Contents

CHAPTER I

INTRODUCTION

Insects have always been man's most serious enemy, even from the prehistorical days. The cave dwellers or pre-historicmen, even in the days of plenty of food, were not spared from the attacks and nuisance of lice, fleas, ticks, etc. Since there was plenty of food to go round in earlier days, the loss due to insect damage was not considered to be of great importance. Due to increasing demands, food was becoming scarcer every day, there came an urgent need to curtail the losses caused by various pests. Reduction of losses by the use of pesticides was one of the many attempts made by man.

Use of insecticides has been known and practised even in prehistoric days. Few suggestions, some of them based even on superstition, like carrying out certain cultural operations at certain phases of the moon to exempt the crop from insect injury, were made and even adopted in those days. From the carvings on some of the stone tablets, it is seen that rat poison was used even earlier than 1000 B.C. Homer has spoken of fumigation with sulphur and also of other forms of pest control. In 900 A.D. the Chinese were reported to have used arsenic for the control of garden insects. Certain early writers have suggested a wide variety of materials used such as hot water, brine, lye, whitewash, soap, vinegar, pepper, soot, tobacco and wormwood as insecticidal agents. During 14th to 18th centuries, mineral oils, arsenic, etc., were used. During 1690, tobacco was used as a contact insecticide and in 1773 nicotine fumigation was also tried to control insects. Even before 1800, pyrethrum (Persian insect powder), mixtures of lime and sulphur and whale oil were recommended for insect control. During the 19th century, much more interest was evinced in the discovery and usage of pesticides. Few others compounds were tried during the period such as fish oil, sulphur, mixture of mercuric chloride and alcohol, whale oil-soap, phosphorus paste, derris, boiled lime sulphur, carbon disulphide, sulphur-tobacco dip, pyrethrum, mercuric chloride (1860),* paris green (1865), phenols and cresols (1867), naphthalene (1882), Bordeaux mixture (1883), rosin-fish oil soap (1886), calcium arsenate (1907) and nicotine sulphate (1909).

Since the last half of the previous century and early part of the - current century, crops in certain areas like Mississippi (potato) and Ceylon (coffee) were either wiped out or about to be completely wiped out by certain pests and diseases. This resulted in the development of an arsenical called *Paris Green* which later

---

* Indicates the year in which they were reported to have been used for the first time.

became extremely popular that people started exploiting it for their personal gains. Though sold under the same name, the composition of this insecticide varied so much, that a law had to be enacted in 1900 to clearly define its purity.

However till the beginning of World War II, pest control in the U.S.A., Canada and most of the European countries had not progressed very much. Potato and tomato blight had wiped out thousands of acres of promising and higher yield fields with disastrous results. Loss in agricultural yields also affect the economy of the concerned countries. So in the earlier years of World War II brought two new synthetic organic compounds, an insecticide (DDT) and a herbicide (2,4-D) into the field of pest control. These two compounds virtually revolutionised pest control and brought new hopes of success in the war against pests. DDT was first used in the U.S.A., by the military for delousing and as a preventive measure against typhus fever. Later in 1944, DDT was successfully sprayed in Naples for the control of locust carried typhus outbreak. However, the housefly which was conquered earlier with DDT has now developed resistance to it and hence other new compounds like the phosphoric insecticides had to be used for controlling them. These organo-phosphorus compounds were developed in Germany as potent pest control agents and introduced as pesticides during 1946. As these compounds are extremely effective against a wide variety of insects and mite pests, an infinite variety of compounds with varying chemical structures have been synthesised and used.

In India, certain chemicals and plant products were being used for controlling the pests in earlier days. However they were not used on a wider scale. The Bengal Famine and subsequent study of the causes of the famine by the Bengal Famine Enquiry Commission brought to light the enormous losses caused by pests and hence, the urgent need for the use of pesticides to prevent further losses. Based on the recommendations of the Commission, State Departments of Agriculture have established Plant Protection Wings in each State. Wide scale use of pesticides in India commenced during 1948-49.

To make the farmers pesticide conscious, subsidy even to the extent of 50% of the cost of the pesticides was given during the first three five-year plans. The use of pesticides continued to grow slowly and reached a market value of 18 crores during 1964-65. So, the subsidy was discontinued in certain States during 1968-69 but Tamil Nadu continued to subsidise the purchase of pesticides to a certain extent for few crops even during 1972-73. According to the official reports, the area covered in India by the plant protection measures increased progressively from 6.4 million hectares in 1956-57, to 17.3 million hectares in 1960-61, then to 36.43 million hectares in 1967-68. The use of plant protection chemicals increased from 460 tonnes in 1954-55 to 20, 900 tonnes in 1967-68 and 29,535 tonnes in 1971-72. The requirements of pesticides for agricultural purposes for, the year 1973-74 is estimated to be around 60,400 tonnes out of a total projected consumption of 68, 890 tonnes.

The introduction of high yielding varieties has resulted with the increased use of fertilisers and pesticides. Fertilisers and pesticides are complementary inputs as the greater use of one leads to greater use of the other. Countries with high agricultural production, the cost ratio between fertilisers and pesticides is 3.5:1 while in India it is around 7:1. During 1963-64, while yields were low, for every kilogram of nutrient used, 19 gram of pesticides was used, while in 1971-72 though the yields considerably increased, the pesticide usage has not kept up but instead has gone down to 13 gram per kilogram of plant nutrient. The per hectare consumption of formulated pesticide during 1973-74 was 336 gram and is expected to be about 570 gram per hectare during 1978-79.

Due to an increase in the acreage under food crops, there was an appreciable decline in the area under cash and commercial crops. This in turn resulted in shortfall of industrial raw materials required for agro-based industries. Consequently some of the raw materials required for these industries had to be imported by spending valuable foreign exchange. Hence, a crash programme for the production of oilseeds, cotton and paddy was introduced. Naturally the introduction of the above programmes led to a considerable increase in the consumption of pesticides in India as could be seen from the following figures:

Year	*Value in Rupees*
1964-65	18 crores
1970-71	31 crores
1971-72	44 crores
1972-73	55 crores
1973-74	75 crores (projected)

In India the basic manufacture of pesticides commenced in 1952 by the setting up of a plant for the manufacture of BHC and DDT. Among the South Asian and African countries except Japan, India is the largest manufacturer of basic pesticidal chemicals. At present, the following 42 compounds are being manufactured in India indigenously and many more are under planning stage.

Insecticides:

1. Aluminium phosphide
2. BHC
3. Chlorinated terpene
4. DDT
5. Dichlorovos
6. Dimethoate
7. EDCT mixture
8. Ethylene dibromide
9. Fenitrothion
10. Lime sulphur
11. Lindane
12. Malathion
13. Methyl bromide
14. Nicotine sulphate
15. Oxydemeton-methyl
16. Parathion-ethyl
17. Parathion-methyl
18. Phosphomidon
19. Pyrethrum extract
20. Thanite
21. Toxaphene

Fungicides:

1. Auerofungin
2. Barium polysulphide
4. Cuprous oxide
5. Ferbam

3. Copper oxychloride
6. Mancozeb
7. Nickel chloride
8. Organomercurials
9. Sulphur-Colloidal, wettable and dusts
10. Streptomycin
11. Thiram
12. Ziram
13. Zineb

Rodenticides:

1. Coumafuryl
2. Warfarin
3. Zinc phosphide

Molluscicide:

1. Metaldehyde

Nematocide:

1. Metham-N-Sodium

Weedicides:

1. Ammonium sulphamate
2. 2, 4-D
3. 2, 4, 5-T

The major raw materials indigenously available for the pesticide industry are benzene, toluene, acetic acid, carbon disulphide, zinc chloride, butyl alcohol, chlorine, caustic soda, ethylene, and acetone. Certain raw materials like zinc, phosphorus, sulphur, maleic anhydride, and mercury are still to be imported until indigenous source of supply is found out or alternate methods of production of the pesticides are adopted.

Additions like chalk, lime, bentonite, and talc are incorporated in order to dilute the active ingredients and thus increase their effectiveness. Suitable solvents like organic hydrocarbons, and alcohols are being used to render the formulation ready for use. The chlorinated insecticides (aldrin, dieldrin, chlordan, heptachlor, and endosulphan) which are based on hexachlorocyclopentadiene (HCCP) are made by the chlorination of cyclopentadiene, which is being produced in the Indian Petro-Chemical plants. Starting with only four primary basic intermediates, viz. maleic anhydride, ethyl mercaptan, metacresol, orthophenylenediamine, none of which are still made in India, seven of the organophosphates like phenothioate, menazon, thiometon, disulphotan, phorate, phosalone, and quinolphos are being manufactured now in India. Of the carbamates, the most popular one carbaryl is produced by reacting α-napthol and methyl isocyanate produced by condensing phosfene with monomethylamine. India is self sufficient in the manufacture of conventional fungicides though most of them are based on the heavy metals, which are still imported.

In India, out of the licenced capacity of 73,171 tonnes, at present (1976-77), the installed capacity is only 52,361 tonnes. There are 18 units manufacturing technical materials and 37 units manufacturing formulations. In addition to these, there are about 110 units in Small Scale Industries Sector which manufacture conventional formulations like dusts and to a limited extent wettable powders and emulsifiable concentrates. It has been estimated that during 1971, about 1,36,000 tonnes of formulations were sold', the market value of which was about Rs. 44 crores.

However in spite of the heavy demand for pesticides, there is under utilisation of the installed capacity of certain pesticides like BHC, copper oxychloride, thiocarbamates, dithiocarbamates and 2, 4-D due to poor offtake and also some of them being narrow spectrum pesticides. At the same time, the installed capacity of many pesticides like parathion, metasystox, dimethoate, malathion, fenitrothion are wholly inadequate to meet the present-day requirements. The demand for the above pesticides will be still greater when the contemplated ban of imports of certain pesticides like endrin comes into force.

CHAPTER II

METHODS OF PEST CONTROL

Losses to national economies resulting from insect damage to agricultural products are enormous but rarely comprehended. Even as early as 1930, Neave computed a 10 per cent loss in crop yields due to the insect damage. Every country makes similar computations every year but most of them have been based on the damage caused by one or two pests on a given crop during one season. Insects and fungi attack the seeds during or before germination and during storage of food crops. Although all the pests are not abundant every year, yet the collective damage is much more than 10 per cent. Recent reports have indicated even 20 per cent loss due to one pest alone (like paddy stemborer). Further with the present-day intensive agriculture, this loss is likely to be as much as 50 per cent since the new high yielding varieties are more susceptible to pests and diseases and also a significant change in the pest-disease complex has been noted recently. The greater use of fertilisers promote greater susceptibility in many crops to various diseases and pests. Added to the above, the introduction of new crops like soybean, sunflower, sugar beets have resulted in their acting as alternate hosts for a number of major and minor pests. Weeds have also added to the problem as they have been estimated to deplete as much as 40 per cent of the available nutrients. The actual loss in yield due to all the above factors cannot be ascertained precisely, though it is likely to be an alarming figure.

The ultimate object of insect control is to reduce the extent of pest damage. A wide variety of control measures have been used to minimise the damage caused by pests. In certain cases, the measures. taken to control major pests have boomeranged, since some of the minor pests have now become major pests. Further the extreme degree of specialisation and remarkable adaptability of the organisms will still enable them to survive.

Pest control broadly includes everything that makes life hazardous for the pests and tends to kill them and prevent their spread from one area to another or increase their population within a given area. Pest control measures can be broadly classified into the following:

PEST CONTROL

- Natural control (through climate, topographic features. insects as prey, disease of insects etc.)
- Applied control
  - Mechanical control
  - Cultural control
  - Biological control
  - Legal control
  - Chemical control

Natural control of pests include all measures that will check or destroy pests and cannot be influenced or modified more than a limited extent by man. Due to the existence of natural control, no species is permitted to increase indefinitely and without limit. Had there been no natural control, one pair of houseflies would have multiplied within six months to 1,910,000,000,000,000,000,000 ($=1.91 \times 10^{21}$). Fortunately enemies of flies like birds, insects, diseases, insecticides and weather take a toll on eggs that are laid, and thus prevent the huge multiplication.

Natural control involves the interplay of factors in the non-manipulatable environment. Climate exerts considerable control on each species by allowing them to survive only under one or two climatic zones and not in all the three zones. Even within one zone, factors like humidity, rainfall, sunshine, temperature changes, wind velocity affect the spread of the pests and diseases. Presence of various barriers and topographical variations control the spread of pests and diseases to a great extent. Natural pests and predators, indigenous bacterial, fungal and viral diseases of the various pests periodically check the population levels.

As opposed to natural control, man completely manipulates the various control measures in the environment and tries to control harmful pests and diseases which have escaped the repressing influences of natural control. Today the increases in both quality and quantity of the crops has created a favourable condition for the spread of various pests and diseases and has led to a stage where successful pest control may often represent the margin between ruin and prosperity of a farmer.

Mechanical and physical control measures aim at controlling pests and diseases through catching them in traps and killing them manually, protecting the trees by tree banding or by special equipments like electrical fences; cutting and burning of diseased plants; flooding the fields to provide unfavourable anaerobic conditions to fungi and pests; or providing light for the control of pests and diseases. Many of these methods are costly and success of these measures are not assured and as such can be undertaken only after the occurrence of the pest and disease and also when costs are of no consideration.

Cultural practices to control pests and diseases include ploughing and exposing the soil to sunshine and natural enemies, to kill various pests and organisms, crop rotations, varying the time of planting, harvesting and carrying out other cultural operations so that the susceptible period of the crop is over by the time the pest or disease attacks, application of fertilisers and manures to the crop to give sufficient vigour to withstand the onslaught of pests and diseases, development of varieties resistant to various diseases and pests. These cultural practices are to a great extent inherent in nature, and as such the exact extent of its effectiveness is difficult to assess.

Biological control of pests and diseases involve the introduction, encouragement or artificial increase of natural enemies of the insect pests which include both insect parasites and insect predators, microbes pathogenic to pests and diseases. The most successful use of biological control has been made by the introduction of new species of insects as enemies of the pest, and is particularly effective when the pest itself is accidentally introduced in new areas. The biological control measures are highly specific, compatible with most pesticides and generally inexpensive as only a low dosage is required. However these are effective only if they are used at that critical time Further because of their specificity, they are of narrow range. In a number of instances, there are certain problems in the production of biological measures and subsequently they require continuous maintenance in the new environment. It is also difficult to assess precisely the effectiveness of these control measures. In spite of the above, the biological control has been successfully used in a number of countries against pests.

Another important method of applied control is the legal control with which countries prevent the introduction of new pests and diseases from other neighbouring countries through agricultural produce. In some countries, even certain States, have, enacted laws to prevent the movement of agricultural produce contaminated with disease and pests from neighbouring States into their areas. Due to the increased movement and speed of international travel nowadays, problems due to accidental introduction must be anticipated well in advance and sufficient quarantine steps should be taken to cover all possible crops on which pests and diseases can spread, and later become a menace.

The next important method of pest control is through chemicals. A chemical to be used for pest control cannot be merely any toxic substance that can kill the pest and/or disease. Many factors like phytotoxicity, effect of the pesticides on animals, pollution effect, degree of persistence of the residues of the chemical on the crops, all need to be considered while selecting any chemical for usage. Chemicals used for controlling pests can be divided into different groups based on the nature of pests to be controlled like:

(i) Acaricides — for the control of mites, ticks etc.;

(ii) Algicides — for the destruction of algae and other aquatic vegetation;

(iii) Antiseptics — for protection of non-metals from microorganisms;

(iv) Arboricides — destruction of undesirable arborell and bushy vegetation;

(v) Bactericides — control of bacteria and bacterial diseases;

(vi) Fungicides — for the control of plant diseases caused by various fungi;

(vii) Herbicides — for the control of weeds;

(viii) Insecticides — for control of harmful insects. Individual groups within insecticides can be divided into subgroups like aphicides based on the nature of insects controlled;

(ix) Limacides or Molluscides — for control of mollusks including gastropods;

(x) Nematocides — for the control of nematodes;

(xi) Zoocides for the control of rodents and other animals.

Chemicals used for stimulating or retarding the growth of plants (growth regulators), for removal of leaves (defoliants), for desiccating plants (desiccants), for repelling insects (repellents), for attracting insects (attractants), and for sterilising insects (sex sterilants) are also included among the pesticides. However out of these, insecticides, fungicides and herbicides in that order are most widely used.

Recent studies have shown that maximum economic benefits cannot be achieved by the use of a single technique in isolation as such a measure is based on the non-consideration of the entire system and concentrates only on a single pest species without considering the effect on other species. So, now emphasis is being laid on integrated pest control which attempts to integrate all suitable techniques of pest management based on the knowledge of all the regulatory forces of the environment and the ecological principles. It is not a mere combination or superimposition of chemical and/or biological control techniques. It aims at the development of a pest control method in relation to the total environment as well as the determination of the injury level from the point of economic crop production and also based on necessity to keep the pests and diseases of a crop below the economic threshold level and prevent its build up to the level of outbreak. The integrated pest control measures vary from crop to crop, number of pests and diseases and even in different locations in the same area.

Insecticides

Insecticides can be divided into the following groups according to the nature of their penetration into the insect: stomach poisons, contact poisons, systemic poisons and fumigants.

Stomach poisons

As the term suggests, it applies to all poisons which act through the stomach. Their use is limited to insecticides like arsenates which control pests like eel-worm and spider mites. As they are stomach poisons, they must be consumed by the insects through the mouth before they can act. Usually, the insecticides are applied to that part of the crop which serves as food for the pest and is ingested in the normal process of eating.

Inorganic metallic salts such as arsenates act as stomach poisons. Ionic or ionisable substances even if organic, belong to this group. The relative speeds of absorption of stomach poisons from the insect gut also influences the efficiency of the formulation, as the insect may regurgitate and throw out the poisoned food before a lethal dose is absorbed. Such poison should be quick in its action as otherwise the insect may not be killed and even if killed, it would have done enough damage before dying.

Chemicals toxic to insects or mites often may be toxic to plant tissues on to which they are applied. For example sodium arsenite although a good insecticide, cannot be applied without considcrable plant injury unless it is diluted. On the other hand, the sparingly soluble basic calcium and lead arsenates can be used without much phytotoxicity. True stomach poisons which are insoluble in plant or animal waxes and are also without effect on plants and pests unless ingested by them, may be selected. When chemicals are ingested by the pests, chemical reactions in the digestive tract may set free secondary compounds which are more poisonous and distribute them throughout the body. These secondary compounds may be lethal by themselves or metabolised to lethal compounds inside the body as in the tissue of the insect gut. Development of increased resistance to some of the modern organic insecticides have revived interest in stomach poisons. On the practical side, frequent applications of stomach poisons are required to be applied as the surface deposits may wither quickly. Some of the important stomach poisons are BHC, DDT, lead arsenate, calcium arsenate, methoxychlor, paris green, sodium fluoride, cryolite, fluosilicate, and certain phosphates.

Contact poisons

In contrast to stomach poisons, these pesticides need not be consumed for killing the pest as they kill by intimate and direct contact with the body of the pest. Usually they kill the pests after they have physical contact with the material. All poisons which act by direct penetration through parts of the body wall, fumigants which enter through the tracheae, and oils which kill the pest by some manner are included in this group. Excluding the fumigants and the controversial petroleum oil mixtures, the so-called contact poisons are either naturally occurring compounds like nicotine, pyrethrum and rotenone or synthetic organic compounds like BHC, DDT and parathion, or mixtures of compounds from either one or both groups.

Solubility of these poisons in insect oils and waxes is a prerequisite for contact action. Generally, contact poisons are soluble in plant and insect oils and waxes, and also animal fats. The lethal action of these compounds in the field are said to be due to the following three mechanisms:

(i) Transport from outside the cuticle to the site of action;

(ii) The primary process or action wherein the insecticide enters and usually inhibits the enzymes' action or other protein interactions;

(iii) Subsequent secondary processes or actions on nervous, respiratory or other systems involving the biochemical effects on the pests.

Most of the substances are rather specific in their action. There are marked differences in their speed of action both against individual pests within a species, and also against a variety of insects. With some of them as with synthetic allethrin, natural pyrethrum mixture, partial recovery of the affected population may take place. Many of them are even more transitory than the inorganic insecticides in their action, for they dissipate or lose their lethal properties quickly after application in the field. Timing and frequent applications are of great importance with such short-lived chemicals. However such chemicals with short life will be of great advantage for application to plants close to harvest time as chemicals less transitory would leave residues dangerous to human beings and animals.

The contact poisons may be instantaneous in their action when they are called "knockdown poisons" like pyrethrum, or may take several hours to kill the pest. Many contact poisons like DDT are also stomach poisons if taken into the stomach of an insect. Some— of the important contact poisons are BHC, DDT, nicotine preparations, lime-sulphur, toxaphene, chlordan, dieldrin, aldrin, methoxychlor, pyrethrum, oil emulsions, rotenone, thiocyanate, TEPP, parathion and systox.

Systemic poisons

This term describes the action of poison in the crop plant rather than in the pest. All poisons are systemic from the point of creature poisoning, because they interrupt the functioning of some of the vital metabolic systems. Systemic poisons are those which are applied to one part of the crop like root or the leaf and translocated to other parts of the plant in the fluid systems for tendering the whole plant toxic to insect or fungi. Some of them are so efficient that as a dust or a dip to seed before planting, will result in sufficient concentration of the active material in the emergent young growing plant to prevent it against the ravages of pests.

Most of the systemic poisons are stomach poisons which attack insect by either chewing the tissue or sucking the juice of the plant before it is killed. Most of the systemic compounds are remote organic derivatives of phosphoric or of thiophosphoric acid esters.

The parent substances as formulated may not be particularly lethal to animal organisms but often they are transformed either by the plant or by the animal organism into extremely poisonous chemicals. The sites of these transformations in the plant are not known, but it is generally accepted that the active materials tend to accumulate in the rapidly growing apical portions of the plant. They help to give the maximum protection to growing and to most vulnerable portions of the plant from the attack by insects and also protect plants which are difficult to protect by residual type materials. The site of these transformations in the animal system appears to be in the liver and in insects, in the gut tissues or other parts of the digestive tract.

The greatest advantage of the systemic poisons lies in that only one part of the plant need receive a dose of the poison for the whole plant to become protected against certain pests. As the natural enemies of these pests do not eat parts of the plant which contain the systemic poison, they are unaffected. Thus, the allies of human beings are unharmed. Insecticides solely relying upon contact for killing the pest can be harmful as they may kill the beneficial insects also. Many sucking pests which cannot be killed by a stomach poison are also too difficult to kill with a contact poison because the pests move very rapidly and have little opportunity to make necessary contact. Of the few systemic insecticides developed till now, the octamethyl phosphoroamide group of chemicals seem to be the most effective against such pests as scale insects, aphids, and mealy bugs.

Fumigants

Certain poisons are transmitted to the insect in the gaseous state. Some of them are applied directly as gases, in contrast to some fumigants such as liquid nicotine or solid naphthalene which operate against the pests in the vapour phase. Fumigants are generally applied to plants or to products in enclosed spaces or in other tight enclosures such as greenhouse, bushes or trees under tents. Some of the important fumigants are liquid nicotine, naphthalene, hydrocy—anic acid, carbon disulphide, chloropicrin, methyl bromide, methyl allyl chloride, ethylene oxide, calcium cyanide and aluminium phosphide.

Majority of pesticides can penetrate the insect organism simultaneously by different routes. Based on their main route of penetration into the insect, they are assigned to one or other subgroup. For example though lindane shows contact, stomach and fumigant action, still it is assigned to the contact insecticides group. For controlling insects with chemicals, the insects have been conveniently divided into two –broad classes based on the type of mouth parts. The chewing insects like caterpillars, leaf eating beetles, and thrips bite, rasp or lap the invaded surface. The damage caused by them can be easily observed. For controlling these insects, a protective insecticide which distributes over the entire surface layers or deposits the chemical on the surface or a systemic insecticide which permeates the plant should be used.

The other group, the suctorial or houstellate insects like aphids, mosquitoes, plant bugs, scale insects have mouth parts modified into piercing and sucking organs to feed on the plant and animal juices. The damage caused by them is not easily apparent but is reflected much later in the yields. A protective surface layer of poison for ingestion into the stomach of the insect is useless against suctorial insects for they take their food from the interior of the plant. So, an insecticide which will kill the pest by absorption into its body tissues by simple contact of the spray with some external part of the body or from contact of sprayed surfaces should be used. Alternatively some local or general systemic insecticides should be used. Number of modern chemicals act both ways.

Fungicides

Fungicides are usually divided into two main subgroups: (i) fungicides used for the control of diseases of growing plants, and (ii) seed disinfectants. Among the fungicides, there are compounds with contact and systemic action, the later group being very small. Some of the systemic fungicides are 1, 4-Oxathiins, chloroneb, terrazole, and benlate.

Herbicides

Herbicides are divided into— (i) non-selective herbicides which act on all species of plants, and (ii) selective herbicides which attack only certain species of plants and are safe for others. This division is very narrow as majority of compounds can either be selective or non–selective depending upon the concentrations used, rate of application per unit area, etc. Based on the external symptoms on plants and the methods used, the herbicides can be divided into four subgroups: contact herbicides, systemic herbicides, herbicides acting on roots system of the plants or on germinating seeds and defoliants and desiccants.

CHAPTER III

FORMULATIONS AND THEIR CHARACTERISTICS

The success of any control measure with chemicals depends on the chemical coming into contact with the pest to be destroyed. This largely depends upon the way the formulation is prepared and the conditions under which the formulation is brought into contact with the pests and plants. Only few modern pesticides like ground rotenone bearing roots, ground pyrethrum flowers, aluminium phosphide, ethylene dibromide, methyl bromide and EDCT mixture could be used in an undiluted form. Generally all the materials are diluted so that more area may be covered or the phytotoxicity may be reduced considerably. Dilution is an effective way of controlling the dosage, as well as securing uniform and thorough distribution over the entire area to be treated. This is usually accomplished by mixing the active material with a neutral carrier or extender, which is either an inert dust or powder or a liquid in which it is to be emulsified. The usual carrier for sprays is water. Those formulations which are insoluble in water, are applied as suspensions of uniform particles and adequate agitation is supplied during the application to ensure an even distribution of the pesticide.

Formulations are classified into the following groups based on the form in which each material is utilised, or the purpose for which it is employed.

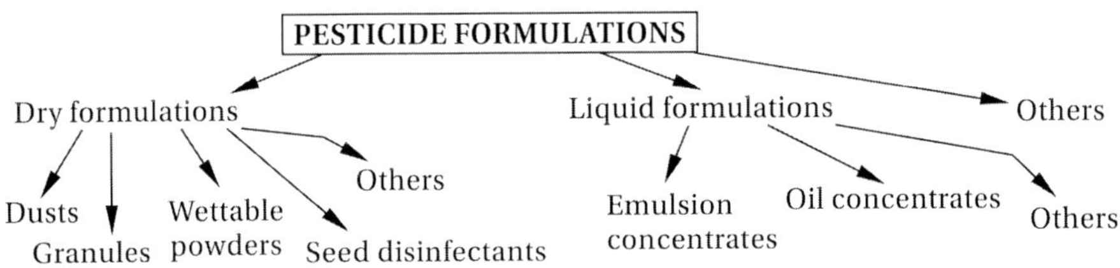

Dusts

Dusts consist of a mechanical mixture of the active ingredients with or without an inert diluent pulverised to a particle size of 3 to 30 μ. Dusts can be classified as follows:

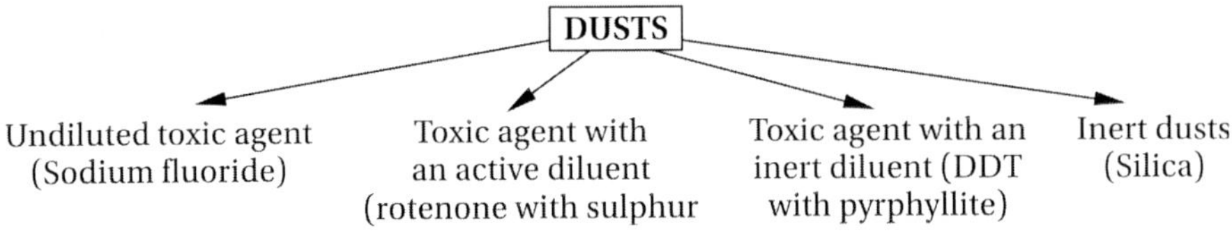

Some insecticides such as calcium arsenates, sodium fluoride or ground pyrethrum flowers may be applied as dusts without any diluent. Some others like rotenone can be mixed with other active pesticides like sulphur and then applied. In other cases, the powdered material of the pesticide is diluted with relatively inert diluents like talc in order to cover more area or to reduce the phytotoxicity or to improve the chemical or physical properties. These diluents are primarily used for extending the pesticide but may function chemically, physically or physiologically.

Some diluents or carriers are utilised for their adsorptive capacity which enables them to take up liquid toxicants or solutions of insecticides. The carriers may be combined chemically or physically with the toxicant or it may be ultimately mixed with it. Often a dust concentrate or master mix having suitable physical properties for blending into a finished dust by adding more diluent is prepared. Chemically active or adsorptive materials such as lime, bentonite or botanical flours are utilised as primary carriers. Relatively inert materials such as talc and pyrophyllite are used as simple diluent.

Dust mixtures are prepared generally by either a solvent-mix method or by the ball mill method. In the solvent-mix method, the toxicant in the form of a liquid is sprayed into the dust mixture during the blending process. The solvent may be allowed to evaporate or it may be a higher boiling solvent of a petroleum fraction. In the ball mill method, the ingredients are intimately ground and mixed together by blending operations of a mechanical mixture. In the grinding process, not only do the pesticide particles get distributed among the diluents but also coat the particles of the diluent. Consequently grinding of the active ingredient together with the diluent gives a more efficient formulation than separate grinding of the ingredients with subsequent mixing. The finer the grinding of the active constituent in a powder or dust type formulation, the more effective it is for dusting, as there is better adherence of more finely ground preparations on plants and insects and also greater uniformity of coverage.

Pesticides can be reduced to the powdered state more or less readily. Compounds with a well formed crystalline structure, high friability and low plasticity grind the best. Technical grade products containing oil as impurities are more difficult to grind as they lump on grinding, and form coarse aggregates which stick to the walls of the mill. Also, lower the melting point of compounds, the more difficult it is to grind them in the ball mill. To prevent lumping, some inert diluent is added during grinding process as well as in storage. The greater the absorptive capacity of the diluent employed as additive, the lesser the quantity of diluent to be used. For example when DDT is ground with silica gel only 5 to 10 per cent diluent needs to be used, while at least 25 per cent of ordinary kaolin will be needed to get the same physical property. If smaller amounts of diluent are used, preparations lump in storage particularly under tropical conditions.

Lumping often occurs in ball mill due to the increase in temperature in the mill and consequent melting of the compound. In grinding low melting pesticides, a large amount of diluent has to be used. Grinding in ball mills is carried out with air separation to prevent lumping. Usually grinding is carried out in two stages—a concentrated powder is first prepared which is then diluted with the necessary amount of diluent and grinding completed in a ball mill. When air jets are used, lumping is prevented but the cost is increased. When a low melting organic compound is ground with a mineral diluent in a ball mill, a dust is obtained that differs slightly from a preparation obtained by impregnation of the diluent with a solution of the compound and subsequent removal of the solvent by distillation, because in the grinding process, coating of the diluent with a thin layer of pesticide takes place.

Several other methods are used for the preparation of dusts: (i) the melted pesticide is atomised in a chamber into which the diluent is introduced in a suspended state, and (ii) the vapours of the pesticide is condensed on the diluent. However these methods are not economical. When dusts are prepared from liquid organic compounds, it is simple enough to mix the diluent, with active constituent. When the content of the active constituent in the dust is very small, the mixing is carried out in several steps.

The best diluent for most preparations of dusts are pyrophyllite and talc which have lameller structure and adhere well to the plant foliage. Further these minerals are quite inert and do not exert an-adverse effect on most pesticides. Alkaline impurities in these minerals may decrease the stability of the dusts that contain esters of the phosphorus acids. Rarely chalk, gypsum, kaolin, kieselgur, tripolic earth, silica gel and various other clays are used. Alkaline diluents are not recommended in the preparations of dusts from esters. Halogen derivatives of hydrocarbons and other esters may be hydrolysed under the influence of compounds of acid nature. In such cases several diluents may have to be used.

Use of hydrophilic minerals like kaolin clay, bentonite type clay are not recommended where there is high humidity, as dusts with these diluents cake easily after even slight damping. To increase the retention of the dusts on plants, hydrophobic compounds of a type of calcium sterate, mineral oils and the like can be added to them. The adherence of powdered formulations and their retention on plants in many cases determine the duration of their action. These characteristics depend not only on the shape of the dust–particles but also on other properties of the additives added to the dusts. Hydrophobic compounds are retained better on hydrophobic surfaces and hydro–phylic ones on hydrophylic surfaces. Hence, it is necessary to consider the crops also.

The important properties of dusts like shape, density, size, hardness, electrostatic charge, adsorptive and absorptive properties and chemical nature of the dust particles affect the quality of the pesticides in storage, application, deposition, adhesion and effectiveness. Certain properties like bulk density, flowability and sticking properties are concerned for the dusts as a whole.

Particle shape: Dust particles may be spherical, polyhedral, tubular, spicular, irregular or other shapes. The effect of particle shape on the performance of insecticides has not been studied much. However highly irregular materials tend to flow poorly. Thin plates tend to adhere well, while materials containing different shapes may not mix evenly and may segregate during application. Particle shape is measured in terms of equivalent spherical particles.

Particle size: The size of the insecticidal dusts fall between macroscopic and colloidal range. The size of a toxic particle is related inversely to its insecticidal activity. For example the toxicity of sulphur and pyrethrum increases with decreasing particle sizes. Small particles are more readily consumed by chewing insects. The effectiveness of particles of a stomach poison is believed to be due to particles less than 10 μ size. Increased activity is due to the large surface area of the product in a finely divided state, in the case of contact dusts, to the greater adherence to insects. Sometimes, fine particles of a diluent may reduce the effectiveness of the mixture by coating the particles of the toxicant. When the size of the particle is reduced, volatilisation and decomposition takes place more rapidly due to the increased surface area.

Particle size affects suspensibility in the atmosphere or in water. Although extremely fine particles may be used for spray purpose, they possess certain disadvantages like slow settling of dust particles, lowering of carrying power, increase in drift, etc. Finer powders are more liable to irritate the operator. Also the finely ground materials cake very badly, but this can be overcome by adding a conditioner.

The uniformity, covering power and adhesiveness of dusts are related to particle size. Small particles of a solid are more adhesive to surfaces and less abrasive to the equipment. In dusts comprising widely different sizes, segregation occurs during application. Particle size is commonly expressed in terms of sieve size. The finest sieve is 325 mesh which has an opening of 44 μ in diameter. Although this sieve is said to have 325 meshes per linear inch (>10, 500 openings per square inch), few of the standard sieves approximate this figure. The proportion of particles that will pass through a fine sieve is inadequate as a measure of fineness of insecticide. Particle size can be determined by microscopic measurements, sedimentation process, and air permeation methods.

Bulk density: Bulk density is the weight of a definite volume of dust measured under specified conditions expressed as grams per cubic inch and is dependent upon particle size, shape, and actual density. Lighter materials do not have such a good carrying power and usually are fed more slowly on a weight basis although they remain in the air for a longer time. Heavy materials do not feed well and fall too rapidly. As the diluent in a mixture has a different density than the toxic particles (sodium arsenate—1.87, bentonite—2.49), all parts of a dust cloud will not have the same toxic concentration. Bulk density is an indication of the fluffiness. A good diluent is one which weighs 300-450 kg/c. metre (30-45 lb/c.ft.).

Particle density: Particle density is the actual density of the solid materials only, as if there were no air spaces between them. Particle density affects the feeding in the duster, carrying power of the dust, segregation of dust components and settling of suspension of the particles. Density is measured by the change in density which occurs when the substance is added to a liquid in a pyconometer.

Electrostatic charge: Electrostatic charges in dusts are produced by friction between the particles themselves and the dusting equipment. Materials high in silica tend to give a positive charge to the blower and a negative charge to the dust stream. Electrostatic charges on particles consi-derably affect the attraction of dusts to the plant surfaces and dust distribution.

Flowability: The flowability of a dust is an indicator of the feed rate of dusting equipment. The angle of slope which is a measure of flowability, is measured by allowing the dust to flow through a funnel upon a disc. Greater the angle the poorer is the flowability. For example, sulphur dust treated with a conditioner has an angle of 40° whereas untreated sulphur has angles > 70°. Dusts with fibrous or needle shaped particles have a slower feed rate than dusts with spherical particles.

Other properties: Hard particles may cause serious abrasion of spraying and dusting equipments. The absorption and adsorption properties of a dust depend upon the nature of the surface, size, shape and other characteristics of the particles. Capacity to absorb moisture affects the caking capacity of the dusts and hence their quality in storage. Their sticking properties depend upon the particle size, shape, adsorptive properties, tendency to form agglomerates or clumps of 25 to 300 individual particles, etc. It may be modified by adding some material like flour or oil and also by the nature of the dusting surface.

Practical considerations in dusts' application: Dusting is more convenient in places where considerable time and labour have to be expended in transport of water. Ground application of dusts to field crops ordinarily range from 10 to 50 lb per acre depending upon the conditions. However dusting cannot be said to be as effective as spraying. Although spray and dust applications are customarily stated in terms of quantities per unit of ground area, it is the amount of deposit on unit leaf area which is important. If a dust diluent was entirely inert, acting solely as a thinning agent, it would make no difference how much the toxic ingredient was diluted, provided same amount of toxic agent was applied per unit area and also if the diluent did not interfere mechanically with the action of poison. At low rates of insecticides the relationship between insecticides content and the rate of application is not important.

Granular Formulations

Granulated formulations are widely used for the control of soil inhabiting pests and also for making plants poisonous to the sucking pests. Granular applications are becoming popular as they are more convenient and leave a smaller residue on the plants.

Granules can be made by several methods. They are prepared by impregnation of prepared granules or minerals of the type of perlite or vermiculite with liquid pesticides or solutions of them or by granulating the powder formulation on a suitable diluent with subsequent screening. The process of preparing granulated pesticide by the latter method is complicated because in addition to preparing the starting powder, the steps of impregnation, granulation, drying and screening have to be carried out. In addition to the diluent and the pesticides, various binding agents based on the synthetic resins or other adhesive materials are also added to the formulation.

Most dusts will aggregate to coarse particles in a fairly uniform manner if tumbled in an oblique rotating drum, while a liquid preferably containing an adhesive substance is sprayed on it at intervals. The formation will be better if the initial mixture contains a proportion of coarse particles to act as nuclei. Pastes containing a self-setting component such as Plaster of Paris are broken down to fairly uniform granules after drying, if rubbed through the sieve before drying. In these methods, the active pesticide is incorporated at the start and is present throughout the bulk of the finished granules

However most of the pesticide granules are made by impregnation of dissolved pesticide into a preformed porous, inert granules or by sticking the powdered pesticide onto the outside of the preformed inert granules which in this case need not be porous. The last method is the most common one and granules of any new pesticide could be made without much preliminary work. Suitable base material is sieved ground mineral matter, like limestone or brick waste. Sieved crushed nutshell or corn cob chips are also used. Walnut shell base is already industrially available. The coating is accomplished by tumbling the base granules in a rotor, together with a suitable admixture of powdered pesticide or mineral mixture during addition of a water solution of a gum, a suitable cellulose derivative, high molecular weight polyethylene oxide or other adhesive.

The most widely used granular formulations are of 0.2 to 1 mm size. The strength of the granules also differs. For the treatment of plants, granules of relatively low strength are used while for the control of weeds in water reservoirs, granules of high strength are used. The rate of release of the pesticide from the granules also varies. If the pesticide has very low solubility in water and is contained within a very close packed mineral matrix, it may take a long time to diffuse out even if the granules are shaken with free water. Equally obviously if the pesticide has a high solubility and forms only a thin envelope around the granular base it will disperse in free water almost immediately. The pesticide in the later case is available to any organism touching the granules. However in moist soils, the difference in action may vary much less.

Granulated formulations of pesticides with fertilisers are also being prepared on a limited scale since many of the pesticides degrade when mixed with fertilisers.

Wettable Powders

Powdered formulations that give stable uniform suspensions with water are called wettable powders. Suspensions usually are more effective than dusts. Spraying is gradually displacing dusting and dusts are being replaced by various formulations for spraying. The use of suspensions is particularly convenient when dust type preparations are poorly retained on the surfaces; emulsions are easily absorbed and as a result, considerable amounts of poisonous residues are retained on fruits. Suspensions are preferred because the particles of a suspension adhere well to the surfaces and also do not penetrate and can be washed out if required.

Important conditions for the wettable powders to be effective are:

(i) That they should be stable in storage and should not cake,

(ii) The suspension should form rapidly but the solid particles should settle slowly,

(iii) The sprayed objects should have good wettability and the sprayed material should spread easily over the surfaces, and

(iv) The sprayed material should be retained on the sprayed surfaces for a long time.

Wettable powders should have a high degree of dispersion of the active ingredient of the preparation. Good wettable powders usually contain not less than 80 per cent of particles with a size below 3 μ.

Wettable powders can be divided into three groups:— (i) Preparations with a high content of the active ingredient content (>60%), (ii) Preparations with a medium content (30 to 60%) and (iii) Preparations with a low content (<30%). The last are usually prepared from liquid or waxy compounds since a high content of such compounds in a wettable powder yields preparations that cake easily.

Wettable powders differ both in their method of preparation and also in their components. A wettable powder will have a diluent, a surface active agent and an auxiliary material in addition to the active ingredient. Sometimes special stickers are also added. For example DDT 50% wettable powder contains DDT technical 50%, kaolin 30-40%, amount of inert diluents 4-5%, dispersing agents 4-5%, and wettable agents 1-2%. The technical DDT of 89-89.5°C is aged for three weeks and then mixed with the inert materials. Then the mix is ground in a hammer mill and mixed with the adsorbents, dispersing and wetting agents in a separate ribbon mixer. The preground mix is collected in drums and set aside for curing for about one week. The product from the hammer mill is coarse with about 25-30 per cent retention on 15 mesh. It is then passed through a disc grinder and ground to 15-20 per cent retention on 200 mesh. Finally, the material is subjected to a high velocity of air at about 12-15°C in a micronizer into which chilled air at 100 lb/sq. inch is fed. The material gets ground to a fine powder. The discharge from the micronizer is let into a chopper containing a series of filters. The product in the chopper, is blended before bagging. The product from the micronizer has only 5 per cent retention on 325 mesh and will have 50-80 per cent suspensibility.

In a high active ingredient content formulation, pure materials without oily impurities and a diluent with a low bulk density and relatively large sorptive capacity are used. Sometimes special kinds of silica gel, hydrated aluminium oxide, synthetic calcium silicate etc., are used. Wettable powders without diluent can be prepared if the active ingredient has a high melting point and grinds well.

With a poor quality diluent, the quantity of surface active agents have to be increased as otherwise unstable suspensions will be obtained. High concentration wettable powders are produced by grinding in a colloidal mill a mixture of the active ingredients with the diluent and the other ingredients. Good results are obtained by grinding in air jet mills of various types. Grinding can be carried out in ball mills but coarser powders are obtained. Satisfactory results are obtained with ball mills by grinding pesticide with the diluent in the presence of water. The paste is then dried and ground a second time.

In case of pesticides obtained by precipitation from aqueous solutions, all materials required for the preparation of the wettable powders are added to the precipitated and washed pesticide and then the mixture is dried in spray driers, thus avoiding the grinding. Wettable powders such as DDT 50 per cent, copper oxychloride, ziram and cuprous oxide are prepared. Wettable powders with a compound action have also been prepared. These may contain several pesticides for different purposes at the same time, so that there may be economy of labour. -

Sprays

In places where water is available, liquid formulations are economical, efficient and convenient to apply. Unless the insecticides are soluble in water, they are usually formulated as emulsifiable concentrates in an aromatic solvent such a zylene or in a petroleum oil. The choice of a solvent is generally governed by the degree of solvency, compatibility with other components of the mixture, phytotoxicity, cost and availability. With easily hydrolysed insecticides, the water content of the solvent must obviously be low. These concentrations are prepared by dissolving the chemical in the solvent with emulsifiers of a type and proportion to provide the desired properties of suspensibility and emulsion stability under conditions of field use. Emulsifiers commonly used are, non-ionic with anionic surface agents. The non-ionic types are generally insensitive to hardness of water and store well. The anionic types produce low interfacial tension thus favouring the formation of small dispersed droplets in an easily emulsified mixture. Most of the surface active agents or surfactants may react with acids to form complexes which are no longer surface active and are therefore devoid of emulsion promoting properties. Soaps may also be used as emulsifiers but only when the water to be added is soft.

Emulsions may be undesirable when two or more formulated sprays (a fungicide and an insecticide) are to be mixed. Here while the actual pesticide may be compatible, either the solvent or emulsifiers may not be compatible. Also when two or more pesticides are used together as emulsions at their respective effective concentrations, the effectiveness of the quantities of solvents or emulsifiers present may be increased sufficiently to make the final spray too stable or even phytotoxic.

Liquid formulations applied to foliage should not only kill the already existing pest but also those which will arrive later. Hence, good coverage of the sprayed surface should be there. In addition to the emulsion stability in spray tank, it should break down quickly as soon as it hits a foliage or other oily or waxy surfaces. As water will run off and also will not wet these surfaces, other components must be added so that they may remain behind. The main disadvantage of dilute-type sprays is the enormous quantity of water needed for spraying, in many cases up to 900 litres. But recently with the invention of special machinery like low volume nozzles, and ultra low volume nozzles, the same quantity of pesticide could be sprayed in 65 to 90 litres of water or even in 10 litres of water over a hectare. However the formulations for LV or ULV have to be slightly different from the regular emulsion concentrates.

Liquid insecticides may be applied as water or oil solutions, or water suspensions of solid particles or water based emulsions. Relatively few water solutions like nicotine or TEPP solutions are used for the control of aphids. Lead arsenate is an example of water suspension of solid particles. Emulsions employed as insecticides are water suspensions of immiscible liquid particles as mineral oil. DDT may be dispersed either by suspending a wettable DDT powder in water or emulsifying an oil solution of DDT in water. Oil base sprays are adapted to household applications as the oil is less likely to injure wood than water. Petroleum oil has the added advantage that many organic insecticides are readily soluble in it. directly or indirectly.

Solutions and solubility: Petroleum distillates and water are the most common solvents of insecticides. Materials which are insoluble in both require some intermediate or mutual solvent often called a solubiliser, blending agent or coupling (bridging) agent. Pyrethrin although insoluble in water, may be dissolved in petroleum and then emulsified in water. Derris is not entirely soluble in oil and hence some solubiliser like dibutylphthalate must be used. The possibility of chemical reaction of the dissolved compounds with the solvent has to be considered. If unsaturated compounds are used as solvents, it is possible that they may react with even such stable compounds as DDT and the duration of the efficiency of the preparation will be less, especially in light and at high temperatures. Sometimes, it will be necessary to mix two or more solvents to increase the solubility, for example the pesticide may be first mixed with small amount of expensive solvent and then the cheaper solvent is added. The first one used is called intermediate solvent. This is often done with DDT and other organochlorine compounds. Intermediate solvents for the preparation of concentrated solutions of pesticides in mineral oils are— cyclohexanone, methylcyclohexanone, mesityloxide, tetrahydrofuran, thiophene, isophorone, methyl ethyl ketone, dimethyl formamide, diethyl carbonate, alkylacetate, xylenes, methyl and polymethyl napthalene and chlorobenzene. Intermediate solvents for halogenated and organophosphorus insecticides are halogenated hydrocarbons of aliphatic series including carbon tetrachloride, methylene chloride, dichloroethane, trichloromethylene and tetrachloroethylene. Most frequently used solvents for preparation of pesticide solutions are petroleum hydrocarbons, dearomatised kerosene, white spirit (turpentine substitutes), mineral oils and diesel or fuel oils. Petroleum solutions containing higher amounts of DDT (>5%) contain an additional solvent such as methylated naphthelene. As the capacity of solvents to hold the pesticides in solution at low temperature varies widely, a formulation should be tested carefully at low temperature.

When the composition or chemical structures are compared, it will be found that the ionisable materials such as inorganic or organic salts are likely to be soluble in water and insoluble in organic solvents. Unreactive organic materials such as hydrocarbons are insoluble in water and soluble in organic solvents. Reactive groups such as —OH, —NH_2, —CHO, and —COOH in organic compounds tend to increase water solubility. High molecular weight organic compounds tend to reduce the solubility in most solvents.

Some of the solvents other than water and petroleum are employed to dissolve the insecticide are ethylalcohol, isopropyl alcohol, acetone, methyl ethyl ketone, ethylene glycomonoethyl ether, diethylene glycol, monoethyl ether, carbon tetrachloride, xylene, cyclohexanol, cyclohexanone, dibutylphthalate, methylated napthalene, pyridine, and pine oil.

Many organic solvents possess a certain amount of toxicity to human beings are also an inflammability hazard. Generally, the aromatic and chlorinated solvents are more toxic than kerosene, when ingested. The vapours of many solvents are narcotic in nature when inhaled in sufficient amount. Solvents of low flash point below 55°C are dangerous from the standpoint of expansions and fire hazard. Normally, refined base oils for sprays possess a flash point between 140 and 170°C. Solvents differ as to their safety to plant tissue when they are employed in formulations. Spray deposition on treated surfaces must be sufficiently continuous to provide a practical degree of protection. For this various materials like spray supplements, adjuvants and conditioners are added. Also other materials like adhesives and stickers to improve the adhesive power, deposit builders to build the spray deposits, deflocculators or protective colloids like gums, casein for stabilising the suspensions of solid particles may be added. Wetting agents and emulsifiers may also be added. A good wetting agent is often not a good emulsifier. Good detergents are frequently efficient emulsifiers but not very good wetting agents. Spreading does not occur without wettability but efficient wetting alone, does not ensure good spreading.

Wetting and spreading: Wetting of a solid occurs when a liquid comes in direct and persistent contact with the surface of the solid and no layer of air separates solid from the liquid. Spreading of a liquid on a solid surface is defined as the ability of the liquid to form a persistent liquid-solid interface, solely by the surface activity over the solid surface. Obviously wetting must occur before spreading can take place. In general this is the only distinction between wetting and spreading.

The smaller the contact angle, the greater is the wetting when used on the same surface and with the same interfacial tension between the liquid and the leaf. A drop of one liquid is said to spread on the surface of another liquid when it thins spontaneously and completely until it forms a layer of few molecules thick. In the case of solids, a moderate increase in the area covered by the drop is referred to as spreading. Dispersing agents or stabilisers are frequently employed to maintain particles of an inscticide in suspensions. Freundlich (1926) applied adhesion tension for expressing spreading ability. Adhesion tension is the product of surface tension and the cosine of the contact angle.

Emulsification: To reduce oil deposits and thus avoid plant injury, petroleum and similar oils are diluted with water. As water is not miscible with oil directly, various processes of emulsification have been developed to disperse the oil uniformly in water. As each oil droplet is surrounded by a more or less tough film of emulsifier, the tendency of droplets to coalesce and separate into a continuous layer of oil is resisted.

Two types of emulsions exist: (i) Oil in water or OW type; water is external or continuous phase and oil is the internal or disperse phase, (ii) Water in oil or WO type; Oil is external and water is internal. Because an emulsion can be diluted only by adding more of the external phase, the OW type of emulsion is the one almost universally employed for insecticidal purposes.

Some conditions which facilitate the formation of emulsion are— (i) the less the relative volume of the disperse phase and the slower that additions of it are made, the easier is emulsification, (ii) increasing the concentration of emulsifiers within certain limits make emulsification easier, and (iii) emulsification is favoured by higher temperatures. When an emulsion breaks, forming separate layers of oil and water, it is practically worthless as an insecticide. On the other hand when an emulsion of the OW type creams, it may be stirred to redisperse the oil. Creaming is due to difference in the specific gravity of the two phases and does not involve separation of the oil into a continuous layer. Petroleum oil emulsion creams towards the top but the tar oil emulsions cream downward.

The properties of oil as well as the characteristics of emulsifiers affect the emulsion. The specific gravity of the oil is one of the factors in determining the type of the emulsion formed with casein oil in water (OW) type. Casein emulsions of petroleum distillates are more stable, as they are of lower specific gravities. The following three methods are commonly employed to distinguish the OW or the WO type emulsions: (i) The indicator method in which an indicator is soluble in one phase and insoluble in the other phase, is used to distinguish the type of emulsion. Sudan III is soluble in oil and potassium permanganate is soluble in water phase, (ii) The drop test, in which a drop of emulsion of the OW type and not of the WO type will mix with water, and (iii) Electrical conductivity test, based on the conduction of an electric current when water is the continuous phase. Many theories have been advanced to explain the mechanism of emulsification. They are Donnan and Pett's theory, Bancroft's ease of wetting theory and the oriented wedge theory of emulsification.

Emulsifiers: Soaps, sour milk, glue and dry skim milk which are colloidal substances have been used as emulsifiers. Alkali (sodium and potassium) soaps are the most commonly used emulsifiers. Soaps form tough interfacial films and hence their emulsion is relatively stable. Under some conditions they protect the plant surface from injury by the pesticide. Part of the soap used in emulsifying a petroleum oil is broken down by hydrolysis, the fatty acid thus set free being dissolved in oil and alkali in the water. Interfacial films are formed from the unhydrolysed soap. The larger the droplets, the less soap films are required for a given volume of emulsion. Whale-oil soap, cresol, phthalic glycerol, alkylresin, sour milk, dry skim milk, egg, flour, glue, various gums and lecithin products have been used as emulsifiers. However the use of a particular material depends on cost. Though clays were tried, they are not suitable for the preparation of spray emulsion which contain free fatty acids, fixed oils and phenols because the WO type of emulsion is liable to be formed. When finely divided solids are used as emulsifiers, the solid must be wetted by the external phase of the proposed emulsion before the other phase is added. Samples of bentonite from different locations revealed significant differences as regards emulsifying capacity. Bordeaux mixture has also been used as emulsifier.

Deposit builders: The addition of a spreader often increases the drain of spray materials from the foliage, but the remaining insecticide forms a more uniform film type of deposit. Although adhesion becomes more effective as it is concentrated during evaporation of the spray fluid, much drain off is prevented by some adhesion. The sticker may act indirectly by causing flocculation of the spray particles. Additives which increase the initial deposit appear to be of a type such as gelatin, glue and casein in preparations which increase the viscosity of the spray.

Adhesives: Where persistence of deposit or residual effectiveness is required, adhesives should replace spreaders. Spreaders usually promote removal of deposits from the foliage during rain. Several distinct mechanisms operate for different classes of adhesive materials which are added to water base suspensions. Proteins and similar colloids tend to adhere and if adsorbed on the particles of insecticides, will promote adhesion. Inorganic materials having hydrophilic properties are adhesives. Oil, if it wets both solid particles and the sprayed surface, will retain the solid mechanically and resist the removal of the water phase. On waxy surfaces, oils tend to soften the wax which then aids in the adhesion process.

Drying oils, plastics and gums form a mechanical film and these bind the particles to the surface. Wamper and Haskins suggested pH of spray suspension must be properly adjusted to permit wetting. An inverted spray mixture is a solid particle like lead arsenate or cryolite initially wetted by water, later become wetted by oil either in the spray tank or upon leaving the spray gun or at impact with the sprayed surfaces. With an inverted mixture it is possible to build up deposits by continued spraying whereas with ordinary mixtures, continued spraying would result in loss by drain off.

Materials of adhesive value include a wide variety of substances like mollasses, sugar, dextrin, starch, wheat flour, soybean flour, gum, tragaconth, gum arabic, gelatin, casein, milk (fresh and dried), egg powder, dried blood, blood albumin, glue, tannins, rosin, soaps, linseed oil, fish oil, saponins, Irish moss, cactus extract, hydrated ferric oxide, bentonite with casein, and aluminium hydroxide gels. Some materials such as lead arsenate are inherently tenacious to most plant specics. Wheat flour was one of the earliest adhesives used extensively. Particulate insecticides such as arsenicals are liable for rapid removal from foliage by wind and other weather factors, unless adhesives are added. Soluble poisons such as nicotines are less rapidly washed from foliage when applied with proper adhesive. Some adhesives require a considerable drying period to set them properly before they are resistant to rainfall. However, the use of large quantities of adhesives complicate residue removal from the crop at harvest.

Tenacity which means the resistance of a spray deposit to removal by weathering, has been used as a criteria of adhesion. Its measurement is based upon the spray residue remaining on a given surface for a given period, after exposure under standardised conditions of artificial rain or sun's heat and light and the dissolving action of fog and dew, the brushing and scraping action of winds, as well as rain. The action of rain is probably most important in most parts of the country. Weber *et al.* studied the effect of different adhesives on the retention of arsenic on apple leaves sprayed with lead arsenate and found summer oil emulsion, bentonite-sulphur and milk to be most effective stickers. Dry skim milk proved to be the most effective of the stickers tested in the deposition and retention of sulphur.

Size of liquid particles: The size of solid particles either in dust or in spray suspensions bears an important relation to such properties as adhesion, suspensibility, and insecticidal efficiency. Similarly the size of oil particles in an emulsion determines the deposit of oil and the effectiveness of the preparation. Small spray droplets provide better control of small insects and spider mites than of larger insects. In any spray pattern there are more smaller sized droplets than larger ones but the greater volume of spray is accounted for by the larger drops. Wetting agents in water based sprays reduce the droplet size from 30 to 50 liquid drops increase their diameter when they strike a surface, those containing water up to 30 per cent, oil droplets up to 10 to 15 times their original diameter, depending upon the surface.

Emulsive Concentrates

The emulsive concentrates upon dilution with water give stable emulsions suitable for spraying plants and surfaces. Two kinds of pesticide emulsive concentrates are known: (i) emulsive concentrates prepared by dispersing an aqueous solution of the pesticides in a water immiscible solvent by means of a colloidal mill. These concentrates are more properly called "concentrated emulsions". Highly dispersed concentrated emulsions can be obtained that are stable in storage, by mechanical dispersion. Homogenised concentrated emulsions withstand low temperatures and prolonged storage without substantial changes. Sometimes concentrated emulsions thicken in storage but after stirring they return to their normal condition, (ii) emulsive concentrates of pesticides known under the name of miscible oils— when the miscible oils are mixed with water, they yield stable emulsions. Concentrates of this type consist of a pesticide, ' solvent and emulsifier. For liquid pesticides, the use of a solvent is not obligatory although the use of organic solvents often permits decreasing the content of expensive emulsifiers in the concentrates.

Hydrocarbons, halogenated derivatives of hydrocarbons, esters, various petroleum products, creolin, coal tar oils and many other compounds may be used as solvents. Emulsifiers are calcium sulfonates, esters of polyethylene and polypropylene glycols, monoesters of sorbitol and mannitol with higher fatty acids, various soaps, and salts of naphthalenic acids. Particularly good results are obtained by the use of mixtures of two or more emulsifiers, one of which is either polyethylene glycol and the other a calcium or ammonium alkyl aryl sulfonate prepared from higher alcohol. A favourable property of calcium alkyl benzene sulfonates is their solubility in organic solvents in contrast to the corresponding alkali metal salts which are relatively poorly soluble in organic solvents, and also crystallise out quickly on standing from emulsive concentrates.

It is convenient to have concentrates that are not too viscous. However for highly toxic organophosphorus compounds, it is less dangerous to use viscous concentrates because if they get on the skin, they are more difficult to absorb and can easily be washed off with water. There is wide use of the so called "inverted emulsion–" that are also prepared by dilution of the appropriate concentrates with water. Enough water is added so that an emulsion of the water-in-oil type is obtained. These concentrates are used for low volume spraying since the presence of an oily film hinders evaporation of the drops. To decrease the evaporation of drops in aerial spraying, special additives like LOVO (butylamine sterate) are introduced. In practical application of sprays, many of the insects' such as houseflies, mosquitoes, gnats are surrounded by spray mist in flight. Conditions for depositing insecticide on the insect are entirely different than when the insects are resting on the water or floors. Any increase in flight activity among insects being sprayed, results in greater pick-up of spray and in consequent mortality, a good argument for incorporating an irritant such as pyrethrins in sprays containing a slow acting insecticide such as DDT. It was found that when houseflies are in flight in a spray mist, three to four times as much insecticide is collected on the wings than on the rest of the body because the wing tips moves far more rapidly during flight than the rest of the insect.

As the pressure of spraying increases, the droplet size, impingement per unit of surface area, time for spraying and mortality of insects are all reduced. As the proportion of droplets below 17 μ in diameter increases, the number impinging on a test surface is less, possibly because the spray follows the air stream around the object. The size of the mist particles is determined initially by the characteristics of the sprayer and the properties of the spray liquid such as viscosity and surface tension. Droplets of oil-base sprays are sometimes measured by collecting the air floated oil particles, on liquid soap which prevents them from coalescing or evaporating. The droplets retain their spherical shape as long as the soap remains in liquid.

Insecticidal Aerosols

Aerosol is one of the new methods in which pesticide is being used in agriculture. An aerosol is a system of particles of ultramicro-scopic size dispersed in a gas. Insecticidal aerosols are produced by either incomplete combustion of tobacco or other organic material containing an insecticide or by spraying a solution of an insecticide such as DDT in fuel oil on a heated surface or by spraying a solution of an insecticide from a container of liquified gas. An insecticide of low volatility remains suspended in the air for a long period if nuclei containing stable smoke particles are provided, upon which the insecticide may condense in minute quantities. Even from early times smoke from smouldering tobacco or pyrethrum has been employed for the control of certain greenhouse and household pests.

When solutions of an insecticide such as rotenone or pyrethrum in petroleum solvents of low volatility are sprayed on a hot surface (say at 375°C), dense clouds of white fog containing insecticide are produced. For efficient formation, there must be good air circulation because without dilutions of the smoke with air, the particles will grow and coalesce. The propellent—insecticide mixture of the liquified gas-type aerosols is usually discharged through an orifice of 0. 013 to 0. 024 inches in diameter. The liquified gas evaporates almost immediately leaving the insecticidal particles suspended in the air. As liquids other than the propellent are relatively non-volatile, the spray particles do not diminish in size. Particles in an aerosol must range from 0.5 to 50 μ in diameter. Particles smaller than 0. 5 μ are not of practical value as lethal dose is collected on a insect very slowly. The most effective range appears to be between 10 to 20 μ. The particle size in liquified gas-type aerosol mist is controlled by the proportion and nature of nonvolatile components. Increasing the velocity of a particle aids it in passing through the stream line layer of air around the insect. Small, slow moving particles are not easily deposited. Very fine particles may remain airborne for periods up to two to three hours.

Aerosols fogs and smoke are normally produced by a toxic aerosol released from a pressurised container or bomb, by means of a propellent gas or by a heat induced volatilisation of a particular insecticide in a suitable petroleum or other solvent. A fog is composed of liquid particles and a smoke is composed of solid particles. If the original insecticide is a solid, aerosol generation through the mediation of a solvent system, will result in a fog if at least one of the solvents is nonvolatile, but a smoke if all are volatile. Although aerosols contain a preponderance of droplets too fine to deposit well on foliage, there are other uses for the type of application in insect and mite control. Many of the resulting small droplets are fine enough to remain suspended in the air for a long time and insects flying through the mist may readily pick up sufficient chemical to be killed. The application of nonvolatile insecticide to flying insects therefore becomes a problem of atomising into droplets, sufficiently numerous, that the victims cannot avoid contracting a lethal dose. The most effective particle size should be between 10 and 30 μ in diameter.

The simplest method of producing pesticide aerosols is burning in special pots, paper and combustible porous materials that have been impregnated with insecticides, fungicides and bactericides. Besides the pesticides, the combustible smoke forming compositions usually contain a combustible substance, filler and an oxidising agent. Nitrates, nitrites, chlorates, chromates, bichromates, persulphates and mixtures of them are used as oxidising agents. Fillers, which passivate combustion include kaolin, infusorial earth. To maintain the necessary temperature, sawdust, waste products from cellulose manufacture, carbon bitumen and other tars are added.

A BHC aerosol contains potassium chlorate 19%, urotropine 8%, diatomite 9% and BHC 64%. Aerosols obtained by atomising solutions of insecticides in volatile solvents are recommended for the control of flies and other flying insects in enclosed premises. Such solutions are placed in metal aerosol cylinder equipped with atomising device. With the aid of carbon dioxide or a low boiling solvent (freon, methyl chloride) pressure is created in the cylinder that facilitates good atomisation of the preparation. When the solution is sprayed into the air, the solvent will evaporate and the pesticide will remain in the atmosphere in a finely dispersed condition. The size of pesticide particles is associated with the size of drops of the solution and depending on the pressure in the cylinder, and construction of the atomising device, it may vary within wide limits.

The use of low volume finely dispersed spray is effective both in sheltered ground and for spraying large bodies of forest and fields with a small expenditure on the preparation. Solvents used for low volume finely dispersed spraying, are petroleum products with a relatively high boiling point. To increase the solubility of pesticide in them, intermediate solvents are added (xylene, polymethylnaphthalene). Usually with higher temperatures, the loss is greater and hence strict observance of temperature conditions are necessary.

Fumigants

Insect fumigants must be sufficiently volatile to produce a toxic concentration of vapour in closed space. Fumigant concentration is usually expressed in weight per volume, for example as milligrams of fumigant per litre, or pounds per 1000 c. ft of fumigated space. The vaporisation of fumigant chemicals seldom approach the point of saturation. The distribution of a fumigant vapour should be uniform throughout the treated surface. Diffusion is faster with gases having lower molecular weights. Adsorption and leakage as well as settling and actual layering of the initially heavy vapours of most fumigants interfere with diffusion to such an extent, that artificial means of stirring the gas mixture are usually employed profitably in a space to be fumigated.

Molecular diffusion must be relied upon to bring about gas penetration of furniture, food and other treated goods. Opposing, the process of diffusion is the absorptive capacity of the product. Hence the amount present in the surrounding atmosphere is the net of the two opposing factors.

Amongst the hazards in handling insect fumigants, two should be stressed because of their insidious nature which may result in unwitting accidents like absorption by skin and inflammability of the fumigant-air mixtures. Substances absorbed through the skin include carbon disulphide, carbon tetrachloride, ethylene dichloride, chlorine, iodine, phenol, formaldehyde, hydrogen sulphide and nicotine. Hydrocyanic acid, ammonia and carbon dioxide penetrate the skin readily. Insect control by fumigation is practised in a number of fields like building fumigation, product fumigation, crop fumigation and soil fumigation. Building fumigation is concerned chiefly with the killing of stored food and fabric pests in infested houses, storage warehouses, factories and transportation facilities. Product fumigation involves the fumigation of such commodities as grains, tobacco flour and other ground cereals. A limiting factor in product fumigation is the difficulty of penetration throughout the material so as to ensure complete disinfection. Crop fumigation concerns chiefly, the gassing of citrus trees enclosed in tents with hydrocyanic acid. Soil fumigation is used for killing of soil infesting insects, nematodes, root and fungi.

INSECTICIDES

CHAPTER IV

INORGANIC INSECTICIDES

ARSENICAL COMPOUNDS

The Greek physician Diszcorides has mentioned about the toxic properties of arsenic even as early as 40 to 90 A.D. The Chinese employed arsenic sulphide, Orpiment (As_2S_3) and Realgar (As_2S_2) for the control of insects on garden plants during A.D. 900's. All the arsenical insecticides act as stomach poisons and are used particularly for the control of chewing insects. All arsenic compounds are highly toxic to all forms of human and animal life. Even some of the organoarsenic compounds that are relatively low in toxicity to warm blooded animals are converted to inorganic compounds inside the animals, in the soil, or in plants and thus become much more toxic. Continuous application of arsenic compounds result in a build up of the compounds in soils to toxic levels. Later these enter the plants. Cases of lungs cancer, liver and other organs in employee who have worked with arsenic compounds for 10 or more years have been recorded. In view of these defects, these compounds have to be used with certain caution especially on food crops. The arsenic compounds have been even banned in some countries like German Democratic Republic and German Federal Republic. Only organic compounds of arsenic at low dosages are showing promise as seed disinfectants and also for growing plants.

Arsenic exists in two states, viz. trivalent and pentavalent and as such forms two oxides, the trioxide As_2O_3 and pentoxide As_2O_5 which in solution become arsenous acid H_3AsO_3 and arsenic acid H_2AsO_4 respectively. In combination with bases, they form two corresponding series of salts, the Arsenites from arsenous acid and Arsenates from arsenic acid. Arsenates form a white precipitate with magnesia mixture (110 g of magnesium chloride dissolved in 700 ml of ammonia of specific gravity 0.90 and diluted to 2 litres) while the arsenites do not form a precipitate. This test could be used to distinguish both the forms.

The insecticidal activity of an arsenical is usually related to the percentage of metallic arsenic, though other metals like lead may also contribute to the insecticidal activity. The phytotoxicity of arsenical is directly related to the percentage of water soluble arsenic that can enter the living tissues. The ideal arsenic compound would be one having a high arsenic content, none of which is water soluble but all of which is soluble in the digestive fluids of insect gut. The arsenites (trivalent) are less stable but more toxic to plants and insects while the arsenates (pentavalent) though less insecticidal are more stable and safer to use.

Arsenic Trioxide (White arsenic, arsenic, As_2O_3)

Arsenic trioxide is a white solid and can exist in three forms— amorphous, octahedral and rhombic. Only octahedral form is stable under ordinary conditions. When heated from 125 to 150°C, it sublimes. This compound is not easily wetted by water but when heated goes into solution quite readily. It goes into solution more rapidly in strong acidic conditions and also in alkaline conditions and as such dilute solution of arsenic trioxide has to be maintained slightly acidic if deterioration has to be prevented. Three series of arsenites—orthoarsenites K_3AsO_3, pyroarsenites $K_4As_2O_5$ and metaarsenites $KAsO_2$—are derived from the corresponding arsenic acids. The pyroarsenites are formed by removal of one molecule of water from the orthoarsenites, while metaarsenites are formed by reacting arsenious oxide with alkali hydroxides or carbonates or bicarbonates.

As mining of arseniferous ores for arsenic production alone is not profitable, it is recovered from flue dust in the copper and lead ore smelters, during roasting of mineral ores. When the flue dust is resublimed, arsenic of 99 per cent purity is usually obtained. Arsenites are more soluble than arsenates and also are more poisonous 'to plant and animal life. Hence, the arsenites are being replaced to a larger extent.'

Arsenic Pentoxide (Arsenic acid, As_2O_5)

This is a white amorphous solid formed by the oxidation of arsenious oxide with nitric acid It is soluble in water and forms arsenic acid H_3AsO_4, which is more correctly called orthoarsenic acid. Arsenic pentoxide combines with alkalies to form arsenates. Due to its high toxicity to plants, arsenic acids are not used as insecticides. However they are used for the preparation of arsenates such as lead arsenates, and calcium arsenates.

Arsenites

Number of arsenites have been used as insecticides of which Paris green, London purple, sodium arsenite ($NaAsO_2$), calcium arsenite $\{Ca(AsO_2)_2\}$, zinc arsenite $\{Zn\ (AsO_2)_2\}$, Scheele's green ($CuHAsO_3$ acid copper arsenite), lead arsenite and ferrous arsenite have been used widely.

Paris green (Emerald green, French green, Mitis green, Schweenfurter green): Paris green is a complex compound of copper metaarsenite and copper acetate (the ratio being 3:1 to 2:1). Acetometaarsenite was first arsenical to be used in large quantities. This compound theoretically should contain an equivalent of 58. 55 per cent arsenic trioxide but commercial samples always contain only 54-57 per cent. Avery found that the ratio between the meta-arsenite and organic portion become less with increasing molecular weight of acids. For example in the butyric acid compound, there were approximately two molecules of copper metaarsenite to one of the copper butyrate. Certain vegetable and animal oils may be used to prepare similar homologues.

Paris green is manufactured by different methods, of which the slow process is commonly followed. In this, copper sulphate, sodium carbonate, white arsenic

(As_2O_3) and acetic acid are reacted as follows:

$$3As_2O_3 + 4CuCO_3 + 2CH_3COOH \rightarrow (CH_3COO)_2Cu.\ 3Cu(AsO_2)_2 + 4CO_2 + H_2O$$

In the other method, sodium arsenite is reacted with sodium acetate and copper sulphate to form Paris green.

$$6NaAsO_2 + 2CH_3COONa + 4CuSO_4 \rightarrow (CH_3COO)_2Cu.\ 3Cu(AsO_2)_2 + 4Na_2SO_4$$

Paris green forms a heavy powder settling in water unless well agitated. Since it is not very adhesive, it is removed rapidly from foliage by weathering. It is gradually decomposed with the liberation of soluble arsenic by continuous percolation with water. It breaks down rather readily even in the presence of water alone due to hydrolysis, forming soluble arsenic. Carbon dioxide and ammonia also decompose the Paris green.

The presence of relatively large amounts of soluble arsenic precludes its use on arsenic sensitive plants. Severe burning of foliage caused by the relatively large amounts of soluble arsenic, its poor adhesive qualities, and its property of settling down rapidly from suspensions have reduced the utility of this compound. Bordeaux mixture may be safely mixed with Paris green whereas lime-sulphur cannot be mixed, as the compounds formed are injurious to plants. It appears that the aquatic flora may convert arsenic to ethylarsine and thus prevent its accumulation in water, although cases of human poisoning from its use have been reported.

London purple (Paris purple, English purple): A purple residue from the manufacture of dyestuffs called London purple, was found to have insecticidal value due to its high arsenic content. Chemically, London purple is a mixture of calcium arsenite and calcium arsenate with considerable quantities of dyestuff and other materials. As the materials prepared have a variable composition, it is difficult to use them due to presence of soluble arsenic, which may cause foliage injury. Hence, its usage has declined.

Other arsenites: Sodium and calcium arsenites are white solid substances practically insoluble in organic solvents but soluble partly in water. In both cases the technical material is mainly metaarsenite with admixture of orthoarsenite. Both arsenites are produced by reacting white arsenic with the corresponding hydroxides or carbonates.

$$AS_2O_3 + Na_2CO_3 \rightarrow 2NaAsO_2 + CO_2$$

$$As_2O_3 + Ca(OH)_2 \rightarrow Ca(AsO_2)_2 + H_2O$$

Sodium arsenite is used for the control of locusts and scabies on domestic animals and also as weedkillers and as fungicides.

Arsenates

Arsenates are generally less active as insecticides than arsenites due to their low solubility. Similar to arsenites, the arsenates form three types of compounds, orthoarsenates, metaarsenates and pyroarsenates—of which orthoarsenates are the important ones. By the replacement of metal ions by one or more of the hydrogen ions or by hydroxyl groups, either neutral or acid or basic salts of arsenic acid are formed. The formation of each salt depends upon conditions such as molar concentrations, proportions between various compounds and upon the temperature maintained during preparation.

Number of arsenates have been prepared for insecticidal use of which those containing lead or calcium have been widely used. Though these arsenates are less active as insecticides than the arsenites, they are safer to use on living plants because of their low solubility.

Lead arsenates: Lead arsenates have been widely used as an insecticide for number of years. This compound is produced in a light, flocculent form, easily dispersible in water. It forms a relatively uniform deposit on plant surfaces and exhibits a high tenacity. It is also not so readily decomposed as many other arsenicals and it causes much less injury to the plants than other arsenicals. Because of the above properties, it is considered an ideal arsenical insecticide. However larger quantities, of this arsenical insecticide have to be applied since this contains less amount of arsenic than the calcium arsenates. Further the cost is likely to be more than calcium arsenates as lead is costlier than calcium. As both lead and arsenic are poisonous to all forms of life, the lead arsenates are potentially more toxic to all forms of life than the calcium arsenates. It is a cummulative poison as it possibly accumulates and is deposited in the epiphysis of the bones.

Depending upon the conditions of preparation, number of lead arsenates like acid lead arsenate ($PbHAsO_4$), basic lead arsenates $\{Pb_4(PbOH)(AsO_4)_3\}$, monolead arsenates $\{PbH_4(AsO_4)_2\}$, trilead orthoarsenates $\{Pb_3\ (As_3O_4)_2\}$, lead metaarsenate $\{Pb(AsO_3)_2\}$, lead pyroarsenates ($PbH_2As_2O_7$ and $PbAsO_7$) have been prepared. Of these the first two are widely used.

Acid lead arsenate (Dilead orthoarsenate, dibasic lead arsenate, diplumbic hydrogen arsenate, diplumbic arsenate, lead hydrogen arsenate, $PbHAs_4$): The acid lead arsenate is usually prepared by reacting soluble lead salts with disodium arsenates (Na_3AsO_4) at low temperature. Commercially, this is prepared from arsenic pentoxide and litharge (PbO). At equilibrium, the product consists principally of dilead arsenate (acid arsenate). With increase in the temperature of reaction, increasing quantity of basic arsenate is formed. To improve the suspensibility of lead arsenate, the bulkiness of the product is increased from a rather compact material (40 cubic inch per pound) to relatively fluffy product.

Pure dilead arsenate is very sparingly soluble in water (0. 002 g As_2O_5 per 100 cc of water after six hours of boiling). Acid lead arsenate which contains 33 per cent As_2O_5 is stable under acid conditions (pH 2 to 5) but tends to change to basic salt containing 23 per cent As_2O_5 under neutral or alkaline conditions. As far as hydrolysis is concerned, acid lead arsenate is stable only in a definite pH range. Outside this range, it will hydrolyse to produce either soluble lead or arsenic compounds depending upon the pH of the solutions. Acid lead arsenates will remain in equilibrium with water if the water is somewhat acidic, like CO_2—dissolved water or water containing small amounts of arsenic acid set free by a slight hydrolysis. When dilead arsenate is heated, it is converted into pyro-arsenates at about 280°C.

Though very slightly soluble in water, soluble arsenic is sometimes formed possibly by hydrolysis, on plant surfaces even to the extent of causing injury to the plant.

$$5\ PbHAsO_4 + HOH \rightarrow Pb_4(PbOH)\ (AsO_4)_3 + 2H_3AsO_4$$

This reaction leading to the formation of basic lead arsenate reaches equilibrium when the concentration of arsenic compound is very low but may proceed to completion if the soluble products are removed.

When acid lead arsenate is applied with other pesticides like lime sulphur, double decomposition may take place readily releasing arsenic acid which may cause serious plant injury.

$$2\ PbHAsO_4 + 2\ H_2S \rightarrow 2\ PbS + As_2O_5 + 3H_2O$$

An excess of lime and other correctives such as proteinaceous materials can be used to reduce the injury.

Lead arsenates are decomposed by the action of dissolved salts in water used in the application of sprays. As the pH of dew on foliage of plants is usually around 5 to 6, the hydrolysis of lead arsenate deposited on plant surfaces by dew and rain is not an important cause of arsenical injury as the soluble arsenic liberated is very small. However, the presence of small quantities of sodium chloride in the atmosphere as in the coastal areas, may. be sufficient to cause decomposition of the deposited lead arsenate, which may be one of the reasons for high incidence of arsenical injury in coastal areas.

The composition of waters used for spray modified the effectiveness of lead arsenates sprays. Waters with hardness caused by sulphates, carbonates and bicarbonates caused rather large release of soluble arsenic from lead arsenate.

Basic lead arsenates: {$Pb_4(PbOH)\ (AsO_4)_3$}: The basic lead arsenates {lead hydroxy arsenates, $Pb_4(PbOH)(AsO_4)_3$ and $Pb_5\ (PbOH)(AsO_4)_4$} are stable in the pH range between 6.5 and 11, below which they are apparently changed to dilead arsenate. Several compounds exist in commercial preparation of basic lead arsenates as there seems to be a whole series of such compounds possible.

The composition of the compound depends upon the concentration and proportions of the reactants used in its preparation. Basic lead arsenate of commerce has approximately the composition $Pb_4\ (PbOH)(AsO_4)_3.H_2O$.

The basic lead arsenate is prepared by slowly adding arsenic acid (H_3AsO_4) to litharage (PbO) with agitation. Small amounts of nitric acid or acetic acid is added to hasten the reaction. The lead arsenate formed is precipitated in an extremely finely divided form. The basic lead arsenate of commerce is safer to use on tender foliage than the acid form although the former has lower arsenate content, less desirable physical properties and lower toxicity than the later form.

The other normal arsenates like monolead orthoarsenates $PbH_4 (AsO_4)_2$, trilead arsenate $Pb_3(As_3O_4)_2$ lead pyroarsenates $PbH_2As_2O_7$ and Pb_2AsO_7, though toxic to insects are not widely used due to their unstable nature.

Calcium arsenates: As lead salts were much more expensive and also scarce during World War I, use of calcium arsenate increased rapidly for the control of potato beetle and similar pests in spite of certain of its undesirable properties like relative instability, serious phytotoxicity that was encountered during the early days (1901). A whole series of calcium arsenates comparable to that of lead arsenates could be produced. However, the solubility of corresponding calcium salts is much more than that of lead salts and also the cost lower because of the higher arsenic content of calcium arsenates.

When calcium hydroxide is allowed to react with arsenic pentoxide (As_2O_5) in solution, several calcium arsenates are formed. Commercial calcium arsenate, a safe preparation produced in 1915 is a mixture of tricalcium arsenate $\{(Ca_3(AsO_4)_2)$ with 37.5% arsenic, calcium arsenate ($CaHAsO_4$), excess of lime and calcium carbonate. These tend to decompose similar to lead arsenate. A safe form of basic calcium arsenate $\{Ca_3(As0_4)_2\}_3Ca(OH)_2$ with 25% metallic arsenic and 0.4 to 0.5% water soluble arsenic has also been prepared. These are of uniform composition and low solubility. As a result, their effectiveness is too low for most effective use on resistant crops. Chemical factors overshadowed the influence of particle size on the solubility of the compounds. Control of temperature and concentration is important in the manufacture of calcium arsenates. Special calcium arsenates containing large particles and a high percentage of CO_2—water soluble arsenic pentoxide have been prepared commercially for use on cotton which is resistant to arsenical injury.

Calcium arsenate has been prepared by a number of methods. One of them is by reacting the water soluble salts of arsenic acid (sodium arsenate) with lime to produce calcium arsenate.

$$As_2O_3 + 6NaOH \rightarrow 2Na_3AsO_3 + 3H_2O$$

$$2Na_3AsO_3 + O_2 \xrightarrow{CuSO_4} 2Na_3AsO_4 \text{ (Sodium arsenate)}$$

$$2Na_3AsO_4 + 4Ca(OH)_2 \rightarrow Ca_3(AsO_4)Ca(OH)_2 + 6NaOH$$

Also calcium arsenate is produced by reacting arsenic acid with lime in aqueous medium of semidry condition (mixing concentrated solution of arsenic acid with lime).

$$2H_3AsO_4 + 4Ca(OH)_2 \rightarrow Ca_3(AsO_4)_2.Ca(OH)_2 + 6H_2O$$

Calcium hydroxide is obtained by slaking calcium oxide with water. An excess of calcium hydroxide is usually present in the reaction mixture and also acetic acid is added to increase the reaction rate. Depending upon the reaction temperature, commercial arsenates contain $CaHAsO_4.Ca_3(AsO_4)_2$, $Ca_5H_2(AsO_4)_4$,

$\{Ca_3(AsO_4)_2\}_3Ca(OH)_2.CaCO_3$ and some unreacted $Ca(OH)_2$. In safe calcium arsenates, the basic arsenates predominates.

Calcium arsenate is a white sparingly water soluble substance. The technical grade product contains a mixture of salts of arsenic acid in which the basic salt of the composition $\{Ca_3(AsO_4)_2\}_3$. $Ca(OH)_2$ predominates. The As_2O_5 content in it is about 40% and the arsenious acid content is not more than 1% and with less than 0. 6% of the water soluble arsenic compound. The phytocidal effect is diminished by the addition of zinc sulphate or oxide.

Calcium arsenates are decomposed much more easily than lead arsenates after application to plant surfaces. Hydrolysis of the calcium arsenates takes place in aqueous suspensions. This reaction

$$CaHAsO_4 + 2H_2O \rightarrow Ca(OH)_2 + H_3AsO_4$$

may continue on the plant surface till the aqueous phase completely evaporates. The nature of reaction proceeding during the drying period of the spray film has not been thoroughly studied.

The presence of excess calcium hydroxide in the "–safe–" formulation apparently prevents the formation of soluble arsenic. However, when calcium hydroxide is changed to carbonate by the action of atmospheric carbon dioxide, soluble arsenic is liberated. S,o the newer safe commercial calcium arsenates are slower in reaction and also slow in releasing soluble arsenic, which possibly accounts for their "–safeness–". The nature of other materials in a spray also influences to a greater or lesser extent the nature of final product. In view of the lower sticking ability of calcium arsenates as compared to lead arsenates, large quantities of calcium arsenates have to be used to obtain the same effect. A combination of calcium arsenate with zinc sulphate and hydrated lime (4:1:2) gave good deposit with low plant injury. The mechanism of action of zinc sulphate in the mixture is not understood.

Comparison of the properties of lead arsenate and calcium arsenate

	Lead arsenate	**Calcium arsenate**
1.	An acid material	A basic material
2.	A single compound	A variable mixture of compounds
3.	Bases release soluble arsenic	Bases react but very slightly
4.	Weak acids decrease soluble arsenic	Acids produce soluble arsenic
5.	Carbonates release soluble arsenic	Carbonates release soluble arsenic
6.	Hard water releases soluble arsenic	Excess lime in the product protects it against action of hard water
7.	Carbon dioxide has little effect	Carbon dioxide releases soluble arsenic
8.	Resistant to weathering	Easily weathered
9.	Reacts with lime sulphur	Compatible with lime sulphur
10.	LD_{50} ca 800	LD_{50} ca 40 to 100

Miscellaneous arsenates: Numerous other arsenates like magnesium arsenates, zinc arsenates, aluminum arsenates, iron arsenates, manganese arsenates, basic copper arsenates have been prepared. Among the four magnesium arsenates known—monomagnesium orthoarsenate $MgH_4(AsO_4)_2$, secondary magnesium orthoarsenate $MgHAsO_4$ and trimagnesium orthoarsenate $Mg_3(AsO_4)_2$ and the basic magnesium orthoarsenates $Mg_3(AsO_4)_2$. MgO or $Mg_3(AsO_4)_2$. 2 H_2O, the last one (basic magnesium arsenate) is the safest probably of its basic nature.

Aluminum, iron and manganese arsenates have not been widely used. Zinc arsenates and arsenites have been used for wood impregnation because of their poor solubility in water and hence preserved from leaching out of the wood for a long time. Basic copper arsenate $Cu(CuOH)AsO_4$ is a stable crystalline substance containing 56.27% copper oxide (44.8% metallic copper), 40.6% arsenic pentoxide, and 3.2% of water of constitution. It is not hydrolysed by water and reacts only slightly with carbon dioxide. It is compatible with lime, calcium caseinate, blood albumin. It reacts slowly with liquid lime-sulphur. Basic copper arsenate is as toxic as lead arsenate.

Compatibility. Though lead arsenate is much more compatible with many of the commonly used fungicides than calcium salt, certain combinations increase plant injury.

Lime sulphur-lead arsenate mixture: When acid lead arsenate and lime sulphur solutions are mixed, the white colour of lead arsenate is changed to brown within a few seconds and the colour darkens with time. Ruther in 1913 found that there was an increase in the concentration of thiosulphate and sulphite salts in solution. As he could not detect any sulphide of arsenic, he concluded that a thioarsenate of some kind was formed. Young and others in 1930 noted an increase in soluble arsenic after mixing and the amount of soluble arsenic formed was in direct proportion to the amount of lead sulphide formed. The decomposition of compounds of lime sulphur— lead arsenate mixture was greater in dilute solutions although at concentrations of lime sulphur solutions greater than 1-50, the degree of decomposition appeared to be constant.

When lime sulphur consisting mainly of calcium polysulphides is mixed with acid lead arsenate, the polysulphides are first broken down to yield hydrogen sulphide which reacts with lead arsenate to form black lead sulphide. As a by-product of these reactions, arsenic acid that is formed, combines with the calcium salts to form calcium arsenates and thioarsenates which are more or less decomposed. Calcium thiosulphate and calcium sulphate are also formed during the course of the reaction. The whole process of reaction is difficult to follow because of the number of components.

Calcium hydroxide-lead arsenate mixture: When calcium hydroxide is added to acid lead arsenate, the amount of soluble arsenic in solution is reduced due to the formation of calcium arsenate and lead arsenate of a more basic nature, or basic arsenate and a very insoluble basic arsenate, the latter compound breaking down in the presence of carbon dioxide to form a less basic compound, and then tricalcium arsenate which may cause the arsenical injury. Possibly the soluble arsenic as such is not free to act upon the plant until all of the lime becomes carbonated. The sequence of events being, acid lead arsenate in the presence of $Ca(0H)_2$ is probably converted to calcium arsenate and lead hydroxide. Calcium arsenate which is relatively stable in the presence of an excess of $Ca(OH)_2$, releases soluble arsenic when the hydroxide becomes carbonated through the atmosphere CO_2.

Multiple mixtures: Various combinations of lead arsenates and other pesticides have been attempted for the control of both chewing and sucking insects as well as fungi. Multiple mixtures containing lead arsenate have one of the two objectives: (i) To decrease the injury of mixture to the plant through the addition of correctives as in the case of lime added to lead arsenate-lime sulphur combination, and (ii) to increase the efficiency of the mixture through the use of wetting, spreading or sticking agents. Lime when added to the mixture, reduced the amount of soluble arsenic set free, preserved the polysulphides content of the fungicide, possibly due to the following reactions:

$$3Ca(OH)_2 + 3S \rightarrow 2CaS + CaSO_3 + 3H_2O$$

$$CaS + PbHAsO_4 \rightarrow CaHAsO_4 + PbS$$

$$PbHAsO_4 + Ca(OH)_2 \rightarrow Pb(OH)_2 + CaHAsO_4$$

$$PbHAsO_4 + CaCO_3 \rightarrow PbCO_3 + CaHAsO_4$$

In the mixtures containing lime, detectable quantities of free hydrogen sulphide were found. Addition of lime however, failed to retain the sulphide—sulphur in solution at a high level as in the corresponding solution of lime sulphur alone.

Proteinaceous materials like gelatin, casein, dried milk powder have been used as correctives more from the point of their spreading, wetting and adhesion properties. The above were also found to increase the soluble arsenic content. However skimmed milk powder lowered the amount of soluble arsenic content. The order of mixing of the ingredients had considerable effect on the composition of the resulting mixture.

Manganese sulphate, ferrous sulphate, zinc oxide, ferric oxide and aluminium oxide have been used as correctives. Ferric oxide was harmless and prevented arsenical injury to apple foliage and also provided greater retention of lead arsenate. Zinc sulphate has been extremely useful as a corrective. The mode of action of these metallic salts as correctives is not well understood, even though it is believed by few that they act to produce elemental sulphur from the polysulphides present in lime sulphur and hence prevent interaction between the polysulphides and lead arsenate.

Toxicity of arsenic

Arsenites are more rapid in their action than the corresponding arsenates. Possibly pentavalent arsenic must be reduced to the trivalent form before it can exert its toxic action. However, if the poison is not lost by excretion before it can affect the tissues, the pentavalent form will presumably exert in the end as great a toxic effect as the trivalent forms.

Arsenical poisoning differs according to the plant age. Older plant foliage are more susceptible to arsenical injury than young leaves. The dew from cotton plants contains large amounts of calcium and magnesium bicarbonates whereas dew from corn grass and various weeds in cotton fields itself is not alkaline. This considerably affects the release of soluble arsenic soluble in suspensions of Paris green, zinc arsenite or calcium arsenate.

Many insects do not consume arsenicals readily. Arsenicals not only kill insects by absorption through the wall of the digestive tract after ingestion of food or water but they are capable of exerting toxic action by penetration through the body wall. Arsenic is a severe gastrointestinal irritant for mammals. Even after application to the eye or injection into the blood vessels, arsenic kills by producing inflammation of the stomach. This is exhibited in insects also. When the poison has been absorbed through the body wall, mammals eliminate arsenic chiefly in the urine and to a less extent by way of the alimentary tract, sweat, bile, milk, tears and hair. Until elimination can take place, the arsenic is stored mainly in the liver and kidneys. Sometimes as much as 10 to 25 times, the quantity of arsenic is to be found in the liver as in the other tissues.

Acute poisoning of man by arsenic is rather easily recognised but symptoms of chronic poisoning are difficult to identify. The symptoms include indefinite disturbances such as loss of appetite, collicky pain, skin eruption and sensory or nervous disorders. Treatment involves emptying the stomach and administering of BAL or 2, 3 dimercapto-proponal, a specific antidote that readily complexes and solubilises trivalent arsenic. Not only the size of a dose of arsenic but its availability for absorption by the digestive system determines the nature and severity of toxic symptoms. The amount of arsenic deposited in the tissues differs when different arsenicals are fed at the same dosage level. There is a marked species difference. Arsenic is not accumulated to nearly the extent that lead is stored. It is eliminated normally much mere rapidly than lead. Elimination from the tissues is normally completed in a few weeks after the removal of arsenic from the diet. Contact with arsenical dusts and solutions should be avoided especially if the skin is broken or irritated.

Mode of action: The arsenates produce regurgitation, turgor and quiescence in insects. The epithelial cells of the midgut of the poisoned insects are generally disintegrated with vacoulised cytoplasm and the chromatin of the nucleus clumped. Arsenical poisoning slowly decreases the oxygen consumption of the insect. This is associated with a reaction among the trivalent arsenic and the sulphahydryl (—SH) groups of glutathione and various enzymes such as pyruvate oxidase.

Arsenical insecticides are general biocides and arsenic trioxide and sodium arsenite have also been used as rodenticides and herbicides. The arsenicals are generally toxic to mammals, the LD_{50} values of arsenic trioxide being 138, calcium arsenate 40 to 100, and lead arsenate >800.

Organic Compounds of Arsenic

Though there are number of active insecticides among the organic compounds of arsenic, they are of no practical significance as their toxicity does not exceed that of inorganic arsenites and arsenates and also, as they are costlier than the inorganics. The most active compounds are those containing one hydrocarbon radical connected with the arsenic. The derivatives of dialkyl and a diaryl arsenic occupy an intermediate position. A disadvantage common to these compounds is that they irritate the upper respiratory passages and also its longer persistence since all the organic compounds of arsenic have to be converted to inorganic form. The latter may also accumulate in the soil and enter into the plants and food products.

FLUORINE INSECTICIDES

Fluorine, one of the members of the halogen family, combines readily with most of the metals and non-metals forming a wide variety of compounds like fluosilicates, fluoborates and fluoaluminates. Also similar to other members of the halogen family, it forms substituted organic compounds. As insecticides, three series of inorganic fluorine salts—fluorides such as sodium fluoride NaF, fluosilicates or silicofluorides such as sodium fluosilicate Na_2SiF_6 and fluoaluminates such as cryolite (sodium fluoaluminate) Na_3AlF_6—are important.

Soluble fluosilicates hydrolyse in pure water to produce corresponding fluorides, the reaction going to complete in the presence of alkali. Inorganic fluoride compounds in solution react with soluble calcium and magnesium compounds to produce insoluble fluoride which is ineffective as insecticide.

$$Na_2SiF_6 + 3Ca(OH)_2 \rightarrow 3CaF_2 + 2NaOH + Si(OH)_4$$

Because of these reactions, calcium compounds such as hydrated lime, calcium carbonate, hard water or alkaline preparations such as calcium arsenate, bordeaux mixture and soaps should not be mixed with fluorine compounds. Barium fluosilicate is incompatible with soluble sulphates such as nicotine sulphate, as barium sulphate is precipitated and also a soluble fluosilicate is formed. Wettable sulphurs, summer oils and lead arsenate are compatible with the fluorine compounds. Most of the fluorides and fluosilicates injure plant life but cryolite is safer than others.

Sodium Fluoride (NaF)

Sodium fluoride was patented in England as an insecticide in 1896. By treating fluorspar or other non-siliceous fluorides with sulphuric acid or other non-volatile acids, hydrofluoric acid is prepared. Some hydrofluoric acid is also produced as a by-product in the phosphate industry. From this sodium fluoride is prepared. In another method, sodium fluoride is produced by reacting sodium carbonate and a silicofluoride:

$$Na_2SiF_6 + 2Na_2CO_3 + H_2O \rightarrow 6NaF + H_2SiO_3 + 2CO_2$$

Sodium fluoride contains 45.2% of fluorine and is soluble in water up to 4%. This prohibits its use in field sprays but it is effective as a dust. Cockroaches walking through powdered sodium fluoride pick the powder on the exudations of their body, from there it may be absorbed through the integuments or be ingested by the insect during the process of cleaning itself. As pyrethrum excites cockroaches causing them to runout of their hiding places, many operators favour a 3:1 combination of sodium fluoride with pyrethrum.

Most of the other simple metallic fluorides like zinc fluoride ZnF_2, barium fluoride BaF_2, calcium fluorspar $CaSiF_6$ have not been used widely but only as a wood preservative.

The Fluosilicates (Silicofluorides)

Sodium fluosilicates (Na_2SiF_6 or 2NaF, SiF_4).

Sodium fluosilicate is a relatively dense material. So, to increase the bulk of the substance for dusting purposes, a light and extra light sodium fluosilicate has been marketed. This contains from 70 to 75% sodium fluosilicate, the balance being alumina. Ordinary sodium fluosilicate is made from hydrofluosilicic acid and soda ash according to the following reactions:

$$CaF_2 + H_2SO_4 \rightarrow 2HF + CaSO_4$$

$$6\ HF + SiO_2\ (\text{Sand}) \rightarrow H_2SiF_6 + 2H_2O$$

$$H_2SiF_6 + Na_2CO_3\ (\text{Soda ash}) \rightarrow Na_2SiF_6 + H_2O + CO_2$$

Alkalies such as sodium carbonate frequently present in commercial sodium fluosilicate as an impurity, or in certain waters used as diluents in spraying, react with it to produce sodium fluoride. As the sodium fluoride is relatively soluble, it used to cause considerable injury to plants.

$$Na_2SiF_6 + 2Na_2CO_3 + H_2O \rightarrow 6NaF + H_2SiO_3 + 2CO_2$$

Attempts to correct the same with lime was not wholly successful. Calcium and magnesia salts react with sodium fluosilicate to form first calcium or magnesium fluosilicate and then the corresponding fluoride. Possibly plant excretions also enter into the decomposition reactions. All the salts of hydrofluoric acid are incompatible with alkalies, particularly lime. When they react with lime or with calcium salts in the presence of bases, calcium fluoride which is insoluble in water and also has no insecticidal or fungicidal activity is formed.

$$Na_2SiF_6 + 3Ca(OH)_2 \rightarrow 2NaOH + 3CaF_2 + H_2SiO_3 + H_2O$$

When heated, sodium fluosilicates decompose into the fluoride salts and silicon fluoride

$$Na_2SiF_6 \rightarrow 2NaF + SiF_4$$

In spite of the high toxicity and low cost, because of the possible phytotoxic injury, sodium fluosilicates have not been widely used.

Miscellaneous fluosilicates: Number of other fluosilicates like calcium fluosilicates ($CaSiF_6.2H_2O$), magnesium fluosilicate ($MgSiF_6.6H_2O$), barium fluosilicate ($BaSiF_6$), double salts of various fluosilicates like zinc magnesium fluosilicate (Zn, Mg, SiF), cryolite ($AlF_3.3NaF$), cryolithionite ($2AlF_3.3NaF.3LiF$), chillite ($3AlF_3.5NaF$), ralstonite ($2NaF.MgF_2.6AlF_3(OH)_3.4H_2O$) and pachnolite ($AlF_3.CaF_2.NaF.H_2O$) have been tested as insecticide.

Calcium and magnesium fluosilicates have been found to be satisfactory against chewing insects. However, little work has been done because of the superiority of barium salt and of cryolite. Barium fluosilicate has low solubility and relatively high toxicity to adult insects. Foliage injury is usually not severe but the cost is higher than the cryolite. Like other fluorine compounds, barium fluoride is not compatible with nicotine, calcium arsenate etc. Common diluents used for this are talc, charcoal, clay or flour. Other fluosilicates have not been widely used. However the salts of heavy metals are all relatively more soluble in water than salts of other metals.

Fluoaluminates

Sodium fluoaluminates (Sodium aluminium fluoride, cryolite, Na_3AlF_6)

Cryolite is a naturally occurring mineral in Greenland. Sodium aluminum fluoride is also made synthetically from aluminium fluoride, ammonium fluoride and sodium chloride

$$AlF_3 + 3NH_4F + 3NaCl \rightarrow Na_3AlF_6 + 3NH_4Cl$$

The natural mineral is in the form of monoclinic crystals while the synthetic compound is amorphous powder. There is very little difference in the insecticidal efficiency of the two forms. Cryolite contains 54.2% fluorine and about 0. 06% soluble in water. The pH of cryolite ranges from 6 to 6. 5. The synthetic form is lighter and fluffier and hence preferred. Because of its low water solubility and low mammalian toxicity, it has been used extensively as a stomach poison. Cryolite although relatively insoluble in water, is soluble in dilute acids and alkalies. This may not be compatible with lime sulphur or Bordeaux mixture. With lime, cryolite forms calcium fluoride:

$$Na_3AlF_6 + 3Ca(OH)_2 \rightarrow Na_3AlO_3 + 3CaF_2 + 3H_2O$$

$$Na_3AlF_6 + 3Ca(OH)_2 \rightarrow 3NaOH + 3CaF_2 + Al(OH)_3$$

Cryolite at 0. 2% aqueous suspension has been used as a stomach insecticide for the control of chewing insects. Systematic consumption of cryolite with food at two to three mg per litre of water damages teeth considerably.

Elemental Fluorine

Elemental fluorine as a gas was tested against a variety of insects and concentrations between 100 and 1000 ppm have been found to be effective.

Insecticidal Action of Inorganic Fluorine Compounds

The relative toxicity of inorganic compounds to insects as stomach poisons compares favourably with that of inorganic arsenicals as insecticides. The fluorine compounds have little value in plant disease control. In the control of cockroaches, when cockroaches run through the dry powder the powder sticks to its legs and antennae. Some of the powder is moistened or dissolved in the exudation about the bases of the legs and on the thinner portions of the outer integuments. This seems to cause some irritation and uneasiness to the insect which begins to clean the moistened powder from the body by licking it. In doing this, thc insect may swallow enough of the poison to result in its death after a period of 5 hours to 10 days. Further absorption of sodium fluoride can take place at the softer, more flexible parts of the body wall at the junction of the head and throat or around the bases of the legs. When cockroaches walk through fluoride, death is more likely to occur from contact than from stomach poisoning. Cockroaches seemed to absorb sodium fluoride much more rapidly when they have been dusted with hydrated lime (not calcium carbonate) and even when the lime has been washed off before the fluoride was applied.

In insects, the fluoride compounds produce spasms, regurgitation, flaccid paralysis and death. Fluorine inhibits the enzymes such as enolase that require magnesium as a prosthetic group, by precipitating a complex magnesium fluorophosphate and thus preventing phosphate transfer in the oxidative metabolism. The insecticidal properties of these fluoride compounds are approximately related to the fluorine content, and the solubility in the digestive juices of the insects. Compounds with higher water solubility such as sodium fluoride and sodium fluosilicates produce severe plant injury. The rat oral LD_{50} value for fluorine insecticides are—sodium fluoride 200, sodium fluosilicate 125, cryolite 13,500 mg/kg. As such cryolite is one of the safest insecticides for mammals.

Soluble calcium salts act as antidotes in fluorine poisoning through the formation of insoluble calcium fluoride. The inorganic fluorine compounds do not possess an acute toxicity as does arsenic. When storage of fluorine occurs in the body, it is primarily in the bone and teeth causing dental fluorosis. Further, the margin between optimum and toxic levels is very narrow. However from one to three pounds of cryolite per animal consumed with food produced no ill effects in horses, cows and sheep.

CHAPTER V

NATURAL ORGANIC INSECTICIDES

Plant materials are important sources of some of the widely used insecticides. Flowers, leaves, roots or other parts are finely ground and used, or the toxic principles extracted and utilised alone or in mixtures, with other toxicants and auxiliary materials. Plant products which are used as insecticides may be classified as follows:

1. Primary toxicants such as nicotine and rotenone.
2. Essential oils and their fractions as pine oil and oil of citronella.
3. Fixed oils as cottonseed oil and fish oil.
4. Miscellaneous, as walnut shell fluor and plant gums.

Tobacco, pyrethrum, derris, hellebore, quassia, camphor and turpentine have been found to be important sources of insecticides. As these are derived from five different families of plants, botanical relationships are not of much use in identifying the potential sources. Further, at least two to four years or more are required to grow a crop of such plants and hence supplies cannot be quickly increased. As most of the insecticides of plant origin have to be imported, the supply will depend upon factors like war, political situation and other considerations. Insecticides of plant origin deteriorate quickly upon exposure to air, heat and also with time. In exploring plants for insecticidal activity, studies have to be made with various solvents, with various parts of the plants, size fractions of the plants etc., and also each one of these fractions have to be tested as stomach poison, contact insecticide and also as fumigant. A toxic principle often does not exhibit its activity until it is separated into its proper chemical form. Number of plant products which are used as insecticides, are all alkaloids (an alkaloid may be defined in general as a naturally occurring heterocyclic, optically active nitrogenous base of relatively high molecular weight and having marked physiological activity.) Of the number of alkaloids, the alkaloids from sabadilla, hellebore, ryania, strychine, atrope and datura, chinchona, delphinium, hap-lephyton, recinus and sophore have not been used widely. Only alkaloids nicotine, nornicotine and anabasine have been widely used.

Nicotine

(l-Methyl-2, 3 (pyridyl) pyrrolidine, $C_{10}H_{14}N_2$)

Nicotine was first isolated by Posselt and Reimann in 1828 and its structure determined in 1893. Nicotine occurs in at least two Solanaceous genera, Nicotiana and Duboisia as well as in Asclepias *(Asclepidaceae),* Equisetum *(Equisebaceae)* and *Lycopodiaceae,* families separated by the primary division of plants into Phanerogams and Cryptogams. Of the 50 or more species of *Nicotiana* of the family *Solanaceae,* the commercial tobacco *(N. tabacum)* alone is an economic source of nicotine. *Nicotiana rustica* also contains considerable amount of nicotine but not much work has been done with that species. The content of nicotine in leaf tobacco varies from 0. 5% in bland varieties to as high as 15% in some Algerian samples, the average ranging between 2 and 3%. Nicotine is present in all stages and in all parts of the tobacco plant, the quantity ranging from a trace in seedlings to a maximum in mature leaves. Mature leaves contained about four times as much nicotine per unit of dry weight as the stalks.

Nicotine in *N. tabacum* is extracted from the woody portions of the plant such as stems and leaf midribs unsuited for smoking or chewing tobacco. Commercially nicotine is extracted with an organic solvent such as benzene or trichloroethylene, a mixture of tobacco refuse, milk of lime and caustic soda. The extract is concentrated in vacuum and again extracted with dilute sulphuric acid, the nicotine then liberated with alkali and extracted with a mixture of ether and petroleum ether. Subsequently by distillation, pure nicotine is obtained. Ordinarily nicotine extracts contain appreciable amounts of alkaloids closely related to nicotine. Most of them are less toxic though some like anabasine and nornicotine are somewhat more toxic than nicotine, to certain insects species. As pure nicotine causes certain ill effects on the operators, nicotine is sold in the form of a 40% solution of sulphate against soft bodied insects like aphids and as contact spray.

Pure nicotine is a colourless, nearly odourless liquid, the alkaloid boiling at 247.3°C with a specific gravity of 1.0093. Naturally occurring nicotine is levorotatory although its salts are dextrorotatory.

Nicotine is soluble in water in all proportions only below 60°C and above 210°C where it forms hydrates. Under room temperature, nicotine is also soluble in alcohol, ether, petroleum oil and chloroform. Nicotine exists in tobacco as salts of such organic acids as maleic and citric acids. Nicotine is appreciably volatile at room temperature and its vapour will burn in the air. Due to the presence of two nitrogen atoms in the heterocyclic structure, nicotine is basic in reaction and readily forms salts with acids. When exposed to air and light, nicotine slowly darkens in colour to a deep brown liquid and becomes more viscous at the same time possibly due to autooxidation, the reaction at first being basic which gradually changes to acidic. Number of position isomers are possible. The α nicotine {2-(1-methyl-2-pyrrolidyl) pyridine} is much less toxic than the racemic *P*-nicotine.

Nicotine as a free base was found to be five to seven times as toxic as nicotine sulphate or hydrochloride. The efficiency of nicotine sulphate sprays increases as the alkalinity of the spray water is increased. Because most of the waters used in the preparation of spray mixture are slightly alkaline and also the soap which is used sometimes with nicotine is alkaline, possibly the nicotine sulphate reaches the insects in most cases as a free base. Of the many nicotinium salts tested, methyl nicotinium iodide, dimethylnicotinium diiodide and dodecyl nicotinium-p-tolune sulfonate were more toxic than nicotine or nicotine sulphate to the insects.

As nicotine is rapidly lost from the mixture applied to plant surfaces, various combinations of alkaloids have been tried to prolong its usefulness. A combination of nicotine and peat, and nicotine and bentonite have been tried. Among the bentonites, bentonites with high swelling properties, good flocculating characters, and with high absorptive capacity should be preferred if the nicotine-bentonite mixture is to be held tenaciously over a considerable amount of time. The nature of union between the alkaloid and carrier is not known, but possibly the base exchange properties of the materials like bentonite affect the retention of the alkaloid. Particle size of the carrier also affects the degree of retentivity.

In studies with 37 different carriers, no clear-cut relationship between pH of the carrier and toxicity could be found. Kaolin, bentonite, pyrophylite, fuller's earth, talc, kieselguhr and other clay mineral sorb the nicotine to such an extent that the mixture becomes less useful as a contact insecticide though may be useful as a stomach poison. Alkaline substances like hydrated lime ($Ca(OH)_2$), calcium carbonate (purified or ground), dolomitic limes and limestone when used as carriers, react with nicotine sulphate to form free nicotine and as such are called active carriers. Such carriers can be used when quick acting contact insecticide dusts are. needed. Based on physicochemical basis, carriers can be classified as: (i) colloidal carriers such as kaolin, fuller's earth which prevent volatilisation of the nicotine by their adsorption; (ii) inert materials such as gypsum which are neutral or inactive towards nicotine sulphate; and (iii) active materials such as hydrated or air-slaked lime, basic or carbonated materials which react with nicotine sulphate to free nicotine and permit it to vaporise. The toxicity of nicotine compounds depends upon the alkaloid content, regardless of physical or chemical combination in which the poison occurs. When applied as contact insecticides, the toxicity not only depends upon nicotine content but also on other factors such as solubility etc.

Insecticidal action of nicotine

Though nicotine is usually used as a contact insecticide, when applied in the form of a dust or spray, it probably acts as a fumigant. One mode of entry of nicotine into the insect is through, the spiracles in the tracheal system. The fumes of nicotine, striking the walls of insects are immediately dissolved, penetrate to vital tissues and bring paralysis of the insects' nervous system. Though vapour pressure of nicotine is low at ordinary temperature, the compound is so active that even low vapour concentration is insecticidal.

Toxicity of nicotine to mammals

Though nicotinc is usually used as dust or spray, from the point of residues, sprays containing nicotine are considered relatively safe because of the volatility of the nicotine and its solubility in water. Nicotine is absorbed rapidly from tongue or eye or even through unbroken skin, more slowly from the stomach probably because of the acidity of the gastric juice. Failure of respiratory mechanism in mammals is primarily due to a peripheral action of nicotine which causes a curera-like paralysis of the endings in the respiratory muscles. Convulsions ordinarily complicate the symptoms by contributing to the development of fatal asphyxia, by increasing oxygen consumptions and carbon dioxide production and by interfering with the respiratory movements by fixation of the respiratory muscles.

Nornicotine

{2-(3-pyridyl) pyrrolidine, $C_9H_{12}N_2$}

For a considerable length of time, nicotine was thought to be only alkaloid in tobacco. However, recent studies have shown that at least 24 substances are present in tobacco, probably as pyridine derivatives. A few like nornicotine, anabasine, oxynicotine, myoamine, 3-acetyl- pyridine, nicotinic acid and 2, 3 dipyridyl have been identified.

Nornicotines are similar to nicotine except that the methyl groups attached to pyrrolidine ring has been replaced by a hydrogen atom. As in the case of nicotine, the form 2-(3-pyridyl) pyrrolidine occurs naturally. The 1-nornicotine occurs naturally in *Nicotiana sylvestris,* a species of tobacco. In this plant, 95% of the alkaloid present is nornicotine. *Duboisia hopwoodii* also contains nornicotine.

Pure nornicotine ($C_8H_{12}N_2$) is a liquid alkaloid resembling nicotine in physical properties and physiological behaviour. All the optical forms, leavo, dextro, and racemic, occur in nature. Pure nornicotine is a colourless, hygroscopic, slightly viscous liquid with an odourless pungent than nicotine. It appears to be more stable than nicotine and is miscible with water and organic solvents in all proportions. It differs from nicotine in being only slightly volatile in steam, which is utilised in separating the two bases. The specific gravity is slightly greater than nicotine (1.07 at 20°C) and its boiling point is considerably higher (266-270°C). It is leavorotatory tests showed that DL-β nornicotine was more toxic than the corresponding DL-β nicotine but almost equal in toxicity to natural α-β nicotine. The DL-β forms of nicotine and nornicotine were of equal toxicity but were far less toxic than the racemic mixtures of the β compounds. The DL-β forms of both nicotine and nornicotine were more phytocidal than the DL-β forms, but in contrast the DL- nornicotine (both α and β forms) were more phytotoxic than the corresponding nicotines.

The differences between the phytocidal effectiveness of the α-β forms was not so great as the insecticidal differences. The 1-nornicotine, dl-nornicotine and α-nornicotine were found to be almost equally more toxic than 1-nicotine to adult winged females of *Aphis fabae* sp. Nornicotine was found to be more poisonous than nicotine to green peach aphid *Myzus persicae.*

Anabasine

{Neonicotine, 3-(2-piperdyl) pyridine}

Anabasine occurs in the seeds of *Anabasis aphylla* Linn. Anabasine closely resembles nicotine in its physical and chemical properties. It is a colourless, somewhat viscous liquid which turns brown on standing in contact with air. It is very stable and possesses an odour similar to that of nicotine.

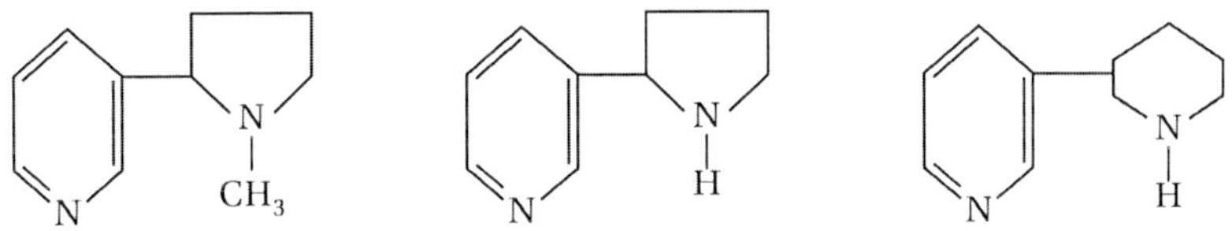

Nicotine Narnicotine Anabasine

Salts are formed in typical alkaloid fashion. It is miscible with water in all proportions and is soluble in all organic solvents. The specific gravity is 1.0481 and boiling point at 760 mm is 280.9°C and has an optical rotation $(\alpha)^{20}\text{-}_D59.66°$.

Anabasine may be distinguished from nicotine by the fact that it is precipitated from methyl alcohol solution and fluoride as the. silico-fluoride, while nicotine remains in solution. An ether solution of iodine precipitates ruby red needles of periodide ($C_{20}H_{14}N_2I_2HI$) from ether solutions of nicotine, while anabasine does not react.

Anabasine occurs in the small green twigs of *Anabasis aphylla* even up to 2.63%, leaves and roots of the tree tobacco *Nicotiana glauca.* Anabasine may be extracted from its plant source by ethylene dichloride or by a water kerosene diffusion process. The crude anabasine sulphate of commerce is a mixture of alkaloids-aphillidine $C_{16}H_{22}ON_2$, aphylline $C_{15}H_{24}ON_2$ and lupenine $C_{10}H_{19}ON$ and methyl anabasine. The crude anabasine sulphate of commerce contains approximately 40% total alkaloids, of which 70% is anabasine and the remainder other alkaloids. For most insecticidal purposes, anabasine has been found to be approximately equal to nicotine in its insecticidal activity and toxicity to warm blooded animals.

Nicotinoids have the highest insecticidal activity have the highest pK_a' and consequently exist largely in an ionised form at the physiological pH. This results in the anomaly that the compounds most highly ionised react most rapidly with the receptor protein, yet are able to penetrate through the ionic barrier surrounding the insect nerve synapse. This has been demonstrated by the lack of toxicity of the completely ionised nicotine methiodide upon injection into Musca. The necessity for nicotinoids to penetrate the insect nerve in the unionised state doubtlessly explains their relative lack of insecticidal action against many insects, whose nervous tissues are protected by a more efficient ion barrier than that of the highly susceptible aphids. Further, the essential requirements of toxicity are a pyridine ring, a highly basic nitrogen at an optimum distance of about 4.2 A from the pyridyl nitrogen and an unsubstituted α-position of the pyridine ring. Many of 3-pyridyl methyl-N alkylamines meet these requirements.

Mode of Action

Nicotine and anabasine affect the ganglia of the insects. Affecting central nervous system, facilitating transsynatic condition at low concentration and blocking conduction at higher levels. The extent of ionisation of nicotinoides plays an important role in both their penetration through the ionic barrier of the nerve sheath to the site of action and in their interaction with the site of action, which is believed to be acetylcholine receptor protein.

Nicotine and to lesser extent anabasine are used as contact insecticide for aphids attacking fruits, vegetables and ornamentals and as fumigants in greenhouse and against poultry mites. Nicotine sprays and dusts commonly contain 0. 05 to 0. 06% nicotine and 1 to 2% nicotine respectively. Nicotine has an oral LD_{50} to rats 30 mg/kg and a dermal LD_{50} (rabbits) 50 mg/kg, thus, it is very highly toxic to man and higher animals. It is very rapidly detoxified and eliminated from the animal body and appears to pose no particular chronic hazard. It's rapid deterioration in light and air favours its use on leafy vegetables within a few days of harvest. Nicotine sulphate is safer and more convenient to use and the free alkaloid is liberated by the addition of soap, lime or ammonium hydroxide to the spray solution. However, nicotinoides have been almost entirely replaced by the organophosphorus compounds which have wider scope of activity and also could be synthesised in large quantities.

Pyrethrum

Pyrethrum has been used for the control of harmful insects for many centuries. The insecticidal principle in pyrethrum is found in the flower heads of certain plants of *Chrysanthemum* genus, family *Compositae.* Only a few species of this large family like *Chrysanthemum* (Pyrethrum) *roseum* web and Mohr., *C. cinerarie- folium* Trev., *C. marshalli* Ach and *C. tamrutene* have been found to be valuable sources of this insecticide. Though pyrethrum was formerly used as finely ground flowers, today it is used as dusts and sprays. The extractives are suspended in a solvent for use as sprays.

Pyrethrum content of flower heads increases up to the time of full opening of the flower. In early days, the flowers were dried in the sun which resulted in a variable product. But now nearly all pyrethrum flowers are being dried at the same temperature, viz. 54.4°C. Then the dried flowers are compressed and exported. These flowers are ground to a fine powder and extracted with the solvents. The solvent is repeatedly percolated through the ground flowers and the weak extracts are evaporated in a vacuum still to recover the solvent. The extracted powder (pyrethrum marc) containing traces of pyrethrins is sometimes used as carriers in dust preparations.

Formerly Japanese held monopoly for the supply of pyrethrum. As their product was low in pyrethrum content and also variable in pyrethrum content, other countries like Kenya which could produce superior quality pyrethrum virtually drove out Japan. Now, pyrethrum is being cultivated in Belgian Congo, Uganda, India, Dalmatia (in Yugoslayia) and the U.S.A. For preparing dusts, a variety of carriers diatomaceous earth, powdered charcoal, pulverised pyrethrum marc, talc, gypsum, bentonite have been used. Bentonite and other highly absorbent materials hold the pyrethrins tenaciously and apparently release it more slowly than carriers like gypsum and talc. Dust concentrates are made from concentrated pyrethrum extracts with a non-volatile solvent and a suitable absorbent carrier and then diluted before use with an inert diluent. Antioxidants such as tannic acid, hydroquinone or some other synthetic organic compounds such as 1, 4 toluidioanthroquinone are used to stabilise the pyrethrins in dust preparations. Concentrated extracts of pyrethrum in acetone, alcohol or a hydrocarbon solvent together with an emulsifier is sold in the market.

Aerosols may be produced by any method which will reduce the particle size of the materials to colloidal dimensions. Pyrethrum aerosol formulations usually have a highly purified pyrethrum extract, a synergist, a diluent (usually a petroleum oil) and a propellant gas. The mixture is compressed into a suitable container. Purification of the pyrethrum extract is necessary to remove substances insoluble in Feron-12, since the insoluble substances interfere with the operation of gas dispenser. In the manufacture of pyrethrum sprays, other pesticides like rotenone, DDT or sesamin are added to increase the toxicity of the pyrethrum extracts.

The structure of the active principle in pyrethrum was established only in 1950's. Staudinger and Ruzicka found two active principles Pyrethrin I and Pyrethrin II which were esters of a keto alcohol (pyrethrolone) and two acids (Chrysanthamum monocarboxylic acid and C. dicarboxylic acid). Two more compounds were identified and the four possible compounds have been named as Pyrethrins I and II and Cinerins I and II. These compounds are all viscous liquids, soluble in a variety of solvents but not in water. The relative toxicities of these compounds are: Pyrethrin I—100, Pyrethrin II—23, Cinerin I—71, and Cinerin II—18. Hydrogenation of the acid component reduces the relative toxicity of pyrethrin I and cinerin I and also the pyrethrolone compound.

Since the pyrethrins and cinerins are esters, they are rather easily decomposed in the presence of moisture and acid or alkalies. Because of this, pyrethrum is not compatible with lime or soaps. Cinerins are more stable than pyrethrins. Points of weakness in these compounds are in the ester linkage which is easily saponified by alkali, due to the unsaturated nature of the alcohol portion (as in the side chain of pyrethrin I) which makes itself susceptible to attack by alkali as well as by oxygen in the air especially in the presence of light, and due to the asymmetry of both acid and alcohol parts of the molecule, resulting in a tendency for racemisation of these optically active components.

As these compounds are not volatile, pyrethrum has no fumigant value. Powdered pyrethrum loses its insecticidal activity rather rapidly on extended storage and this is further accelerated on exposure to air in thin layers. Certain antioxidants like pyrogallol, hydroquinone retard the loss of activity. Neither acids nor the cyclopentenolones are toxic to insects. Mild catalytic hydrogenation destroys the greater part of the toxic action of the pyrethrins.

The insecticidal properties of the pyrethrum are due to five esters (Pyrethrins I and II, Cinerins I and II and Jasmolin II) that are present mostly in the acheness of the flowers, ranging from 0.7 to 3.0%. These esters are:

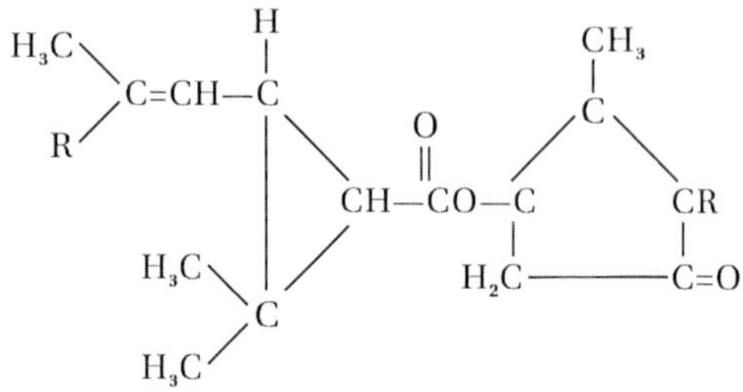

where

	R=	R_1=
Pyrethrin I	— CH_3	—$CH_2CH{=}CH.\ CH{=}CH$
Pyrethrin II	—$COOCH_3$	—$CH_2CH{=}CHCH\text{-}CH_2$
Cinerin I	—CH_3	—$CH_2CH{=}CH.\ CH_3$
Cinerin II	—$COOCH_3$	—$CH_2CH{=}CH.CH_3$
Jasmolin II	— $COOCH_3$	—$CH_2CH{=}CH.CH_2CH_3$

These esters have asymmetric carbon atoms and double bonds in both alcohol and acid moieties.

Insecticidal activity of pyrethrins: They are more toxic than cinerins and variations in their activity depends on the insect species and the technique employed in any particular test. Pyrethrum affects the nervous system of insects such as cockroaches, resulting in muscular excitation, convulsion and paralysis. The affected cockroaches emerge from their hiding places, fly and become paralysed. The solubility of pyrethrum in oils promote their penetrations of the insect integuments. Because of the residual life of pyrethrins, insects are paralysed much more rapidly so that the insects are usually incapacitated before they are able to eat it. Some insects although rapidly paralysed by pyrethrum, may recover entirely within a few hours or a day as the pyrethrum may decompose inside the insects and lose its activity. Insecticidal activity of pyrethrum is reduced by 49 per cent and 68 per cent when the treated surfaces are exposed to a sunlamp for 10 minutes and to air and sunshine for 5 hours respectively. On the other hand, the activity is reduced only by 5 per cent in darkness even after 7 hours. However antioxidants reduce the effect of exposure to air and light especially during storage. The finer the pyrethrum powder, the more rapid is its action on insects, but it rapidly deteriorates also upon exposure to light.

The most active isomers are the α trans acid esters of the d cis alcohols. Possibly the biochemical activity is associated with a specific three point contact between pyrethroid and biological receptor at the isobutylene group, the dimethyl cyclopropane ring of the acid and the unsaturated group of the keto-alcohol.

Pyrethrins when applied to the insect nerve produce arhythmic spontaneous discharge at 0.01 to 0.1 ppm and a reversible blocking of conduction at 1 ppm. Paralysed insects exhibit characteristic vacuolisation of the nerve tissue. A threefold increase in dosage is required to produce the same percentage of mortality as that of knockdown since there is a substantial recovery after rapid knock-down when pyrethroids are applied. Recovery from paralysis is the result of rapid enzymatic detoxification in the insect apparently by mixed function oxidases. However the detoxification enzymes are inhibited by a number of compounds like methylene dioxybenzenc derivatives such as piperonyl butoxide, propyl isome and sulfoxide. Pyrethroids at about 0.03 to 0.1 per cent with 5 to 10 times of synergists have been used as household fly and cattle sprays.

Pyrethroids are of low mammalian toxicity because of their rapid detoxification following oral ingestion. Oral LD_{50} values for rats are pyrethrum 820, allethrin 920, cyclothrin 1,400 mg/kg. Synergists and combinations are also of low toxicity with oral LD_{50} values to the rats. Piperonyl butoxide 11,500, propyl isome 15, 000 and sulfoxide 2,000 mg/kg. Pyrethrums are practically non-toxic to warmblooded animals. However they appear to be toxic to all animals when introduced into the circulatory system. In certain cases, pyrethrum dust produces allergic attacks in many sensitive persons. Injury to plant life by pyrethrum insecticide is not due to the pyrethrum but to a base oil or to some other ingredient.

The unstable nature of the pyrethrins to light and oxygen coupled with the establishment of the structure of pyrethrins stimulated research on synthetic pyrethroids. From 1964, a group of compounds like K-othrin, Kodethrin, properthin and enprothrin, were synthesised but they could not excel the natural pyrethrins as all of them were unstable to light and temperature. The first synthetic pyrethroid was Permethrin- which had high activity against insect, low mammalian toxicity and at the same time greatly increased stability as described in 1973. This group of compounds were more photostable than many of the common organophosphorus and carbamate insecticides and were also insecticidally more active than others by at least one order of magnitude.

With the structural requirements for the photostable pyrethroids established, many new compounds which retained the chemical and steriochemical features of parent materials and their biological activity were synthesised during 1973-1977. In 1974, Okno and others described highly active compounds like Cypermethrin, Fenvalerate, Decamethrin. These compounds are viscous lipophylic liquids having high boiling point and low volatility. They are practically insoluble in water, highly photo-stable, biodegradable, have low mammalian toxicity, and do not leave residues in the biological systems. Their movement from the site of application is limited in the environment due to their low volatility and in soils and in plants as they are non-systemic in nature because of low polarity. Hence, they will cause less contamination in the environment. However, the synthetic pyrethroids are likely to be expensive on weight basis than many other insecticides as their synthesis is more complex. However as they are found to be effective even at 1/10 to l/50th of the normally recommended rates of conventional pesticides, the synthetic pyrethroids are likely to be competitive.

Allethrin

The isolation of cinerins and the determination of the structure of cinerolone by LaForge and his associates led to further work on the synthesis of closely related compounds of possible insecticidal value. One of the compounds, Allethrin (divinyl pyrethrin I, DVPY—I, allylrethrin-allyl analog of cinerin I) is an ester made from alcohol having the allyl side chain 2-allyl-4 hydroxy-3 methyl- 2-cyclopentene-l-one (allethriolone) which is chemically identical with naturally occurring cinerolone except that the side chain contains one less carbon atom. Allethrin is a mixture of four racemic forms or eight optical isomers—four of which are esters of the cis- acid and four are esters of the trans acids—because commercial allethriolone is the DL-form and it is esterified with the DL-cis and DL-trans chrysanthemum monocarboxylic acids.

Commercial allethrin is a clear, brownish liquid having a specific gravity between 1. 005 and 1. 0 i 5 at 20°C and refractive index of 1.504 at 20 to 40°C. Allethrin is formulated in the same way as the natural pyrethrums. Allethrin shows the same general toxicity as natural pyrethrums against many insects but at the same time is reported to be without harmful effects on warm-blooded animals. Number of analogs of allethrin like ethythrin, furethrin and cyclothrin have been prepared but they have been found to be less toxic than natural pyrethrums.

Rotenone

A crystalline substance was isolated by Nagai in 1902 from *Derris chinensis*. As this plant was known locally as Rohten, the crystalline substance was called Rotenone. Substances isolated from other plants were given different names like Nicouline, Derri, and Tubatoxin but apparently these substances were all identical and so the name rotenone was adopted.

Several fish poison plants belonging to *Galegeae* and *Dalhergieae* of the subfamily *Papilionatae* and family *Leguminosae* possessed considerable value as pest control insecticides. Important species are— *Derris elliptica* and *D. malaccettsis, Lonchocarpus nicou* and *L. utilis*. Selection and breeding have raised the content of the active principle rotenone in several of the commercially developed species. The rotenone and related principle often occur chiefly in roots but may occur in other parts of the plants like leaves, stems or seeds. Rotenone containing roots are rated commercially in terms of standard rotenone content of 5%.

Among the 30 species listed, *Derris elliptica* followed *D. malac-censis* are the most important sources of rotenone. The roots of the above species contain the active principles, even as much as 13% in certain clones, with total ether extractives of over 30%. Of the twelve species of *Lonochocarpus, L. nicou* grown in Puerto Rico has been reported to contain as much as 20.6%. The rotenone content of *Tephrosia* (Cracca) and *Millettia* has been found to be much lower than other plants. For commercial purposes, only *Derris elliptica* and *D. malaccensis* are grown. Optimum time of harvest is from 18 to 27 months after planting. Yields vary from 400 to 1800 lb of dry roots per acre if trailing system is followed.

Extracts known as derris or cuberesins contain rotenone with certain rotenoid compounds of lesser activity than rotenone. Rotenone usually makes upto 30 to 40% of the total extractives. Unlike pyrethrum extractives which may be mixed directly with a mineral oil base to provide a homogeneous product, the formulations of petroleum base products containing derris cube extractives requires an additional material which will dissolve resins in oil. Dibwylph-thalite, comphsassafras (oil camphor sasasfrass), methyl-n-amyl ketone, 2-(4-tert-butyl phenoxy) ethanol and candinol (cashewnut shell oil) have been used as mutual solvents (also called cosolvents, solubilisers, blending agents, coupling agents, bridging agents).

Rotenone is extracted from finely ground plant material with such solvents as ether and chloroform. Number of closely related rotenoid compounds like deguelin and elliptone which occur along with rotenone have less toxicity. Derivatives of rotenone, deguelin and elliptone having a hydroxyl group in the 15th position are sumatrol, toxicarol and malaccol respectively.

Pure rotenone crystallises from alcohol in white six sided plates belonging to the orthorhombic system with m.p. 163°C. Rotenone is relatively soluble in chloro-organic solvents with the exception of carbontetrachloride. It is practically insoluble in water and only slightly soluble in petroleum oils. The rotenone molecule contains three asymmetric carbon atoms. However only one optical isomer is known though eight are possible.

Rotenone because of its complex structure, is more easily decomposed than either nicotine or pyrethrum. The colourless solution of rotenone in organic solvents becomes successively yellow, orange and finally deep fed when exposed to light and air, as a result of oxidation and may deposit crystals which contain dihydrorotenone and rotenone which are non-toxic to insects. Thin deposits of rotenone on plants exposed to sun and light change colour in about 10 days. This change may be retarded by the addition of lamp black to the spray thus excluding light from the deposit.

Rotenone is unstable in alkali which splits the molecule at the central pyrone ring or when dilute, racemice the compounds and promote oxidation. The first stage in decomposition results in the formation of a hydroxyl group, the second in the production of hydrorotenone and the third in a split molecule. Racemisation of rotenoid derivatives during extraction and isolation has created uncertainly as to the nature and relative toxicity of the compounds as they occur normally in the plant materials.

Dihydrorotenone is produced by the catalytic hydrogenation of the side chain of rotenone with the elimination of double bond. This compound is somewhat more resistant to detoxification by sunlight and is not much less insecticidal inder some conditions than rotenone itself.

Some solvents form crystalline solvents with rotenone, a fixed proportion being held as solvent of crystallisation. Rotenone when crystallised from certain solvents contains definite amount of the solvent in the crystal similar to water of crystallisation in inorganic salts. These are called solvates and contain one molecule of solvent of crystallisation, when crystallised from chloroform, benzene and carbon tetrachloride, when crystallised from acetic acid two molecules of solvent of crystallisation and none when crystallised from acetone, ethyl alcohol and ethylene dichloride.

Most of the rotenone from an extract of derris (or cube) roots can be separated in a crystalline form. The uncrystallisable residue however is approxi-mately as toxic to fish as crystalline rotenone. It contains large quantities of deguelin, toxicarol and other rotenoid compounds which can only be separated as crystalline entities by weak alkaline treatment. Although pure crystalline rotenone, deguelin and toxicarol are almost insoluble in petroleum ether when derris root is extracted with the solvent, a large proportion of these compounds is dissolved, presumably because of the presence of contaminant resinoid materials which is also extracted. The rotenoids appear to be far more soluble in solvent plus resin than in solvent alone, because of the toxicity of nonrotenone extractives, both rotenone and total extractives content must be considered in the insecticidal evaluation of roots. Rotenone is crystallised in carbon tetrachloride from the rotenone containing material. Then the method involves the separation of rotenone carbon tetrachloride solvate transforming it to rotenone by treatment with alcohol, weighing and calculating the rotenone content.

Dusts are prepared by extracting dried roots with a solvent such as chloroform and then mixing the extract with a finely divided diluent or carrier. This enables an even distribution of the toxic material throughout the mass of dust. Walnut shell flour, pyrethrum marc, clays and other carriers and absorbers have been used in the preparation of coated dusts. Sprays are prepared by mixing the finely ground roots with water and the suspension is applied directly. Solvents used for the preparation of such extracts in order of their dissolving power are: chloroform, ethylene dichloride, trichloroethylene, chlorobenzene, ethylene chlorohydrin and benzene. Carbon disulphide, ethyl formate and ethylacetate are also good solvents for rotenoids.

Rotenone Derivatives

Deguelin

There is a great similarity between the composition of deguelin and rotenone. In rotenone, one benzene ring is a substituted dihydrofuran nucleus while in deguelin, it is a 2-substituted α pyron ring. Since there are two asymmetric carbon atoms in the deguelin molecule, four optical isomers are possible, although none is known to exist in the crystalline form. Deguelin like rotenone contains two hydrogen atoms which are readily removed by mild oxidation. A double bond is thus introduced in the molecule between carbon atoms, shared by rings B and C and the product is dihydrodeguelin $C_{22}H_{20}O_6$. Deguelin itself in less toxic than dihydrodeguelin and rotenone when tested.

Other derivatives

Tephrosin and toxicarol have been found along with rotenone and deguelin in derris and cube roots. Tephrosin has been isolated from the leaves of *Tephrosia vogelii.* Tephrosia does not occur as such in derris but is formed by the oxidation of deguelin. The toxicity of tephrosin to insects is apparently less than rotenone and deguelin but greater than toxicarol. Sumatrol is similar to tephrosin but differs in being phenolic. Sumatrol has roughly the same toxicity to the *Apis rumicis* as toxicarol.

The above rotenoids plant constituents are only about 0.1 to 0.2% as active as rotenone in insecticidal activity. With the exception of hydrogenation of the isoproponyl side chain of rotenone, which produces dihydrorotenone of nearly equal toxicity, all other chemical alterations greatly decreases the toxicity of the rotenone molecule. Dehydrorotenone and tetrahydrorotenone are non-toxic as are derivatives with major alterations in rings A, B, C, D and E.

Insecticidal activity of rotenone

Rotenone is extremely potent against many insects, acting both as contact insecticide and as stomach poison. Rotenone is rather specific in its action upon insects. A considerable proportion of the insecticidal activity of derris and cube extractives is due to the nonrotenone portion, principally to levodeguelin. Deguelin occurs in inactive form in the plant. The optically active form dihydrodeguelin is about 10 times as toxic as the inactive form. Derris and other rotenone materials deteriorate through the action of light and oxygen. Solutions of rotenone are photosensitive, turning from colourless to yellow on standing. The change takes place rapidly in pyridine, less so in chloroform, only slowly in acetone and impreceptibly in benzene or alcohol. It is accelerated by access to air and is accompanied by change in optical rotation and less of activity. Although dry crystalline rotenone is relatively stable under appropriate storage conditions, it loses its activity upon exposure to the sun for some weeks.

Rotenone and other related compounds are considerably more toxic to insects, fish and other cold blooded animals than to mammals. The LD_{50} oral of derris is 75 mg/kg for guinea pig, 600 mg/kg for rabbit, and 132 mg/kg for rats. However the rapid decomposition of spray or dust residues exposed to air and sunlight renders the possibility of health hazard rather remote. Rotenone is readily converted to non-insecticidal derivatives by oxidation in the presence of light and air with the formation, first of a hydroxy derivative at C—7 and then dehydration to form dehydrorotenone with a C=C bond between rings B and C. These reactions render rotenone residue innocious after 5 to 10 days in normal light. Rotenone compounds cannot by any means be considered non-poisonous. Derris dust when inhaled over extended periods, can be fatal to experimental animals. Ground plant materials containing rotenone cause dermatitis in a high percentage of persons when applied next to the body inside the clothing. Rotenone dissolved in oil is much more toxic when given orally than as solid or in suspension.

Insects poisoned with rotenone exhibit steady decline in oxygen consumption and the insecticides have been shown to have a specific action in interfering the electron transport involved in the oxidation of DPNH to DPN by cytochrome b. Thus, rotenone at 1×10^{-7} to 1×10^{-6} M inhibits the mitochondrial oxidation by *Periplanata americana* tissues of α-glutamate, α-ketoglutarate, malate, fumarate, oxaloacetate and citrate all of which are catalysed by DPN.

Miscellaneous insecticidal plants

Among thousands of species of plants tested for insecticidal use. Hellebore *(Veratrum album* or *V. viride)* and Sabadilla *(Schoeno- caulon officinale)* have been found to be toxic to certain species of insects. Number of other species have also been studied for their insecticidal value but none of them have been extensively tested. However the following plants have shown some promise:

1. *Aesculus* sp. (Buck eye)
2. *Amorpha fruticosa* (Indigo bush)
3. *Anacardium occidentale* (Cashew tree)
4. *Ananiirtus cocculus* (Fish berry)
5. *Annona species* (Custard apple)
6. *Celastrus* sp.
7. *Croton tiglium*
8. *Dryopteris* (Male fern)
9. *Heliopsis longipis*
10. *Mammea americana* (Balsam family)
11. *Phellodendron amureuse* (Velvet tree)
12. *Quassia amara* L.
13. *Rhododendron* sp.
14. *Tripterygium* sp.
15. *Vetivera zizanioides* (Vetiver)
16. *Zanthoxylum clavaherculis*

Essential oils: Certain essential oils have been found to be of practical value as insect repellents and attractants:

Oil of citronella—Citronellai (17 to 34%) and Gran oil (16 to 45%)

Lemon grass oil—Citral (65 to 85%) Citronellal

Oil of cassia—Cinnameic aldehyde (85 to 90%)

Oil of pennyrayol—Isomenthane (50%), pullgone (30%)

Oil of thyme—Thymol or carvacrol

Clove oil—Eugenol (85 to 95%)

Pine needle oil—Bornyl acetate (0 to 49%)

Turpentine or pine oil

Camphor.

CHAPTER VI

OILS

Goeze in 1787, first mentioned petroleum oils as insecticides. He had also indicated that there was a possibility of phytotoxicity if adequate precautions are not taken during its application. Only in 1865, a petroleum distillate was tried against scale insects. Later kerosene was mixed with soap suds to form emulsion of varying degrees of stability. During 1880-1900, kerosene emulsion appeared to the entomologists as an ideal contact insecticide for general use. Emulsions of crude petroleum were tested in New York during 1896. Though crude oil was suited for dormant use, it was not suited for summer applications due to its volatility and its persistence on foliage.

At about this time, a different method of preparing oil emulsion came into use. When certain compounds or mixtures of compounds mutually miscible in petroleum and water are added to the oil, an emulsifiable oil results which is called "miscible oil" or "soluble oil". The earlier mixture of this sort was largely oil solutions of soap and carbolic acid, the latter compounds serving as emulsion stabiliser. When an emulsifiable oil is added to water, a stable milky emulsion is formed with relative ease.

The crude oils produced were not uniform and contained impurities of varied nature and in varying quantities. Lubricating oils came into wide use from 1906 after they were successfully tried by Yothers in that year. Yothers recommended the use of boiling water in the preparation of oil-emulsion (Lub oil 2 gallons, water 1 gallon, potash fish oil soap 2 lb). These were mixed thoroughly and brought to boil as heat reduced the surface tension of the ingredient. The stock emulsion was stable for at least one year. After 1923, various cold pumped emulsions became popular. Generally, a non-soap emulsifier such as calcium caseinate or bordeaux mixture was added to the water.

The earlier whale oil emulsions contained up to 10% oil but with the development of quick emulsion, oil concentration could be reduced from levels of 4 to 10% to as low as 1.5 to 2% and still be effective. Quick breaking, non-soap emulsions tend to deposit more oil than the more stable, soap emulsions both being applied at the same rate. The effect of oils on both insects and plant life are generally governed by the physical and chemical properties of oils and by the oil deposit on the insect or plant, a factor which depends on the emulsifiers and is not easily regulated. This was overcome in the tank mixture method where uniform mixture is kept up by high speed agitation.

Petroleum oils used as insecticides consist largely of aliphatic hydrocarbons-but may be paraffinic or isoparaffinic. Four chief types of hydrocarbon structures are present in petroleum products. They are as follows:

(i) Aliphatic compounds of the paraffinic series including saturated, straight and branched chain structure like methane CH_4, butane C_4H_{10}.

(ii) Naphthenic compounds having saturated ring structures like cyclohexane.

(iii) Aromatic compounds of the benzene series having unsaturated ring structure like benzene C_6H_6, naphthalene $C_{10}H_{10}$.

(iv) Unsaturated compounds largely of the olefines and acetylene series like ethylene C_2H_4, acetylene C_2H_2.

All petroleum compounds are subjected to one or other kind of separation and purification. The process of distillation separates the viscous fractions composed of larger molecules and having greater viscosity. After distillation, the various fractions of oil are refined, the degree of refinement depending upon the intended use of the oil. Certain unsaturated compounds may be present in the original crude oil or may be formed during the distillation process also. These are removed by treating the distillate with sulphuric acid or sulphur dioxide, thus sulfonating the unsaturated compounds. Petroleum sulfonates designated as mahogony soaps in trade are valuable detergents, wetting and emulsifying agents. Higher fractions are used occasionally as summer spray oils in household sprays, as solvents and carriers for pyrethrum, DDT and other synthetic insecticides. When crude petroleum is subjected to fractional distillation, the following six classes of products can be separated:

(i) Naphtha, gasoline, petroleum ether and benzene (b.p. 70-150°C)

(ii) Kerosene (150-300°C)

(iii) Gas oil, distillate fuel oil (250-350°C)

(iv) Lubricating oil, petroleum (>300°C)

(v) Paraffins separated from (iv) by chilling.

(vi) Asphalt, tar, pitch coke remaining as residues.

Petroleum products may also be classified according to the paraffinic compound which they contain:

1-4 carbons — Gas

5-10 carbons — Naphtha and gasoline

11-16 carbons — Kerosene

14-18 carbons — Lubricating oil

17-21 carbons — Petroleum

21-27 carbons — Paraffin wax m.p. 40-60°C.

Each of the primary fractions may be distilled further if narrower cuts are desired. Now the trend in development of spray oils is towards more highly refined narrow cut fractions, with low viscosities, which are safer on plants. Petroleum oils to be used as insecticides must meet certain specifications depending upon the purpose for which the oil is intended. Some of the important characteristics of spray oils are their viscosity (or resistance to flow), low volatility or distillation range, purity or degree of refinement as measured by the unsulfonated residues. In general, the higher the boiling point, the more toxic that oil is to the insect. The practical problem is to use the heavier oil without phytotoxicity.

(i) *Viscosity:* Viscosity varies with molecular weight as well as with composition. Penetration of the tracheae is related inversely to viscosity. Oils of low viscosity are more active in this respect because of their greater mobility. On the other hand oils of high viscosity are less likely to be expelled from the tracheae.

(ii) *Volatility of oils:* It is important since rapid evaporation reduces exposure of the treated insects to oil and increases survival. On the other hand undue persistence of oil on plant surfaces may result in the absorption of injurious quantities by the plant.

(iii) *Flash point:* It is the minimum temperature at which a heated oil will give off inflammable vapour in sufficient quantity to produce a combustible mixture. This will give the fire hazard of the unemulsified oils and their mixtures either when handled in bulk or when sprayed.

(iv) *Sulfonation:* This even expressed as percentage, indicates the degree of refinement of the oil in terms of the percentage of oil which does not react with sulphuric acid under NTP. The sulfonation test is in terms of the unsulfonated portion or the residue of oil which does not react with acid. The oil which reacts with acid is made up largely of aromatic compounds and unsaturated aliphatic compounds. Sulfonation has been used as a measure of safeness for application to plant and animal life. The degree of plant injury caused by an oil could be estimated with considerable accuracy, by means of the sulfonation test. The higher the unsulfonatable residue percentage in an oil, the less likely it is to produce plant injury. The mechanism of phytocidal action is that the hydrocarbons of spray oils have little toxic action on plant leaves until they are oxidised under normal conditions with sufficient rapidity, to cause plant injury but the unsaturated hydrocarbons under the influence of light, particularly UV radiation rapidly absorb oxygen and thus become increasingly dangerous. The exact mechanism of this reaction is not thoroughly understood but it appears that peroxides are first formed which are later oxidised to acids. In addition to the direct effect of these acids upon plant tissues, their presence is equally undesirable in oil mixtures containing dissolved or suspended insecticidal materials such as the pyrethrum which may be subjected to decomposition. Dormant spray oils commonly have a sulfonation test 70-85%, summer foliage oils 90-98%.

(v) *Iodine value:* It is the absorption of iodine by the unsaturated aliphatic compounds in oils. This is used in the place of sulfonation test.

(vi) *Oxidation:* Oxidation of saturated hydrocarbons which make up a large part of a highly refined oil, takes place particulary when oil films are exposed to the action of air and sunlight. Oxidation is measured by the titration of acidity which is formed when hydrolysed with KOH. Neutralisation number is an expression of acidity in terms of mg of KOH required to neutralise the acid in 1 g of oil under prescribed conditions. It is useful particularly as a measure of breakdown in oils. The effects of oxidation upon petroleum toxicity are little known. It has been said that the oxidation of kerosene increases both toxic action and spreading power, so that these properties may depend upon the same substances in oils. Unrefined oils possess greater ability than refined ones to spread upon the surface of water.

(vii) *Sulphur:* Sulphur is present in small amounts in crude petroleum. Some oils have been reported to contain as much as 4% sulphur. Even though the use of oil sulphur combination sprays often caused severe foliage injury, the addition of free sulphur in combinations of hydrogen sulphide, organic sulphides, disulphides and mercaptans to the oil did not increase the oxidation of the oils used. Addition of sulphur in any of the forms did not however result in severe leaf injury.

Properties of Petroleum

(i) Crude oil

Crude oil was apparently first used by Smith in 1897. The oil was emulsified with soap, a hot soap solution being mechanically mixed with oil. Later "miscible oil" appeared, in which the emulsifier was dissolved in oil with the aid of phenol, cresylic acid and naphthalene. However the variability of crude oils made it difficult to use and the impurities present often caused severe damage to plants. This led to their replacement by more refined lubricating oils.

(ii) Gasolenes and naphtha

The lowest boiling petroleum fractions are seldom used for insecticidal purposes because of their high volatility and fire hazard. They possess considerable toxicity towards insects. For spraying work, it is generally agreed that the flash point should not be less than 125°F. Some of the higher boiling naphtha, such as Stoddard solvent which approaches the properties of kerosene is used by pest control operators as solvent in the control of insects in furnitures, mattresses and similar articles.

(iii) Kerosenes

Kerosenes although used as early as 1865, was first used in the form of an emulsion with soap by Cock in the 1877. Other materials such as sour milk, condensed milk, resin, clays, bordeaux mixture and others have been used as emulsifiers, Kerosene is available in a number of broad boiling range from about 182°C to about 266°C. The low boiling fractions of kerosene probably because of their low viscosity and high volatility, have been found to be less effective than higher boiling fractions as emulsions, and have largely been replaced in agricultural uses by more viscous and less volatile oils. The properties of kerosene vary over a rather wide range. Kerosenes are composed of hydrocarbons having from 10 to 16 C atoms per molecule. The viscosity of kerosene is relatively low, too low to be measured by the saybolt viscosimeter.

All kerosenes are subjected to a number of chemical treatments depending upon the purity required. Pure, bland, water white products are widely used in sprays containing pyrethrum. DDT or synthetic contact insecticides or combinations of these. Since kerosene itself is toxic to insects, it assists in the killing action of the incorporated toxicants. Deodorised kerosene is also widely used in the preparation of DDT residual sprays. The toxicant is usually present at a concentration of 5% by weight, solution is frequently accomplished by the addition of a neutral solvent such as methylated naphthalene, xylene, toluene. When applied to an exposed surface, the solvent rapidly evaporates leaving a microcrystalline deposit of DDT.

(iv) Light and medium oils

Oils with a viscosity between 40 and 65 seconds saybolt are called light oils. These oils are highly refined and contain more than 90% unsulfonatable residues, i.e. they are composed almost entirely of saturated hydrocarbons, presumably made up to 14 to 18 carbon atoms per molecule. The so-called medium oils have characteristics similar to that of light oils but are more viscous from 65 to 85 seconds, saybolt. These oils are also highly refined and contain but small amounts of unsaturated hydrocarbons. Both the light and medium oils are applied as emulsions and are usually emulsified by users.

Paraffinic oils are more toxic to insects than the corresponding naphthenic or aromatic oils. Low molecular weight fractions of all oils exhibit low insecticidal toxicity. Fractions of paraffinic oils boiling below 670°F and having a viscosity of less than 55 seconds saybolt, were practically non-toxic to eggs of the oriental fruit moth. For naphthalene base oils, the limits were higher— a boiling point of 690°F and a viscosity of 110 seconds saybolt being required before appreciable toxicity was apparent.

The so-called superior type oil having a higher content of paraffinic hydrocarbons and small amounts of aromatic and naphthenic constituents have become very popular because of their high efficiency as insecticides and because they are less phytotoxic than the aromatic or naphthenic oils. For orchard work, superior oils are preferred by growers. These oils are efficient insecticides and are low in phytotoxic properties. Oils containing a high percentage of paraffinic hydrocarbons are more efficient insecticides than those containing aromatic or naphthenic compounds.

(v) *Heavy oils*

Oils having a viscosity greater than 55 seconds saybolt are usually classified as heavy and are used for insecticidal purposes principally as dormant sprays. For this purpose these oils need not be highly refined. In many cases, oils intended for lubrication have been used successfully. These oils are also applied as emulsion containing upto 85% oil. For the control of certain insects, other materials such as dinitro-o-cyclohexylphenol are added to increase the toxicity of the dormant spray and thus allow a reduction in oil deposit and still maintain a satisfactory insect control. Combinations of petroleum and tar oils are also widely used as dormant sprays.

Highly refined whale oils of the type commonly used for medicinal purposes and having a viscosity of 150-250 seconds saybolt at 100°F have been used for the protection of sweet corn ears. Oils for this purpose must be exceptionally pure. Heavy oils are mainly used for application to the dormant vegetation and hence do not need to be highly refined. These are frequently used in combination with other materials such as dinitro compounds.

Petroleum oil is often utilised as an emulsifiable solvent and carrier of such water insoluble insecticides as DDT. It may also increase the effectiveness of insecticides through improved spreading and penetration. One or two per cent of petroleum oil may be added to dust formulations to increase the adhesion and the killing efficiency of such materials. Not only any addition of oil to an insecticide reduce the quantity of an expensive ingredient necessary to obtain effective insect control but addition of a toxicant to an oil spray makes possible a reduction of oil content in the interest of safety to plant life.

Insecticides which are soluble in oil but practically insoluble in water will not pass into the water phase of an emulsified oil solution or vice versa. Base nicotine (not the sulphate) is mutually soluble in water and oil phases when an oil solution of nicotine is emulsified in water. At the low concentration utilised in spray practices, the nicotine is about equally distributed between oil and water. This means that if the distribution ratio is "1" indicating equal concentrations in the two phases, a nicotine oil emulsion when diluted so as to contain 99% water and only 1% oil will release 99% of its total nicotine content to the water phase. Since much of the water in an oil emulsion drains off during and immediately after the application and a larger percentage of oil remains, a distribution of nicotine such that a greater proportion of it is dissolved in the oil phase would tend to extend the period of effectiveness of the nicotine. A very low distribution ratio would be required if most of the nicotine were to be concentrated in the oil phase of the diluted emulsion. Alkalies above 0.1 N tend to drive nicotine into the oil phase. Acids tend to extract nicotine base from the oil through the formation of water soluble salts which are not soluble in oil.

Most vegetable oils were found to retain more nicotine than mineral oils in competition with water. This is due to the greater polarity of vegetable oils. Some materials added to oil may be intended to kill one pest while the oil is applied against another pest. For example when a mixture of oil and cryolite is used against citrus pests, cryolite is intended to kill orange worms, and the oil—the scale insects and mites. Further, oil also improves the action of cryolite by increasing its spread and adhesion.

Insecticidal Action of Petroleum

Highly refined petroleum oils are often not very effective as insecticides. A high content of unsaturated hydrocarbons seemed to hasten the time of death while 2 or 3% unsaturated hydrocarbons seemed to affect scale insects very little. Thus there was an idea that the aromatic and other unsaturated structures in an oil confer insecticidal potency, but however, in certain studies it was found that the efficiency of killing certain pests by spray oils of the lubricating type high in paraffinic compounds, was much superior to that of naphthene in spray oils.

Efficiency is increased as the aromatic content of the oil is decreased. The belief that the aromatic content of an oil results in a lessening of the insecticidal efficiency probably stem from the relationship known to exist between these structures and the safeness of oils to plants and foliage. Contrary to the situation with eggs of fruit tree leaf roller, aphid eggs are quite resistant to petroleum oils but very susceptible to low concentrations of tar oil which are predominantly aromatic. It therefore appears that some insects are more effectively controlled by oils of high paraffinicity while others are more susceptible to oils of high aromatic content.

Regarding viscosity it is doubtful if it is of importance with respect to the killing action of oils. When a petroleum product is applied as a contact insecticide, it may actually kill by the toxicity of its vapours entering the respiratory 'system of the insect. The fumes of low boiling fractions, such as gasoline, kill insects readily but the higher boiling fractions in the lubricating oil range have so little volatility as to possess no apparent fumigant action. The lower the boiling point of a petroleum product, the higher the possible concentration of toxic vapours but below a certain point, the increase in concentration is of no advantage because for homolgous compounds, toxicity is reduced as boiling point is lowered. When oil emulsions are applied to aphids, the liquid phase of oil appears to be the active agent and higher boiling point fractions are thus more toxic.

Oils are being used extensively as scalecides and mosquito larvicides. The mechanism of oil penetration of the tracheae has been considered chiefly with respect to scale insects and mosquito larvae. Oil penetrates very slowly but apparently it reaches the spiracles of an insect readily if the insect covering is composed of rather loosely woven waxy threads.

De Ong observed that when a drop of kerosene was placed on a scale insect, it penetrated throughout the tracheal system but was frequently expelled almost as rapidly, apparently by the efforts of the insects. Penetration of the tracheae is related inversely to viscosity, oils of low viscosity being more active in this respect because of their greater mobility. On the other hand, oils of high viscosity are less likely to be expelled from the tracheae. Suffocation is an extremely slow process in an insect immersed in a non-toxic liquid, probably because an inactive insect carries sufficient air within its tracheal system to supply its tissue oxygen for considerable time. Potato tuber moth larvae become motionless much more rapidly in either nitrogen or water than they do in the oil. The oil treated larvae however become discoloured if left in oil or dried and shriveled if exposed to the air in much less time than is required for death in nitrogen or water. Thus it can be inferred that even the most highly refined oil kills the larvae by virtue of its toxicity many hours before death could be assumed to take place by suffocation.

Petroleum oil has been extensively employed for the control of mosquitoes in their larval haunts. Many theories have been advanced to explain the action of the oils upon mosquito larvae. The earlier theory said that oil lowered the surface tension of water and this prevented the larvae from clinging to the surface of the liquid environment and breathing. The other theory postulated that oil layer on the surface of water acted as a mechanical barrier between the larvae and the atmosphere thus resulting in suffocation. However, it has been found that oil killed the larvae while simple suffocation took place only after long submersion. According to the other theory, toxic compounds from oil are dissolved in water. However it has been found that larvae were not killed by water which had been exposed to the oil. Ross suggested that the oil enters the siphon and blocks the tracheae. Observations showed that mere blocking even with oil is not important. It was later suggested that oil entering the tracheae act directly on the tissues as a contact poison. However in tests the larvae were found to die before the oil had an opportunity to penetrate. At present the most reasonable explanation seems to be that the vapour from kerosene penetrates this tracheae rapidly with lethal effects. This theory applies in the case of relatively volatile kerosene but with heavier oils, action in the manner would be progressively less as the boiling point of oils becomes higher. Even the larvae of the petroleum fly is killed by vapours which enters the tracheal system from the more volatile fractions of petroleum.

Similarly several theories have been proposed as to the mode of action of oils when used as ovicides. Some of them are: (i) oil may coat the egg in such a manner that the young insect cannot get energy, (ii) The oil may soften the chorium (or shell) of the egg to interfere with its embryonic development, (iii) The oil may act by contact with the delicate integument of the young insect when the latter emerges from the egg. (iv) Oil may prevent respiration of the developing embryo by interfering with the passage of oxygen or carbon dioxide through the chorium, and (v) the oil may actually penetrate the egg and coagulate the protoplasm within. Oil deposits on a surface may prevent the development of newly hatched larvae. Sublethal contact of a scale insect with an oil deposit may result in a reduction in the emergence of the young scales or an appreciable percentage of young may be dead when born. The aphid and psyllid eggs are more effectively killed by the aromatic structure such as that occurs in tar oils, but geometrid and capsid eggs are destroyed more readily by what appears to be the physical environment and stifling of the eggs.

Horticultural sprays

Oil sprays can be roughly divided into winter or dormant sprays and summer sprays. Winter sprays contain heavy oils and are applied to deciduous trees in the dormant stage. Oils need not be highly refined for this purpose and combinations of petroleum and tar oils are often used. They are used largely to kill aphids and the overwintering eggs of any insects and mite pests. Since winter oil sprays alone are often unable to control the overwintering eggs of some of these pests, incorporation of well-known lime sulphur or dinitrophenol insecticides into them increases the ovicidal effectiveness of dormant sprays, yet allows a reduction in oil deposit while maintaining satisfactory pest control.

Summer spray oils are much more carefully prepared, for they are lighter and are highly refined. They may be toxic by themselves against many pests or essentially non-phytotoxic and suitable solvents or carriers for other pesticides and other crop control agents. These petroleum sprays are widely used to control sap feeding insects such as scale insects and mites on fruit trees. On deciduous trees in full foliage and on green house plants, much use is made of combined sprays, the insecticidal additives being as diverse as DDT, lead arsenate, nicotine, organophosphorus compounds or rotenone. Various fungicides may be added to these combinations to provide a complete pest control treatment. Studies have shown that there is a high probability that spray oil induced reduction of insoluble solids of orange juice and this can be avoided if the application is made in August and September and according to the recommended dosage and formulation.

Effect of Petroleum on Plants

Adverse effects from petroleum sprays restrict the use of insecticides and acaricides that are otherwise, often satisfactory. Control of pests with oil without causing injury, required a balance among at least three major interdependent factors: nature of the oil, concentration of the oil in the spray mixture and amount of oil remaining as a residue on the plant surface.

The emulsifiers used with an oil may sometimes increase its toxicity to plants. Oil emulsion made with soap may be injurious even though either soap alone or oil emulsified with another emulsifier is not injurious. If dormant spraying is practised, it is desirable to increase moderately the oil content of an emulsion whenever full strength bordeaux mixture is used with it, as bordeaux mixture reduces the efficiency of oil emulsion. The sulphonation test is the principle test for predicting the safety of oil for foliage application but it is not always the best one. Oils extracted with sulphur dioxide are more injurious than those treated with sulphuric acid, possibly due to the differential separation of the toxic chemicals from oil.

Hydrocarbons of spray oil have little toxic action towards the leaves until they are oxidised to oil-soluble acids. Saturated oils are oxidised too slowly to become harmful under ordinary conditions of use and slow oxidation is also characteristic of unsaturated hydrocarbons if kept in the dark. Upon exposure to sunlight however, the latter rapidly absorbs oxygen and becomes increasingly dangerous. Oils composed primarily of aromatic hydrocarbons absorb oxygen and develop acidity at a more or less uniform rate, when exposed to light and oxygen. Whereas in oils chiefly paraffinic in nature, these processes are delayed during the latent period. Following this slow start, the processes proceed at an accelerating rate. Greater changes may occur with paraffinic than with aromatic oils. This induction period may be long enough for of the oil to disappear from the leaf. The greater tolerance of plants towards paraffinic oils may be partly due to it.

As in the case of foliage injury from other spray materials, injury to plants may be either acute or chronic and both involve deep seated physiological changes. Acute injury often appears within a matter of days after application and is marked by foliage burn followed later by leaf or fruit drop, but the use of highly refined oils, minimum effective dosage and better timing of treatment will minimise this type of injury. Injury to fruit or wood seldom occurs except with oils having a high percentage of unsaturated hydrocarbons such as commonly found in untreated oils or those\slightly refined. Contact of these oils with the roots may cause the death of a tree within a relatively short time. Unsaturated hydrocarbons are responsible for practically all leafy drop and any other obvious injury occurring within a few days after spraying.

Chronic injury, on the other hand, may not develop until after long periods of repeated applications. It may result due to plugging of the intercellular spaces and surrounding of the plant cells with oils, thus preventing normal physiological functions of the cells. The foliage may become yellow slowly and subsequent gradual defoliation may extend over a period of weeks. Fully mature leaves particularly those which are senile are more susceptible to injury from neutral oils than younger ones, still in the growing stage. However very young leaves are much more susceptible to injury from unrefined oils than mature but not senile leaves. Even this varies greatly according to the degree of refinement of oil used.

Under certain conditions, too frequent applications of oil are a source of injury. The respiration rate of dormant apple twigs was reported to be accelerated more by a light unsaturated oil than by a heavy unsaturated oil. On the other hand no significant difference in respiration rate was exhibited by the twigs treated by light and heavy saturated oils. Oils differ in their effect upon respiration depending upon the stage or seasonal development of the plant. Green and Johnson concluded that oils with 16 per cent or more sulphonatable residue increased the respiration of bean leaves whereas oil less than 16 per cent sulphonatable residue decreased respiration.

Oils have two main effects on leaves, viz. a physical effect tending to decrease respiration by plugging the stomata and intercellular spaces, and a chemical effect in which toxic chemicals increase respiration. Accordingly the effect of oil on respiration depends on the predominance of physical or chemical effect. Spray oils were found by Kelley to retard the transpiration rate of apples, pear, cherry, peach and plum shoots. The rate of transpiration of sprayed shoots ranges from approximately 1.4 to 2.3 times that for checks. Oil sprays retard transpiration in apple foliage only when they come in contact with lower surfaces where the stomato are located. Transpiration of old leaves is reduced much more than that of young ones. The degree of saturation does not affect the disturbances of transpiration. The length of time required for recovery is dependent also upon the amount of oil applied.

Photosynthesis is also affected by oil sprays, the starch content of the foliage dropping off markedly for a considerable period. In some cases the starch content returns to a point far in excess of that in the check. Oil sprays intensify the green colour of foliage of sprayed trees. Gunsburg found an increase in the chlorophyll content of apple tree leaves sprayed with oil, the increase being as high as 47 per cent. After application of summer oils, foliage may drop later in the season, than normal.

When citrus foliage and foliage of others is sprayed with unemulsified oil, the oil passes through the cuticle on either surface of the leaf, through the stomata on the lower surfaces, through the cuticle and lenticles of smaller stems and through the corky bark of larger branches and trunk. Oil apparently enters any or all intercellular spaces as it reaches them, but advances only as long as an oil supply remains on the surface. The disappearance of an oil sprayed on a leaf surface appears to be due more to absorption by the plant than to volatility, at least in oils as heavy as those of the lubricating type. Penetration of leaves by an oil depends upon the plant species. On immersion in kerosene the leaves of some species may be penetrated throughout in 13 minutes while others such as the succulent, xerophytic cactus like stonocarop are not penetrated even at the end of 8 hours. Apparently these characteristics which protect from water are also effective against oil penetration. However Rohrbagh in 1941 criticised earlier research based on penetration of unemulsified oil into the plant tissue, because it did not relate closely to the use of oil emulsion under practical conditions.

Evaporation of oils from the interior of living orchard foliage apparently is negligible. Much of the oil is retained by the leaves and this is removed from trees by defoliation. Removal of oils and the resistance of apples to oils helps to explain why there is usually no cumulative injury by oils sprayed on apple trees.

Oil emulsions are commonly mixed with other insecticides. They are usually compatible with arsenicals, bordeaux mixture, and various plant and synthetic products. The emulsifier is often the limiting factor with respect of compatibility. Combinations of oil with lime sulphur although recommended have caused serious injury to foliage. Severe burning may result if an oil emulsion is mixed with a sulphur spray or is applied within 30 days of lime sulphur or other sulphur sprays. Combinations of sulphur with oil emulsion are utilised in certain treatments of citrus against mites but they are hazardous in hot weather. Spraying with petroleum should not be done at temperatures of 90°F or higher nor below 40°F. Injury may result if rainy weather follows an oil spray owing to interference with respiration at higher humidities.

Determination of oil deposits on plants

Oil deposits were determined by English who cut discs of known area from orange leaves which had been sprayed with a three per cent emulsion. Swain and

Green called attention to the distinction between insecticidally active deposit (surface oil) and total deposit (surface oil plus oil absorbed by the leaf). They developed a comparatively simple method for an approximate determination of surface oil which involves the washing of leaf surfaces with methylene chloride, evaporating the solvent and then weighing the residue consisting of plant wax and petroleum oil. This procedure appears most useful when determinations arc made immediately after the water phase of oil emulsion has evaporated and the oil has had little opportunity to penetrate the leaf. Since petroleum oils and leaf were known to be mutually soluble, uncertainty exists as to the error of determination of the surface oil.

Another method is to dry and grind the citrus leaves before making extractives with petroleum ether. However Gunther and Ebelhay pointed out that petroleum oils become located intercellularly and intracellularly. Hence all methods up to that time had failed to remove the oil quantitatively. They devised a method involving selective chemical disintegration of leaf cells, accompanied by a steam distillation of the stable oils present. These methods do not apply to oils other than petroleum. Glass or waxed surfaces have been employed in studying oil deposits because of the ease of weighing the deposit and because even animal and vegetable oils can be used.

Effect of Petroleum on Man and Domestic Animals

Unemulsified petroleum based insecticides have fire hazard associated with them. Petroleum fractions used as insecticides are poisonous when taken internally. Oil sprays may affect farm animals by inflamming the skin even causing excessive rise in body temperature and burning the hide severely to produce visible lesions. The best index of effect on skin was sulphonation of the oil. The viscosity of oil was found to be unreliable as an index of skin reaction. The chemical reaction of oils with the skin caused inflammation of the peripheral capillaries, thereby reducing water dissipation from the skin and impairing the cooling power of animals to critical limits, at high atmospheric temperature. The milk yield might also be reduced over twice as much as a result of the application of oil base sprays as by the activities of the most serious of common fly pests of cattle, the stable fly and others.

Vegetable and Animal Oils and Their Derivatives

Earlier a number of animal fats and vegetable oils have been used as insecticides and acaricides but only under limited and specialised conditions and most often as insecticidal soaps. Several of these are still widely used in Africa, China and Europe against a variety of insects and mite pests, usually as spray emulsions.

Chemistry of vegetable and animal oils: The terms, fat and oil as applied to substances of animal and vegetable origin are interchangeable although a fat is usually considered to be solid at ordinary temperature while an oil is a liquid. Chemically all such naturally occurring fats and oils are esters or organic salts of the trihydric alcohol, glycerine and fatty acids. Because the glycerine molecule contains three hydroxyl groups, it is possible to have a molecule of fat in which there are one, two or three different fatty acids. A wide variety of fatty acids have been found to occur in natural fats and the characteristics of fatty acid molecules composing, the fat, determine to a large extent the properties of the fat. With the exception of margaric acid, all naturally occurring fatty acids are composed of an even number of carbon atoms, although they may be either completely saturated ($C_nH_{2n}O_2$) or unsaturated ($C_nH_{2n-2u}O_2$) where u equals the number of double bonds. All even numbered members of the saturated series of fatty acids from C_2 to C_{30} have been found in biological material but unsaturated acids with less than 18 carbon atoms occurs only rarely.

Vegetable and animal oils differ from petroleum oils in that they are less stable, being hydrolysed without difficulty in the presence of acid and forming soaps in the presence of alkalies.

Vegetable oils

As insecticides, vegetable oils are not often used because of the availability of cheap petroleum and tar oils which are generally more effective insecticides. Castor, sesame, cotton seed, linseed, olive, coconut, corn oil, hemp, palm, groundnut, soybean and rape oils were tested as sprays and it was found that rapeseed oil was the most satisfactory as an insecticide, killing certain insects at a concentration of 0.5 per cent. This investigation used oil emulsions both at the pink stage and as summer sprays on apples. Rape oil contains esters of rape and erucic acids as well as acids of the linoleic and linolenic series. Rape oil stands between the drying and non-drying oils. In other words the presence of unsaturated fatty acids permits combination with oxygen to a limited extent. Sulphur is present in cruder grades of rapeseed oil which may account in a measure for its effectiveness as an insecticide. Though vegetable oils have been found to be highly efficient ovicide, their cost was prohibitive when compared with petroleum and tar oils. Vegetable oils are better solvents than petroleum for such materials as Derris resins.

Animal oils

Whale oil and certain varieties of fish oil such as manhaden and herring oils are the principal animal oils with insecticidal applications. Forsyth is reported to have been the first to advice the use of whale oil against scale insects. Bone oil is derived from the destructive distillation of bones in the preparation of bone charcoal. It is the source of pyridine and other nitrogenous bases. However animal oils have seldom been used as insecticide. Fish and whale oil soaps are rather commonly used as emulsifying and wetting agents.

Fatty acids: The component fatty acids making up oil molecules are known to possess insecticidal value. Fatty acids produced by the hydrolysis of animal or vegetable fats are extremely varied in chemical composition.

Siegler and Popenoe tested a number of fatty acids as contact sprays against insects, particularly aphids. The toxicity of fatty acids increased with molecular

$$
\begin{array}{lcl}
H_3C{-}O{-}CO{-}R' & & H_2C{-}OH \\
| & & | \\
H\ \ C{-}O{-}CO{-}R'' + 3HOH \rightarrow & & H\ \ C{-}OH + HO{-}CO{-}R' + HO{-}CO{-}R'' + HO{-}CO{-}R''' \\
| & & |\qquad\qquad\qquad \text{Fatty acids} \\
H_2C{-}O{-}CO{-}R''' & & H_2C{-}OH \\
\text{Neutral fat} & & \text{Glycerine}
\end{array}
$$

weight at least up to a certain point C_{10} and the free acids were more toxic than their corresponding soluble neutral or alkaline salts. Against tests, a preparation of long chain fatty acids derived from coconut oil were found to be highly toxic at a concentration of 1 to 800-1200 parts of water. Acids present in the mixture were said to be mostly caprylic, capric, lauric, myristic, palmitic, oleic and stearic. Among fatty acids, the most toxic acids were those containing 10 to 12 carbon atoms with toxicity to insects increasing with molecular weight up to this point and decreasing markedly with molecules having more than 12 carbon atoms.

Although molecular size is a factor in determining the rate of cell penetration by fatty acids, lipoid solubility may so dominate the situation that penetration becomes actually more rapid with increasing size of the molecule. Double bonds in fatty acid chain as in oleic acid, change the physical characteristics and toxicity. Toxicity seemed to be correlated with the degree of hydrolysis. The peak of toxicity lies near the 10 carbon capric acid. Siegler and Popenoe recommended the use of the coconut fatty acids for insecticidal purposes, since they contain large proportions of the more toxic acids like capric, lauric and myristic acids. The toxicity of acids increases with molecular weight from acetic acid to undecylic acid (11 carbon acid). Acids heavier than tridecyclic (13 carbon atoms) exhibit decidedly less toxic action. Formic acid is exceptional, being as toxic as butyric acid.

Soaps

Goeze in 1787 appears to have used soap as an insecticide. Among the soaps which were widely used as insecticide earlier, the most important ones were made from animal fats, particularly whale and fish oil soaps. Soaps are alkali salts of fatty acids and are most commonly sodium salts (hard soaps) or potassium salts (soft soaps). The low surface tension of soap solution is of importance in the performance of soaps as insecticides and as spray adjuvants.

Siegler and Popenoe advanced the hypothesis that the toxic action of soaps on insects was due to the action of fatty acids produced by hydrolysis rather than to the soap molecule itself. Previously, it had been held that a combination of soap with small amount of free alkali present was responsible for its insecticidal action. Later it was confirmed that the toxic properties of soaps made from fatty acids parallelled the toxicity of fatty acids themselves and increased with the increasing molecular weight up to carbon chain of 10 to 12 carbon atoms, above which the toxicity fell off. Comparing soaps and series of saturated fatty acids, it was found that soaps in general were less toxic than fatty acids alone, the toxicity of acids and soaps were in the same general order, although the most toxic acid tested was with capric $C_9H_{19}COOH$, i.e. a C_{10} acid, the most toxic soap was that of lauric acid $C_{11}H_{23}$ COOH in the saturated series. The soap of oleic acid was more toxic than any of the soaps of the saturated acids. The toxicity of potassium soaps to plants decreased as the size of the molecule increased. Certain authors found little difference between the toxicity of Na and K soaps while certain others found that sodium soaps were more toxic than potassium soaps. Various authors tried to correlate physical properties such as surface tension and angle of contact to toxicity but were not successful.

Reactions of soaps with other insecticides: Like all salts of weak acids, soaps hydrolyse in water, the extent of reaction being dependent upon the weakness of the fatty acid and strength of the base. Hydrolysis limits the use of soaps as insecticides. Free alkali released from sodium and potassium soaps by the reaction of water is usually undesirable from the insecticidal standpoint. The alkali would react with arsenical such as lead arsenate to produce soluble arsenic. The fatty acids thus would combine with the metallic ion of the arsenicals to form an insoluble soap. Such metallic soap are worthless for wetting or emulsifying purposes because of their gummy nature and their facility for forming inverted (water in oil) emulsion. Formerly, the toxicity of soaps to insects' life was attributed to one of the two causes: (i) The mechanical stoppage of body openings particularly spiracles, and (ii) The caustic properties of free alkali for the soaps. Only in recent years, the action of soaps as insecticides have been found to depend upon the toxic action of fatty acid as well as upon the wetting and film forming properties conferred upon the soap by the fatty acids present.

In combinations with inorganic insecticides such as lead or calcium arsenate, a double decomposition takes place resulting in the formation of insoluble lead and calcium soaps and soluble arsenic. Therefore, such combinations should be avoided. As pyrethrums are themselves esters, the action of alkaline solution such as soap solution is to breakdown the esters with subsequent formation of alkaline salts of the organic acid portion of pyrethrum molecules. Hence soap and pyrethrum preparations should not be combined, particularly if the mixture is to be stored for any length of time. In most of the materials proposed for the combination, soap is present as an agent to assist in the wetting and spreading of the liquid.

Coal Tar Derivatives and Their Classification

The use of coal tars (sometimes called carbolineums, tar cresols, tar distillate, tar oils) as insecticide is of comparatively recent origin. Tar preparations for this purpose had their origin either in Germany or Holland, shortly before 1920. Though the use of coal tar products increased during the period 1930-40, it declined very rapidly with the advent of newer organic compounds.

Coal tar is the term used to describe the distillate obtained from the destructive distillation of coal and is a byproduct in the production of coke and illuminating gas. The composition of this varies very much as it depends upon the type and composition of coal used, temperature and method of distillation. Usually the crude tar is obtained from the coking process and is subjected to fractional distillation during which process it is separated, with their approximate boiling point ranges. The general composition of various fractions are given below:

< 210°C—Light oil, benzene, toluene, xylene

210-240°C—Middle or carbolic acid, phenols, naphthelene

240-270°C—Heavy or cresoate oil

> 270°C—Anthracene oil

In spite of the fact, that relatively little is known about the toxic properties of tar distillate fraction, certain workers have set up certain standards: (i) It should be derived from tar secured from bituminous coal, the latter having been heated to high temperature in gas retorts or byproducts of coke ovens in the making of illuminating gas, (ii) Tar oil shall distill within the following limits: less than 410°F not more than 1 per cent, 410 to 445°F—not more than 10 per cent, 445-671°F not more than 65 per cent, (iii) The tar distillate should not contain more than 3 per cent water, (iv) It shall not contain more than 10 per cent tar acids and less than 5 per cent is to be preferred, and (v) It shall be free from crystals on standing for three hours and with occasional stirring at 41°F.

Oils, comprising one of the most useful groups of insecticides, of these petroleum oils are used most frequently. Animal and vegetable oils are seldom used for pest control purposes, since most of them are in steady demand and command higher prices for other uses as in food production and soaps. Because of their irritating and phytotoxic properties the use of tar oil for insecticidal purposes is no longer practical in this country.

Coal tar produced in the manufacture of illuminating gas and coke is a mixture of hydrocarbons and differs from petroleum oil in that compounds of the aromatic series predominate rather than the paraffinic and naphthenes. The tar is distilled and several fractions are separated which resemble petroleum fractions. In general, it is the neutral fraction of the highly boiling distillation that is most useful for spray purposes. Tar distillates must not only be fractionated as to boiling point but also with respect to acidity. Unrefined tar distillates contain both tar acids and tar bases along with the predominant neutral oil.

Tar acids are phenols (hydroxy benzene), cresols (methylphenols), xylenols (dimethylphenols) and higher hydroxy compounds. They are the principal factors in plant injury and are irritating to the human skin. Their removal however is said to reduce the effectiveness of higher boiling tar oils. They can be extracted with caustic soda by the formation of water soluble sodium salts. A certain minimum content of the acids is an aid in the emulsification of the tar distillates. The tar bases are mainly nitrogenous heterocyclic compounds related to pyridine and quinolines. They are soluble in acids.

Neutral oil used as an insecticide is derived from the creosote and anthracene fractions. It is chilled to remove crystalline naphthelenes and anthracene neither of which appears to add much to the effectiveness of the oil. The degree of effectiveness varies with the kind of pest against which it is applied. Due to this, ccrtain dinitrophenols have to a large extent displaced tar distillate emulsion as orchard ovicides.

Not only are the higher concentrations of tar oils injurious to dormant trees in normal health, but after severe winters tar oils are likely to cause greater injury. Coal tar creosote is standard wood preservative but one disadvantage is that freshly treated wood is easily ignited and burnt. The liquid itself may cause burns and as such protective clothing must be worn while working with it.

Wood tar products: Raw pure tar oil obtained in the process of destructive distillation of pure wood is high in organic acids and liable to injure plant foliage although processed pine tar oils have been prepared which are said to be safe for foliage applications. Pine tar oils are sometimes recommended for use in animal life. Wood creosote may be derived from pure wood or from hard wood such as birch, beach or maple. Hard wood creosote is likely to be undesirable because of its odour.

Coal tar or tar oils are still being used as dormant sprays in Europe but not in the U.S.A. These strongly irritating and phytotoxic products are obtained from the destructive distillation of coal or from coke ovens, from coal tar or from wood. These are extremely variable and complicated in composition. Aromatic hydrocarbons dominate these mixtures but tar oils (phenols and cresols) and tar bases are also present. Coal and wood tars are incorporated into the emulsions and are used for the control of overwintering aphids and psyllids. However, they do not control redspider mites.

CHAPTER VII

ORGANOCHLORINATED HYDROCARBONS

DDT (Dichlorodiphenyl trichloroethane: 2,2,-bis (p-chlorophenyl)- 1,1,1-trichloroethane: 1,1,1-trichloro-2, 2-bis (p-chlorophenyl) ethane, pp'-DDT)

The synthetic organic group of insecticides is of very recent origin. Many of the synthetic organic compounds are so new that their properties have not been fully studied. This compound first synthesised. by Zeidler in 1874, was tested as an insecticide by Swiss workers during 1939-1940 and was found to be very effective. At that time as the natural insecticides like pyrethrum and rotenone were extremely scarce due to war and as DDT possessed a long residual activity, it was quickly adopted as an insecticide. Its extreme toxicity towards insects particularly those affecting man is practically harmless nature against human beings or animals helped its rapid spread. However, DDT is being used much less at present for housefly control since the fly has developed resistance to DDT.

DDT is prepared by the original Baeyer condensation method used by Zeidler. It is produced by reacting 225 parts of chlorobenzene with 147 parts of chloral in the presence of 1000 parts of monohydrated sulphuric acid.

$$Cl_3CCHO + 2C_6H_5Cl \rightarrow Cl_3CCH(C_6H_4Cl)_2 + H_2O$$

After cooling the mixture to room temperature, the mixture is poured into a large excess of water where the solid product separates out. This can be filtered off. This reaction takes place in the presence of condensing agents like concentrated sulphuric acid, oleum, chloro-sulphonic acid, fluorosulphonic acid, hydrogen fluoride, anhydrous aluminium chloride.

The composition of crude DDT varies with the conditions of reaction. The average commercial product contains 70 per cent of pp'-DDT (l, l, l-trichloro-2, 2-bis(p-chlorophenyl) ethane) and other related compounds even as many as 13 compounds. As DDT is a mixture of compounds, U.S. Specifications have stated that the technical material should have an organic chlorine content between 43 and 51% by weight and a pH between 5 and 8.

Pure pp'-DDT is a white crystalline compound occurring in long tubular needles that are biaxial. It has a melting point of 108. 5 to 109°C and a density of 1. 556. Pure DDT is virtually insoluble in water but is soluble in many organic solvents (xylene-60%, carbon tetrachloride 47%, ethylacetate 68%, kerosene 8%). The best solvents for DDT are ketones, esters of lower fatty acids, aromatic hydrocarbons, halogen derivatives of the aliphatic and aromatic series. The higher the aromatic hydrocarbon content of petroleum products, the more DDT they will dissolve. In most solvents, the effect of temperature on the solubility of DDT is pronounced.

Pure DDT samples are relatively stable and decompose only when heated to a temperature of 195°C. However impure samples containing iron or aluminium salts and alkalies as impurities, catalyse the thermal decomposition of this compound resulting in the formation of 1, 1-dichloro-2, 2-bis (p-chlorophenyl) ethylene, an interior insecticidal compound due to dehydrochlorination. Even as little as 0. 01 per cent of anhydrous ferric chloride will catalyse this reaction.

$$(ClC_6H_4)_2CHCCl_3 \rightarrow (ClC_6H_4)_2C = CCl_2 + HCl$$

In the presence of strong alkali, DDT is dehydrochlorinated to DDE (2, 2-bis p-chlorophenyl)-1, 1-dichloroethylene) which is oxidised to pp'-dichlorobenzophenone. The thermal decomposition point of pure DDT is 125°C. The tropical

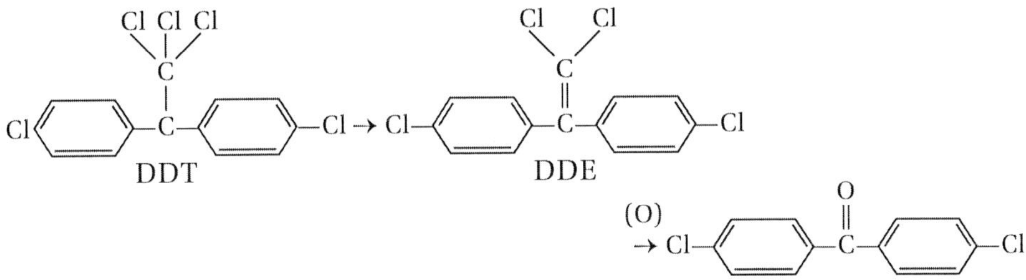

Light pp'-dichloro benzopheonon

and subtropical temperature may be sufficiently high to decompose technical DDT rapidly causing complete loss of activity in a short time, although long exposure to ordinary temperatures in temperate zones may not cause appreciable changes. UV light and in a few cases sunlight may cause decomposition of DDT. Many solvents like dioxane fuel oil-2, kerosene, motor oil, octadecyl alcohol, soybean oil, tetrahydronaphthalene and methylated naphthalene have an inhibiting action on the catalytic decomposition reaction caused by anhydrous ferric chloride. DDT was readily decomposed in solvents containing chloro- and nitro- groups such as mono- and dichlorobenzene, ethylene chloride and nitrobenzene and also by basic organic compounds such as methylamine and dimethylamine.

A detailed study of the compatibility of a number of diluents, insecticides, fungicides and fertilisers with DDT using recrystallised DDT and measuring the amounts of HCl evolved after heating to 115 to 120°C for one hour, has revealed that the compounds can be grouped into three groups, viz. (i) materials resulting in severe decomposition of DDT (aluminium chloride, chromium dolomite limestone, fuller's earth, iron filings, kaolin, nicotine and stainless steel); (ii) materials causing some decomposition of DDT-(bentonite, bordeaux mixture, cupric chloride, ferric dimethyl dithiocarbamate, pyrophyllite, sulphur and talc); and (iii) materials causing practically no decomposition of DDT (alumina, ammoniated superphosphate, ammonium nitrate, ammonium sulphate, calcium arsenate, calcium oxide, copper powder, cryolite, cyanamide, 2,3-dichlor-1,4-naphthoquinone, lead acetate, lime sulphur, Paris green, potassium chloride, potassium sulphate, pyrethrum, tin, rotenone, sodium fluoride, sodium fluorosilicate, sodium nitrate, steamed bone meal, uranium, zinc chloride, zinc dust). However in certain cases like pyrophyllite and talc wide differences were observed in samples from different sources. Sprays containing only DDT in solution should be used as residual sprays for maximum effect. However, when a combination of immediate contact action with residual action is desired, pyrethrum extract or its equivalent is mixed with DDT.

Compounds related to DDT: Number of closely related compounds with insecticidal action have been synthesised:

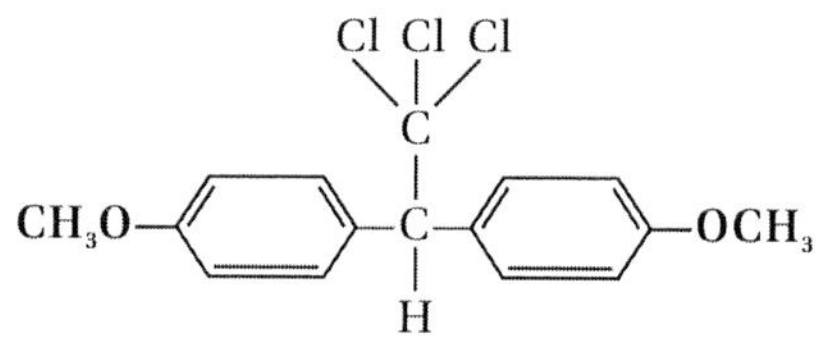

Methoxychlor

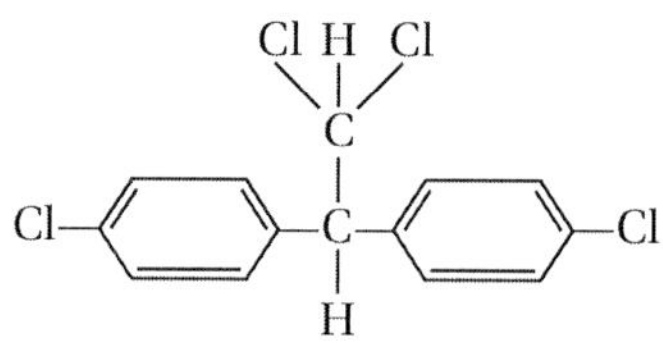

DDD

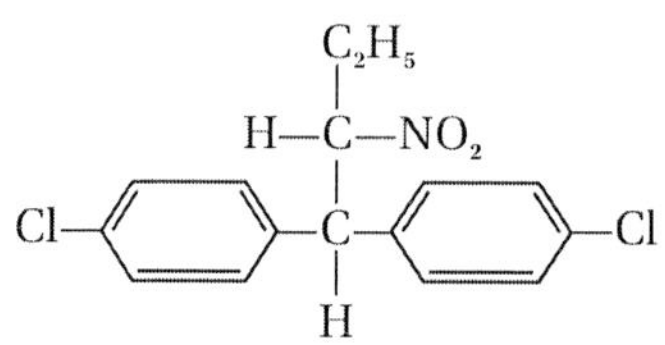

Bulan

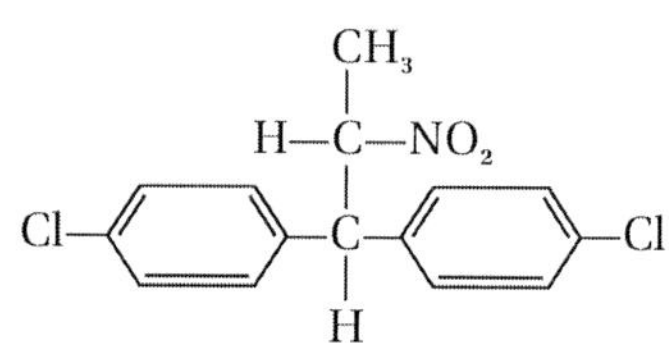

Prolan

Dilan = Mixture of Bulan and Prolan

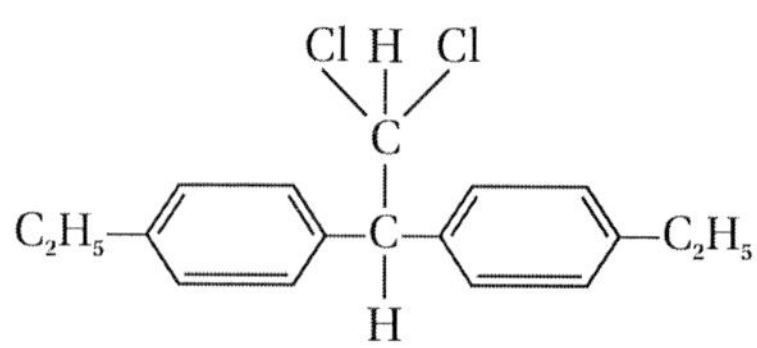

Perthane

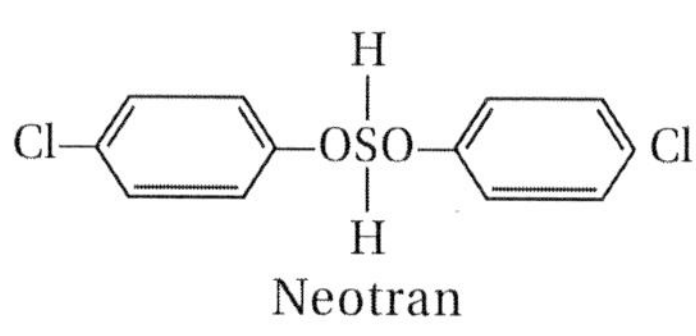

Neotran

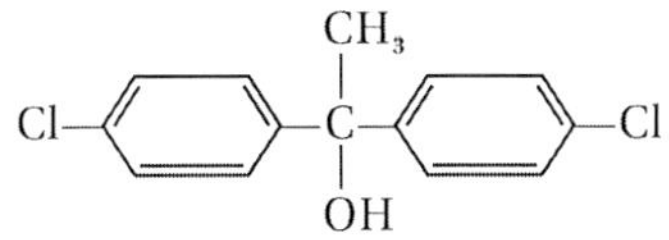

Dimite

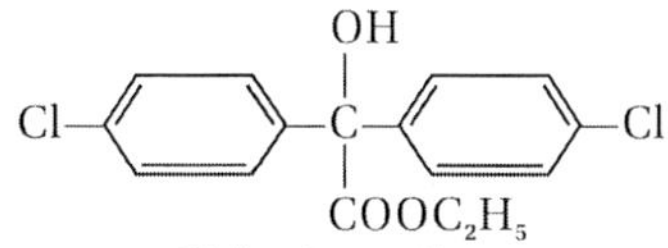

Chlorbenzilate

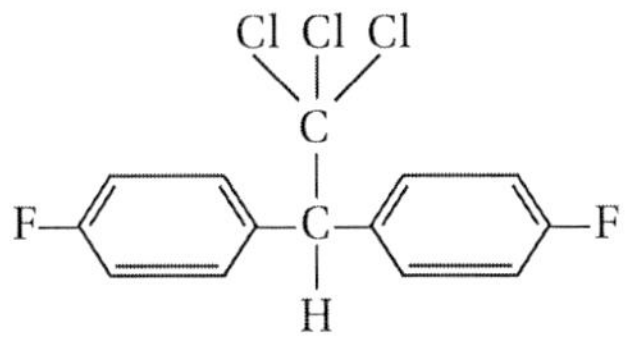

DFDT (Gix)

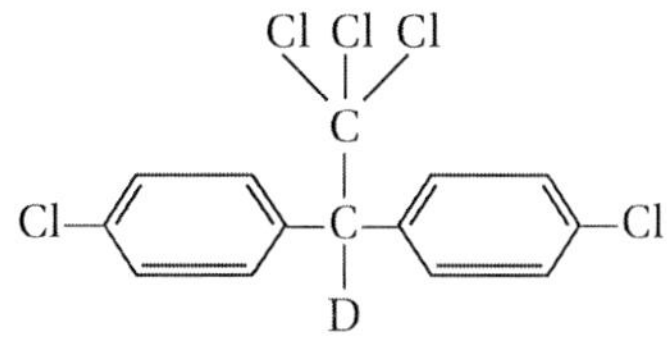

Deutro - DDT

Methoxychlor (DMDT, Methoxy analog of DDT, dianisyl DDT, 2, 2-di-p-anisyl-1,1,1-trichloroethane)

Methoxychlor was prepared by Elbs in 1893 by reacting anisole and chloral in the presence of sulphuric acid or glacial acetic acid. Commercial methoxychlor

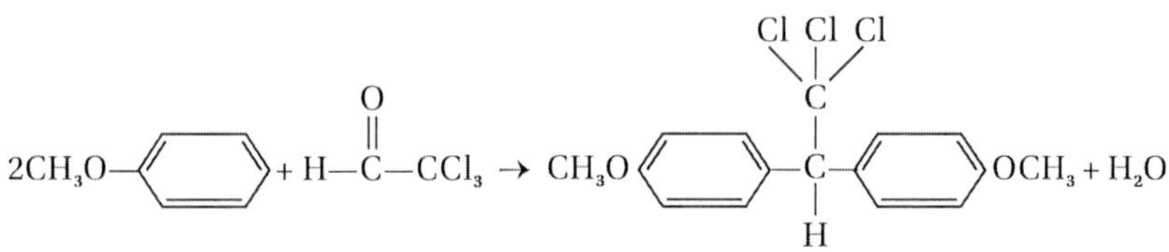

contains about 88% of 1,1,1-trichloro-2, 2-bis (p-methoxy phenyl), ethane and 125 related materials. It has generally the same physical and chemical properties as DDT but is somewhat more resistant to decomposition in the presence of alkalies. It is a crystalline substance, insoluble in water but soluble in many organic solvents. It has a relatively low toxicity to higher animals, with little residue hazard and is also not concentrated in animal fat. However because of its high cost and low effectiveness towards a number of insects, its use is limited. Methoxychlor in petroleum base however has been found to have good knockdown effect on houseflies. Methoxychlor is often used as a third component in the preparations containing DDT and lindane.

DDD {TDE, Dichlorodiphcnyl dichloroethane, tetrachlorodiphenylethane, l,l-dichloro-2, 2-bis (p-chlorophenyl) ethane}

DDD differs from DDT only in the absence of one chlorine atom in the ethane moiety. Commercially, TDE is made by reacting mono-chlorobenzene with the products of chlorination of ethyl alcohol at 25-30°C. The crude products contain a number of related compounds of which o,p'-isomer is important.

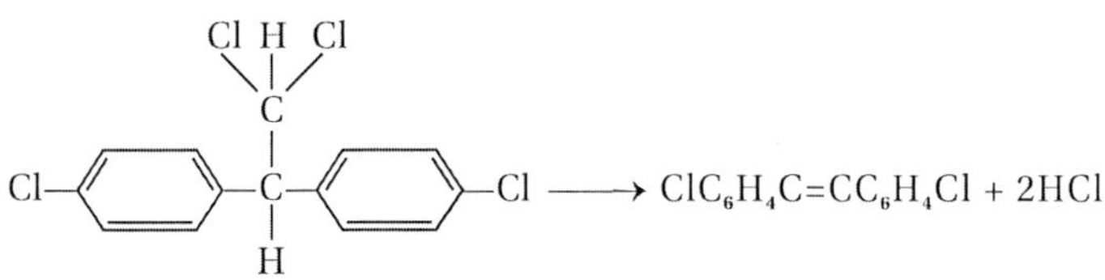

Chemically, DDD is similar in most properties to DDT but is more slowly dehydrochlorinated than DDT. However when heated with ferric chloride, it releases HCl and dichlorotolan. It has considerably less toxicity than DDT.

Dilan

Dilan is a mixture of two compounds, viz. 53.3% Bulan (DNB) {2-nitro-1,1-bis (p-chlorophenyl) butane} and 26.7% of Prolan (DNP) {2-nitro-l,1-bis (p-chlorophenyl) propane}. In addition to the above, the commercial product contains 20% of related compounds (chiefly the o,p'-isomer). The technical product is a brown sticky plastic semisolid at room temperature and is practically insoluble in water. It is slightly soluble in oils and ethyl alcohol but soluble in methyl alcohol and aromatic solvents. Like DDT, dilan is unstable in the presence of alkali and is also oxidised rather easily.

Perthane {2,2-Dichlor-1,1-bis (p-ethylphenyl) ethane, di-(p-ethylphenyl) dichloroethane}

Perthane is a selective insecticide with low toxicity to higher animals and is 1/5 to 1/10 as active as DDT for different insects. It is a white crystalline substance insoluble in water, but highly soluble in organic solvents. The LD_{50} value for mice is 6600 mg/kg.

Neotran {DCPM, Bis (p-chlorophenoxy) methane}

Neotran is an acaricide, produced by reacting p-chlorophenol, dichloromethane and sodium. It is almost insoluble in water and oil but is readily soluble in ether and acetone. It is stable in the presence of alkali and is compatible with practically all the common spray materials.

Dimite {DCPC, DMC, 1, 1, -Bis (p-chlorophenyl) ethanol, di-(p-chlorophenyl) ethanol}

Dimite is colourless and crystalline compound made by reacting 4, 4-dichlorobenzophenone with methyl magnesium bromide and later treated with water. Similar to the other chlorinated hydrocarbons, dimite is also practically insoluble in water but soluble in most organic solvents. On heating, it gives di-(p-chlorophenyl) ethylene. The same decomposes in acid solutions but however this compound is generally compatible with most common spray materials. Its toxicity to mammals is apparently low.

Chlorobenzilate {Formerly Geigy 328, 2-hydroxy-2, 2, -bis (4-chlorophenyl) ethylacetate}

The pure compound is a yellow viscous liquid, slightly soluble in water, soluble in most organic solvents. Being an ester, it is decomposed by strong alkalies and acids and hence not compatible with alkaline compounds. The technical material contains about 90% of 2-hydroxy-2, 2-bis (4-chlorophenyl) ethylacetate and is a brownish liquid with a specific gravity of 1.2816 at 20.4°C. It is primarily an acaricide and has an acute oral LD_{50} of 3200 mg/kg to mice.

Sulphenone (4-chlorophenyl sulphone: Para-chlorophenyl-phenyl sulphone)

Sulphenone also exists as colourless crystals with a slight aromatic odour. Two forms with different melting points, 90 and 94°C are known to exist. It is practically insoluble in water, slightly soluble in oil and freely soluble in aromatic and polar organic solvents. At ordinary temperatures, it is stable to acid and alkalies and hence is compatible with all common spray materials. Sulphenone possesses considerable toxicity to mites but does not appear to be toxic to other insects.

Ovotran (4-Chlorophenyl-4-chlorobenzene sulphonate)

Ovotran is a white tan flaky solid, insoluble in water, slightly soluble in alcohol and oil, and readily soluble in acetone and aromatic solvents. Because of its ester nature, this compound is decomposed by alkalies but is compatible with neutral or acid spray materials.

Aramite {2-Chloroethyl-1-methyl-2-(p-tert-butyl phenoxy)-ethylsulphite: Butyl phenoxy isopropyl chloroethyl sulphite: 2-(p-teri-butyl phenoxy) isopropyl-2-chloroethylsulphite}

Though aramite does not possess the characteristic DDT type structure, there is chemical similarity between these compounds. The pure compound which is a colourless liquid with boiling point 175°C at 0.1 mm Hg is prepared in 2 stages:

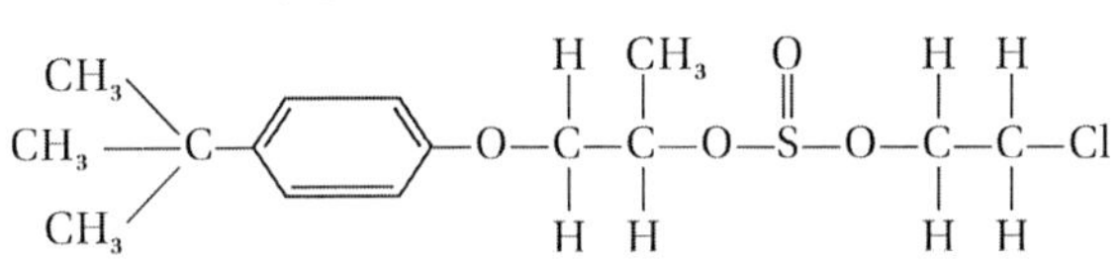

(i) the 2-chloro-ethanol is reacted with $SOCl_2$ to give 2-chloro-ethylchlorosulphonate. (ii) This is then reacted with the condensation product of propylene glycol and p-tert-butylalcohol. Being an ester, aramite is readily decomposed by alkalies and hence not compatible with lime or other alkaline spray materials. It is also slowly decomposed by light. It is used principally as an acaricide.

DFDT {Gix: 1, 1, 1-trichloro-2, 2-bis (p-fluorophenyl) ethane}

The fluorine analog of DDT was prepared during World War II by reacting chloral with fluorobenzene. The purified p, p'-isomer crystallises from alcohol as white needles and has a fruity odour. Like DDT, DFDT is easily dehydrofluorinated by alkali but unlike DDT it has a very high solubility in many organic solvents. Like DDT it is accumulated in the body fat but somewhat less toxic to warmblooded animals.

In view of the remarkable insecticidal properties of DDT, hundreds of compounds with similar constitution have been synthesised and tested. Of the impurities presents in DDT, only o,p'-DDT {1, 1, 1, -trichloro-2(0-chlorophenyl)-2-(p-chloro phenyl) ethane} was found to be from 8 to 21%. o,p'-DDT is a solid, crystallising in hard rhombic, is dehydrochlorinated by alcoholic alkali like p, p'-isomer but at a much slower rate. It is a reasonably stable compound and does not eliminate hydrogen chloride upon heating at 115 to 120°C for 2 hours although the reaction is catalysed by traces of anhydrous ferric chloride. The toxicity of o,p'-DDT is considerably less towards insects than the p,p'-isomer. Only about 0.1% of another isomer {O,O'-DDT, 1,1,1-trichloro-2,2-(O- chlorophenyl) ethane} is found but it is not highly toxic to insects.

Structure activity relationship

Of the large number of DDT analogs, only a few like 1 DE, methoxychlor and Gix have been developed commercially. For maximum activity, the DDT-type molecule must be substituted in p, p' position with halogen or short chain alkyl or alkoxy groups. The nature of substituents for the ethane moiety is also critical and this molecule needs to have an optimum complementary to hypothetical protein receptor. This is expressed by the activity in the following order $CCl_3 > CHCl_2 > CH(NO_2)CH_3 > C(CH_3)_3$. The substituent on the α carbon not only affects the optimum configuration of the molecules but also, influences detoxification in the insect which can be reduced by α-D or α-F analogues.

Mode of action

DDT is particularly effective against many species of insects like leafhopper, flies, mosquitoes, lice, fleas and bedbugs, beetles and plantbugs and is ineffective against Mexican bean beetle, red spiders, other spider mites, cotton bollworms, cotton leafworm and many aphids. Although DDT inhibits the development of orchard mites, as number of predators are eliminated, mites increase very rapidly after an application of DDT.

The effectiveness of DDT depends often upon the type of formulation employed. DDT was found to be 4 to 6 times more toxic against the aphid *Macrosiphum soalanifolii* when dispersed as an emulsion in the benzene solution than when dispersed as a suspension of fine solid particles made by stirring an acetone carbitol solution of DDT into water.

When houseflies walk on a film of crystalline DDT deposited on a surface such as glass, they become incapable of flight, and later are unable to walk. For a considerable time, they exhibit the tremors characteristic of DDT poisoning in insects. When mosquitoes come in tarsal contact with DDT deposits, the following symptoms appear in the order: excitation, ataxia (loss of coordination), convulsion movement, knockdown (general paralysis) and death. The insect may become restless and excited within a few seconds of contact with DDT. Mosquitoes excited by contact with DDT show unusual preference for light and fly out of hiding places. Such DDT excited mosquitoes bite readily until they are knocked down. The sensitivity of insects in general to external applications of DDT is correlated to adsorption of DDT by the chitnous cuticle. Such cuticle may act to concentrate selectively DDT from the surrounding medium. In insects, the major site of action of DDT appears to be the peripheral nervous system including both sensory and motor nerves.

Two general theories based on the Overton-Meyer theory on lipoid solubility and physiological activity have been advanced to explain the relation of chemical structure of DDT to insecticidal activity. Langer and his associates in 1944 postulated that the trichloromethyl part of a molecule (similar to the structure of chloroform which confers upon DDT, its lipoid solubility) is a requirement to bring about penetration to the site of action and that the p-chlorophenyl groups linked by a carbon atom constitute the toxic component of the molecule.

The theory of Martin and Wain is based on the opposite view that the p-chlorophenyl grouping is responsible for the lipoid solubility of DDT and trichloromethyl radical furnishes toxicity in relation to the amount of hydrochloric acid which is split off by dehydrochlorination of vital centres of the organism. However, considerable evidence refutes these theories, because when hydroxyl radicals replace the chlorine atom of the trichloromethyl group, and thus prevent dehydrohalogenation, the resulting compound is no less toxic than DDT. Many compounds which split off hydrochloric acid are not insecticidal. Speed of splitting is not a vital factor because DDT yields acid rapidly while methoxychlor does not.

Malachite green and others which exert insecticidal action, contain linked phenyl groups and dehydrochlorinate to form the corresponding leuco compounds. The process of DDT degradation *in vivo* appears to involve dehydrochlorination to dichloroethylene derivative, then to di-(p-chlorophenyl) acetic acid (DDA) which is excreted in the urine of higher animals. Both the dichloroethylene derivative and di-(p-chlorophenyl) acetic acid are relatively inactive insecticidally, yet they contain the linked chlorophenyl structure unchanged.

Cambell and West pointed out that the mode of action of DDT is clearly very complex, and the mechanism of action on insects has not yet been determined. When chlorine in the aromatic radical of 1,1,1, -trichloro-2, 2-bis (p-chlorophenyl) ethane is replaced by hydrogen, bromine, iodine, hydroxyl, a higher hydrocarbon radical, amine, thiocyanide, carboxyl, nitro or cyano groups, the insecticidal activity of the compound is substantially decreased. A lowering of insecticidal activity also occurs when several methyl or alkoxy groups are introduced into the aromatic radicals. Replacement of chlorine by fluorine, methoxyl, methyl or ethyl does not substantially change the insecticidal activity of the compound but it lowers toxicity for the vertebrate animals and man.

Replacements of one chlorine in the trichloromethyl group by hydrogen lowers the activity for insects by 2 to 4 times and the toxicity by 5 to 15 times. When a second atom of chlorine is replaced by hydrogen a five-fold decrease in toxicity for animals is observed when compared to DDT and the insecticidal activity decreases 5 to 50 times. Generally, it has been found that all the trichloroethane studied having p, p substituents like DDT are mostly insecticidal but have no acaricidal activity. Alterations in the tri-chloroethane portion of the molecule, the p-chlorophenyl radicals remaining unchanged, result in reduced insecticidal action but an increased acaricidal action. Bis-(p chlorophenyl) methane is unique in being effective both as an insecticide and as an acaricide. Langer's theory of lipoid solubility is conferred on the DDT molecule by the trichloromethyl group is not applicable to acaricides, probably because there are distinct physiological differences in the insect and mite integument and/or nervous system and also there exists a marked specificity in acaricides.

With regard to the chemical mechanism responsible for toxicity, Campbell and West suggested the evolution of hydrogen chloride as only a secondary effect accompanying the more fundamental alteration in the electromagnetic nature of the group from which it is evolved. The effect of chlorine atoms therefore may be entirely steric and that the three α atoms of chlorine are necessary for conferring optimum toxicity on the compound due to their providing the required steric orientation and that this reaches a maximum when the remaining chlorine atoms are in the para-positions. Prill *et al.* pointed out that the active substituents in general must be in the para-positions (4, 4 positions) for the compound to have a high degree of toxicity. The presence of additional substituents in other positions appear to cause a diminution of activity.

The insecticidal activity of DDT is related to the ease with which it is absorbed by the insect cuticle. DDT affects the peripheral sensory organs to produce violent trains of different impulses that produce hypersensitivity and then convulsions of the insect. The paralysis and death that ensure are thought to occur from the metabolic extraction or from elaboration of a naturally occurring neurotoxin.

Insecticides containing DDT affect the growth of certain crops like cucurbits by producing chlorosis, stunting and consequent reduction in yield. Injury of cucurbits may be avoided by using purified DDT as the injury is caused by isomers other than p, p'-isomer. In most applications, DDT does not break down readily in soil and under certain conditions causes serious damage to plant life, if it became sufficiently concentrated in the soil.

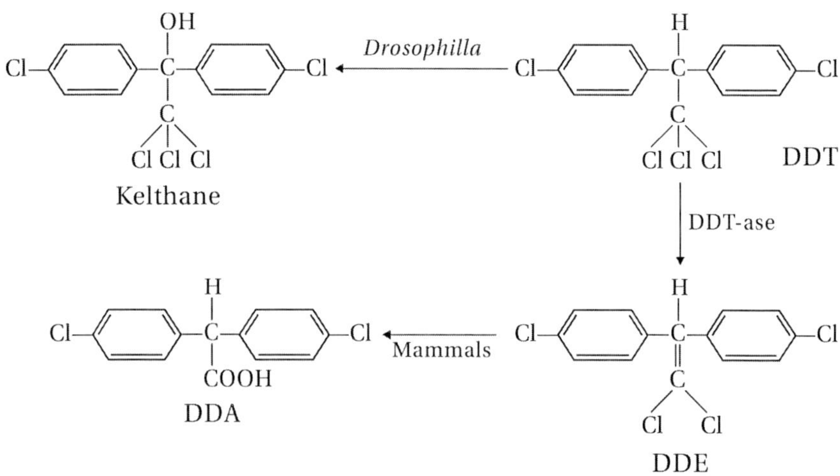

Though there are considerable losses of DDT spray deposits on citrus, much crystalline material may still persist on the leaf surfaces. Studies have shown that dehydrohalogenation is the key reaction in the field decomposition of DDT, and this can be catalysed by traces of iron. This is largely dependent upon temperature.

In insects, DDT is metabolised by dehydrochlorination to form DDE by α-hydroxylation to form 2,2,-bis-(p-chlorophenyl)-2- hydroxy-1,1,1-trichloroethane (kelthane). In mammals DDT is dehydrochlorinated to DDE and then to DDA {bis(p-chlorophenyl) acetic acid} which is excreted in urine.

Insect resistance to DDT has become widespread in many species possibly due to the selection of naturally occurring genotypes that possess enhanced amounts of the enzyme DDT-ase which detoxifies DDT to the inactive DDE. Comparative inhibitors of DDT-ase such as 1,1-bis (p-chlorophenyl) ethanol, DMC, N,N-dibutyl-p- chlorobenzene sulphanamide. Antiresistant have sometimes been used successfully as synergists with DDT to restore activity against resistant races to the housefly. DDT exhibits affinity for lipoid tissue and as such is concentrated and stored in fatty tissues at a rate of 10 to 20 times that of ingestion and these storage deposits are only slowly converted to DDE and ultimately excreted as DDA. However, DDT is the most permanent and durable of all the contact insecticides. Because of its insolubility in water, its has low vapour pressure and its resistance to destruction by light and oxidation.

DDT is relatively non-toxic to higher animals with an oral LD_{50} of 200 and a dermal LD_{50} value of 3000 mg/kg to the rat. The comparable oral LD_{50} values for the other DDT analogs are TDE 3400, methoxychlor 6000, perthane 8170, prolane 4000 and bulan 330 mg/kg. Methoxychlor is favoured because, it does not accumulate in animal tissues and is also not concentrated through the food chains. Another defect of DDT is the absence of acaricidal effect on plant-infesting mites, necessitating application of acaricide on a number of crops. Chlorine comprises almost 50% by weight of DDT materials. The other compounds mentioned earlier contain an even greater proportion of chlorine. Hence the compounds of this group can be said to come under the group called "polychlors".

BHC (Benzene hexachloride, Hexachlorocyclohexane, Gammexane, HCH, 1,2,3,4,5,6-Hexachlorocyclohexane, $C_6H_6Cl_6$)

Benzene hexachloride was first synthesised by Michael Faraday in 1825 and its structure was determined in 1836. In 1942, the Imperial Chemical Industries Laboratories in England found this compound to have considerable insecticidal properties but did not give considerable encouragement since consistent results were not obtained. This inconsistency was later found to be due to the presence of isomers of different toxicity.

BHC is prepared by a number of methods of which the following are important:

(i) Chlorination of cyclohexane— $C_6H_{12} + 6Cl_2 \rightarrow C_6H_6Cl_6 + 6HCl$

(ii) Chlorination of cyclohexene— $C_6H_{10} + 5Cl_2 \rightarrow C_6H_6Cl_6 + 4HCl$

(iii) Chlorination of benzene— $C_6H_6 + 3Cl_2 \rightarrow C_6H_6Cl_6$

The first one is the most widely used among the three methods. The chlorination of benzene is a chain reaction and proceeds through the stages of formation of dichlorocyclohexadiene and tetrachlorocyclohexene. The chlorination of benzene is carried out in the presence of different initiators like organic peroxides, unsaturated compounds like ethylene terpenes, irradiation with actinic light or hard gamma radiation from radioactive cobalt Co^{60}.

In industry, chlorination is carried out with irradiation from the medium UV light and in the presence of excess of benzene or other organic solvents, like methylene chloride. In the chlorination of benzene, number of isomers are obtained. At present 8 isomers are known and these have been named as α, β, γ, δ etc. in order of their discovery. Of these γ-isomer is highly toxic to insect pests.

The following conditions have been found to favour the formation of BHC with the largest possible content of γ-isomer: (i) Low temperature chlorination from 20 to 30°C. However, the chlorination at low temperature decreases considerably the reaction rate and consequently productivity, (ii) A high concentration of chlorine in the reaction medium also favours an increase in the γ-isomers content of the technical grade product, (iii) Concentration of chlorination products in the final reaction should be as low as possible. Impurities such as oxygen of the air and iron compounds affect the formation of BHC negatively and also help in the formation of side products. Hence, the use of iron containers or iron compounds should be avoided. After chlorination the product can be separated from the reaction solution by various methods like distillation and fractional distillation.

The molecule of 1,2,3,4,5,6-hexachlorocychlohexane may exist in 16 possible sterio isomeric forms. The BHC molecule has to be thought of as 3-dimensional structure. All the six carbon atoms do not lie in one plane. Three of them lies in one plane while the other three lie in another plane as in the given figure below.

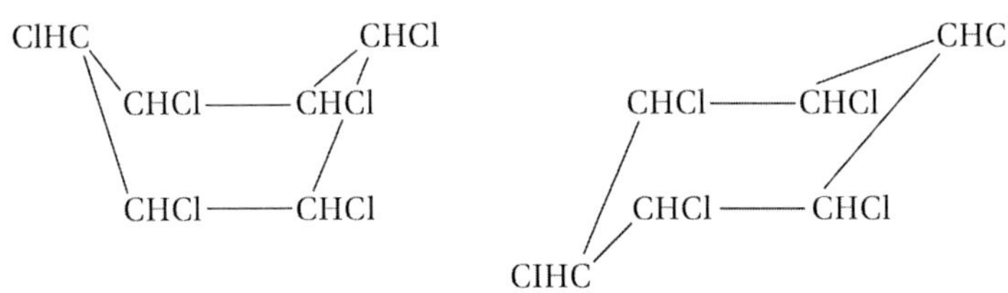

Cis (Boat form) Trans (Chair form)

The crude product of BHC is a greyish or brownish amorphous solid with a musty odour, melting at 65°C. It consists of 10 to 18% of the active γ-isomer, 55 to 70% of α-isomer, 5 to 14% of β-isomer, 6 to 8% of δ-isomer, 3 to 4% of ε-isomer, traces of η-isomer, BHC is considerably more volatile than DDT which accounts for the shorter residual action of BHC. The γ-isomer is by far the most toxic of the isomers to insects, being from 500 to 1000 times as active as the δ-isomer. The β and ε isomers are non-toxic. The γ- isomer has a slight aromatic odour and is stable to the action of heat, fight, and oxidation and can be burnt without appreciable decomposition. The musty odour of BHC limits its use, as it gives off flavours on the harvested crop of fruits and vegetables. The nature of the compound responsible for this odour has not been identified.

The isomers of BHC are stable to light, heat, air, carbon dioxide and strong acids. In the presence of alkalies the α, β, γ, λ and ε isomers are readily dehydro halogenated, forming 1,2,4-trichloro- benzene (main product) and 1,2,3 and 1,3,5 trichlorobenzene.

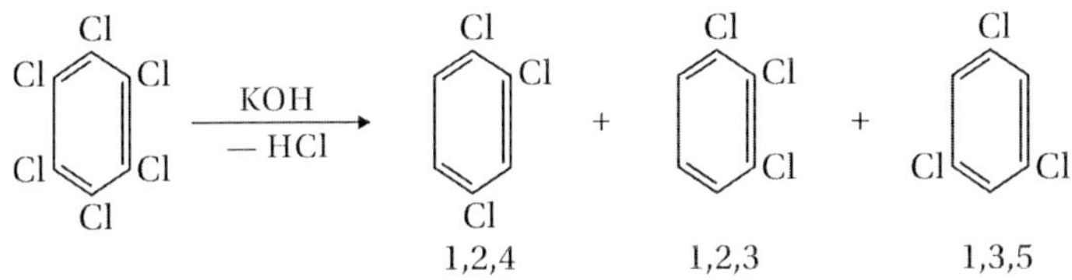

The dehydrochlorination of BHC is accelerated by water, light and bases. It is assumed that the first product of the splitting out of HCl is pentachlorocyclohexene which further goes to tetrachlorocyclohexadiene and trichlorobenzene. Impurities of pentachlorocyclohexene and tetracyclohexadiene possibly give the characteristic odour. Splitting out of HCl also occurs when reacted with lime,

$$C_6H_6Cl_6 \xrightarrow[\text{-HCl}]{} C_6H_5Cl_5 \xrightarrow[\text{-HCl}]{} C_6H_4Cl_4 \xrightarrow[\text{-HCl}]{} C_6H_3Cl_3$$

ammonia and organic amines and also at elevated temperature of 250 to 350°C. By heating to 170°C, it is possible to isomerise the individual isomers to the other ones and in the presence of ferric chloride, the reaction proceeds to not more than 50%. BHC is reduced by zinc dust in acid medium to benzene and this is utilised for the quantitative determination. Comparatively BHC differs little in its chemical properties from other halogen derivatives of hydrocarbons that have halogen atoms on the adjacent carbon atoms. Water acts slowly at room temperature on BHC and only on heating to 100°C an appreciable amount of HCl is formed. Thus when steam acts on BHC at about 102°C for an hour, about 0.13% HCl is split out from the γ-isomer. When γ-isomer BHC is heated with water at 200°C in sealed tubes, 1,2,4-trichlorobenzene and its hydrolysis products are obtained.

The crude BHC has been extensively used as a soil poison and for seed treatment. Great care should be taken in applying BHC to edible crops or to soil where tuber and root crops are to be grown as severe off-flavours may be produced. Highly purified product containing over 99% gamma isomer has recently been prepared and this has less objectionable odour and plant injury.

Gamma BHC is a colourless crystalline solid, practically insoluble in water but soluble in organic solvents. It is practically odourless or has slightly aromatic odour. It is very stable to the action of heat, light and oxidation but is readily decomposed by alkaline materials to form principally 1,2,4-trichlorobenzene and 3 moles of hydrogen chloride. The isolation of pure γ-isomer from the technical product is an important problem. To obtain γ-isomer the crude product is treated with calculated amount of methanol at a particular temperature, and the un-dissolved α and β isomers are filtered off and the filtrate is cooled. By this process the technical γ-isomer is isolated after removal from the mother liquor and dried. It contains more than 90% of the γ-isomer. To obtain Lindane (99% of pure isomer) this preparation is again recrystallised and γ-isomer obtained. The efficiency of the extraction and yield of pure Lindane depends upon the temperature, properties of various compounds and size of the crystals in the technical materials. The pure compound is named as Lindane after van der Lindane, the discoverer of gamma isomer. Lindane is appreciably volatile at room temperature and sublimes most rapidly. Lindane is widely used because it has no odour and also leaves no residues on food and forage products. Because of its cost, lindane has been widely used in household sprays, livestock dusts and seed treatment.

BHC is used to control various harmful insects, plant pests and animal parasites. As it may accumulate in the body of animals and enter into meat and milk, its use in animal husbandry is decreasing. BHC and Lindane have been used in combination with seed disinfectants such as ethylmercury chloride (Mercuran), hexachlorobenzene (hexagamma), TMTD, copper trichlorophenolate (Phenthuramal). Usually such seed disinfectants contain a diluent up to 10 to 15%. Lindane is also combined with other insecticides like DDT but present compatibility problems when alkaline materials are involved. When mixed with BHC, calcium carbonate (powdered limestone) did not cause decomposition but commercial calcium arsenate when mixed decomposed BHC and pure γ-isomer. However, the presence of delta isomer reduced the amount of dehydrochlorination of the gamma isomer.

Lindane is moderately toxic to mammals with an oral LD_{50} of about 100 and a dermal of 1000 mg/kg to the rats. The mode of action of Lindane is not known. However the specific toxicity of γ- isomer suggests that like DDT, it may interfere with the pores of lipoprotein structure of the insect nerve, causing distortion of pores and consequent excitation of nerve impulses transmission. The γ- isomer is slow in its killing action and is not an effective knockdown agent. Like DDT, BHC application results in increased red spider population on crops like cotton. Foraging honey bees are killed within few minutes of contact with surfaces treated with BHC.

Chlordane (Chlorondin, 1,2,4,5,6,7,8,8-octachloro-4,7-methane-3,4,7,7a-tetrahydroindene, $C_{10}H_6Cl_8$)

Chlordane was prepared first by Hayman by combining hexachloropentadiene with cyclopentadiene which results in a Diels-Alder addition product.

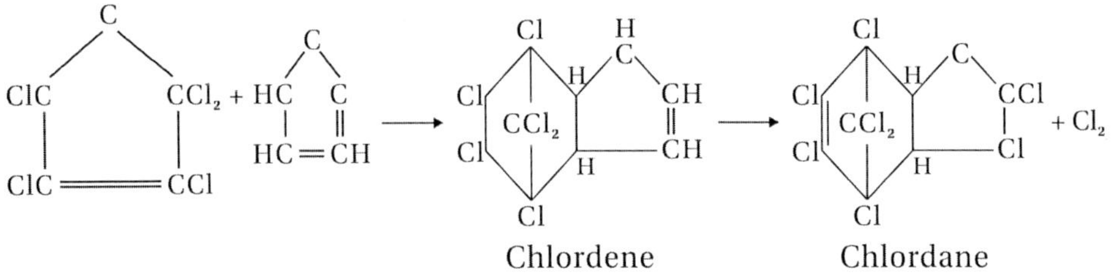

Though the chemical name is 1,2,4,5,6,7,8,8,-octachloro-4,7- methano-3a,4,7,7a,—tetraindane, commonly it is written as Octa-chloro-4,7-methano-tetra-hydrolindane.

Technical chlordane is a dark, viscous liquid with a specific gravity of 1.56 to 1.63 at 25°C. It is insoluble in water but is soluble in most of the organic solvents and other chlorinated insecticides like DDT and BHC. Chlordane is considerably volatile and in this respect it is a intermediate between DDT and γ BHC. Technical chlordane splits off chlorine as HCl when reacted with caustic alkalies but it is not possible to split out all the eight chlorine atoms under ordinary conditions. Two atoms of. chlorine are released when chlordane is acted upon by caustic alkali in alcoholic solution, or by zinc dust in an acid medium. Moist chlordane strongly corrodes the containers as a small amount of HCl is released when chlordane is reacted with water. Epichlorohydrin is added to stabilise the chlordane.

Chlordane is slow in its action upon insects. However its residual treatment is not so lasting. Chlordane is being replaced by the more effective heptachlor and other related compounds. As technical chlordane is decomposed by alkalies, alkali diluents should not be used nor should it be mixed with alkaline pesticides such as lime sulphur, Bordeaux mixture or calcium arsenate. Chlordane at concentrations normally used is relatively non-toxic to plants except cucurbits. Certain amount of injury to germinating seeds have been reported. As chlordane has been found to cause severe liver damage in chronic toxicity tests, it is normally not recommended in food and forage crops except as early season treatments.

Heptachlor (1,4,5,6,7,8,8-Heptachloro-3a,4,7,7a-tetrahydro-1,7-methanoindene)

Heptachlor may be prepared by the action of sulphuryl chloride on chlordene (condensation product of cyclopentadiene and hexachloro-cyclopentadiene in carbon tetrachloride) in the presence of benzyl peroxide. Heptachlor is produced by substantive chlorination of chlordene at low temperatures like 0 to 5°C, in carbon tetrachloride solution and in the presence of catalysts like infusarial earth, silica gel and other similar material. After chlorination the solvent is distilled off and the technical grade product is recrystallised from methanol. Recrystallisation permits the removal of most of the oily impurities.

The pure compound is a white crystalline solid with m.p. 95-96°C. The technical grade material contains about 72% heptachlor and 28% related compounds like chlorodrin, nonachlor and octachlor. This is a soft waxy solid, light tan in colour and it melts between 46 and 73.9°C. It has a density of 1.57 to 1.59 at 90°C. It is insoluble in water but soluble in most organic solvents like kerosene and various aromatic hydrocarbons and halogen derivatives of hydrocarbons. Heptachlor is more inert chemically than chlordane. Contrary to chlordane, it is not readily dehydrohalogenated by alkali and water. Heptachlor splits of one chlorine atom under the action of silver nitrate in 80% acetic acid. This reaction is specific as neither DDT nor toxaphene releases chlorine atoms when heated with silver nitrate in acetic acid. However octachlor gives off two chlorine atoms.

When heptachlor is oxidised, it is converted into the corresponding epoxide. This reaction takes place readily under the influence of soil microorganisms in animals and probably in insects.

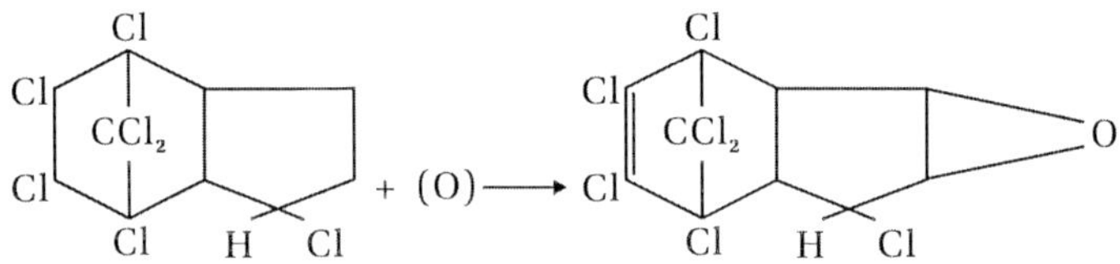

Heptachlor   Heptechlor epoxide

Heptachlor epoxide is the main metabolite of heptachlor and is more toxic to all species organisms. Practically all of the heptachlor in an animal organism and in the soil go to the epoxide. Heptachlor lasts for a long time in soil in the form of epoxide as it takes two years for 90% of the compound to decompose when 25 kg/ha was applied. Further decomposition takes place slowly when heptachlor is oxidised with nitric acid, chlorondic acid {3,4,5,6,7,7-hexachlorobi-cyclo (2,2,1) heptene-4,5-dicarboxylic-1,2-acid} is formed in small quantities.

The LD_{50} for different animals ranges from 60 to 300 mg/kg and chronic toxicity of heptachlor is higher. Heptachlor is used for the control of soil inhabiting pests and also as an addition to seed disinfectants. It has an advantage over lindane, in that it is stable in alkaline soils and also considerably cheaper. However like lindane, heptachlor protects the seedlings from damage by insects and also stimulates the germination of seeds. Number of compounds related to chlordane and heptachlor such as diheptachlor, hexachlor, nonachlor (or ennoachlor), Trichlor 237 have been described. However as these have very weak insecticidal properties they have not been widely used.

Aldrin (1,2,3,4,10,10-hexachloro-l,4,4a,5,8,8a-hexahydro-1,4-endo,exo-5,8 dimethano naphthalene, HHDN, $C_{12}H_8Cl_8$)

Many of the variations of Diels-Alder reaction with hexachlorocyclo-pentadiene have produced useful insecticides. Aldrin is one of them and has been defined as an insecticide having not less than 95% of its principal constituent (HHDN)

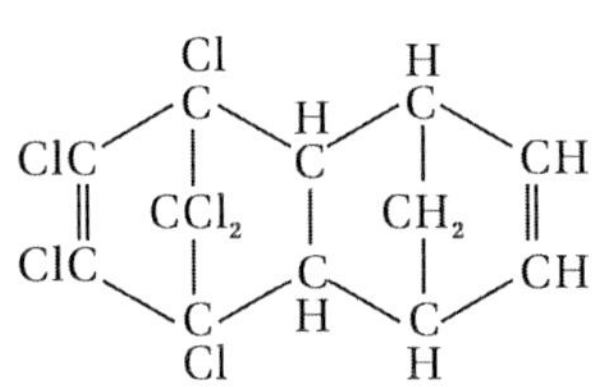

Aldrin

Aldrin is produced by reacting hexachlorocyclopentadiene with an excess (2-3 moles) of bicyclo (2,2,1) heptadiene-2,5 at 100°C. If excess of heptadiene is not there, other compounds like dodechlo-rohexacydoheptadecadiene will be formed.

Aldrin (95% HHDN) has a m.p. between 100 and 102°C while the pure material melts at 104 to 104.5°C. Aldrin is a white, crystalline solid without appreciable odour at room temperature and a pine like odour at higher temperatures. Like other chlorinated compounds, it is insoluble in water but soluble in organic solvents. Water and caustic alkalies at room temperature do not break the compound. However the slight evolution of HCl which is produced when the impurities reacted with boiling water, peroxides, peracetic acid and other organic peracids convert aldrin to dieldrin in the soil, plants, insects and vertebrates. Aldrin is rapidly oxidised in the liver. Zinc dust in acetic acid reduces aldrin. It readily reacts with phenylazide with the formation of aldrin phenyldihydrotriazole. This compound when acted on by HCl releases a molecule of nitrogen and goes to chloroaldrinaniline which readily combines with the diazo compounds to form bright coloured compound.

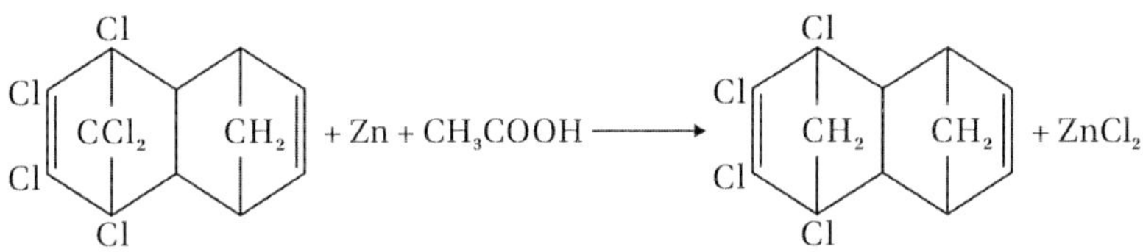

Structural isomers are possible in the bridging ring. The active principle of aldrin is the endo-exo-isomer. These isomers are named from the relative positions occupied by the two rings in relation to each other.

The technical grade compound melts at 45 to 60°C and is practically insoluble in water. Technical preparations are a brown mass containing 82% HHDN and containing not more than 0.1% water and not more than 0.1% free acid calculated as HCl. Technical aldrin contains not less than 85% of aldrin. Aldrin is formulated as emulsifiable concentrates, wettable powder and dusts.

The toxicity of aldrin to insects is very high. Upon exposure to the air on plant surfaces, aldrin disappears in less than three weeks, so that there is no great danger of residue hazard if sufficient time elapses between the last application and harvest. The LD_{50} for experimental animals is 40-50 mg/kg. With the continuous introduction of aldrin, its toxicity increases although after a certain time an equilibrium is reported. Using C^{14} labelled compound it has been found that the aldrin entering was equal to the amount of its conversion products that were excreted. Aldrin is used for the control of grasshoppers and soil inhabiting insects. As this is highly stable, the use of aldrin and dieldrin with fertilisers into the soil has been forbidden in England in 1965.

Number of analogs of aldrin have been synthesised but none of them have good insecticidal value. Removal of double bonds by the additions of different atoms and groups like hydrogen, halogen leads to a sharp decrease in the insecticidal activity of the compound. The activity is also lowered when chlorine atoms are replaced by fluorine atoms.

Dieldrin (1,2,3,4,10,10-hexachloro-6,7-epoxy-1,4,4a,5,6,7,8,8a-octa-hydro-1,4,5,8 endo, exo-dimethanonaphthalene, HEOD)

Keran described this compound in 1948 under the name "Compound 497". Like aldrin the name was chosen since this compound is made by the Diels-Alder condensation reaction. Dieldrin is defined as an insecticide having not less than 85% of the principal constituent HEOD. Dieldrin (85% HEOD) is a buff coloured crystalline solid without appreciable odour and melting at approximately 180°C. The melting point of recrystallised dieldrin is 175-176°C. It is less volatile than DDT at temperature below 43°C and more volatile at higher temperatures. It is insoluble in water but soluble in most of the organic solvents. However its solubility is considerably less than aldrin in methanol 49%, acetone 54%, benzene 75%, hexane 7.7%, deodorised kerosene 4.3%. Pure dieldrin is relatively stable compound in spite of its epoxy linkage. Dieldrin approaches aldrin in its chemical stablity. Its reactions are determined to a considerable extent by the presence of the epoxy group. On heating with monoethanol amine for 12 to 16 hours at 140°C, two chlorine atoms are completely split up from the endomethylene bridge. When

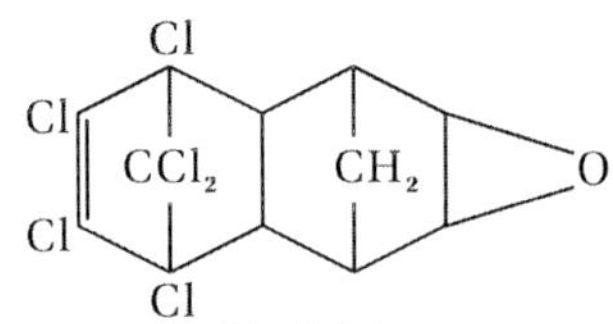

Deildrin

dieldrin is dissolved and refluxed with halogen acids, it will yield halohydrin With mineral acids and alkalies under normal conditions, dieldrin is not decomposed and as such may be used with a variety of other pesticides.

Dieldrin is used in different ways such as seed disinfectants with fungicides, and aerosol bombs with pyrethrum and DDT for a prolonged residual activity. Dieldrin acts as a contact and stomach poison. Against many insects, it is more toxic than DDT, BHC and aldrin. It is not phytotoxic but is highly toxic to warmblooded animals as its LD_{50} value is 25 to 50 mg/kg. Due to its persistence for a long time, it may present a residue hazard on food and feed stuffs unless applied at the recommended dosage and intervals before harvest.

Endrin (1,2,3,4,10,10-hexachlor-6,7-epoxy-1,4,4a,5,6,7,8,8a-octa- hydro-l,4,5,8-endo-endo dimethanonaphthalene)

This is a sterioisomer of dieldrin. Endrin is produced by the oxidation of isodrin with hydrogen peroxide in acetic acid at the lowest possible temperature. When the oxidation is carried out at 100°C, the endrin rearranges to a ketone that has no insecticidal effect. Such isomerisation also occurs upon prolonged storage in the presence of compounds of an acid nature and special stabilisers like epichlorohydrin are added to the technical grade compound. Endrin is a white crystalline solid, melting at 200°C with gradual chemical rearrangement.

Endrin is not decomposed by alkalies but acids cause a rearrangement into a compound without insecticidal activity. It is compatible with most pesticides except those of acidic nature. It is insoluble in water but is soluble in most organic solvents The solubility is greater in aromatic hydrocarbons such as benzene and xylene than in paraffinic solvents such as hexane and kerosene.

The technical grade product is brown and contains not less than 85% principal compound. Endrin exceeds many other compounds of this series in toxicity for man, domestic animals and insects. The LD_{50} for rats is 17 mg/kg. An advantageous property of endrin is its low persistance since it is relatively isomerised to the non-toxic ketone on exposure to light. Phytotoxicity tests indicate that some injury may result from the use of endrin on sensitive plants. Endrin has considerable specificity.

Isodrin (1,2,3,4,10,10-hexachloro-1,4,4a,5,8,8a-hexahydro-1,4,5,8- endo-endo-dimethanonaphthalene)

Isodrin is obtained by the condensation of cyclopentadiene with 1,2,3,4,7,7-hexachlorobicyclo-2,2,l-heptadiene-2,5. It is also produced by the condensation of hexachlorocyclo-pentadiene with acetylene at 5 to 18 atmosphere pressure and temperature below 150°C. Isodrin is a white crystalline substance m.p. 240-242°C. It is reminiscent of aldrin in its chemical properties but is not used much in agriculture. However it is employed for the production of another insecticide, endrin. The LD_{50} for experimental animals ranges from 7 to 15 mg/kg.

Endosulfan (Thiodan, 6,7,8,9,10,10-hexachloro-1,5,5a,6,9,9a-hexa- hydro-6,9-methano 2,4,3-benzodioxathiepin-3-oxide)

Endosulfan is the adduct of hexachlor cyclopentadiene and 1,4- dihydroxy-2-butene subsequently chlorinated with SO_2Cl_2 to produce the endosulfan. The technical material is a brownish solid, m.p 70 to 100°C consisting of about 4 parts of α-cis isomer m.p. 108°C and one part of β trans isomer m.p. 206°C. The α-cis isomer is slowly converted to the more stable p form at high temperature. Both the isomers are slowly oxidised in air and in biological systems and are rapidly oxidised by peroxides or permanganate to endosulfan sulphate with m.p. 181°C. In acidic media the isomers hydrolyse to form endosulfan alcohol m.p. 203°C which has the endo cis configuration.

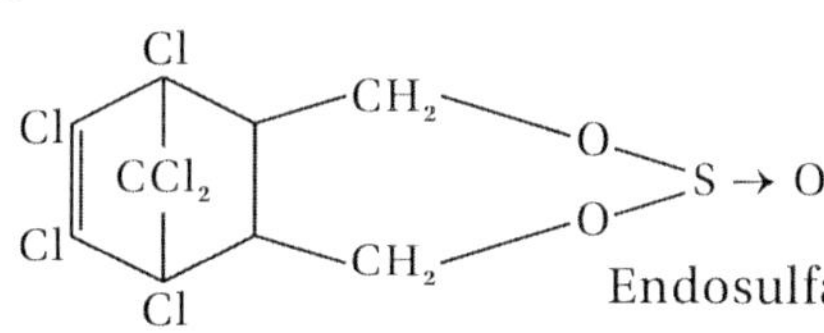

Endosulfan

Structure Activity Relationship

Variations in the steriochemical configuration of the rings like endo, exo, endo-endo, have a substantial effect upon the spectrum of insecticidal activity. The highest activities of the adducts of hexa-chlorocyclo pentadiene is shown by the compounds that contain a polychlorinated centre together with another electronegative site such as a double bond or Cl, O, S or N atom.

The symptoms of poisoning in insects are hypersensitivity, hyper activity with violent bursts of convulsions and finally complete prostration with convulsive movements. The site of these disturbances lies in the ganglia of the central nervous system rather than in the peripheral nerves as with DDT. The precise biochemical lesion responsible for the toxic action is unknown. Perhaps the most promising theories related to the absorption and critical fit of the active compounds into the pores of nerve membranes to produce convulsant action.

Chlordane is used for the control of cockroaches, ants, termites and soil insects. Heptachlor has much the same spectrum of effectiveness and is also used for grasshopper control. Aldrin is a broad spectrum insecticide. Dieldrin is more stable and residual and is used in addition for the control of grasshoppers, insects of public health importance, termites and mothproofing. Endrin is especially used for the control of lepidopterous larvae attacking cotton fields and vegetable crops. Thiodon is a broad spectrum insecticide for pests of fruits, vegetables and ornamentals.

Oxidative conversion of heptachlor and aldrin to their much more stable epoxide is an important reaction in both plant and animal tissues. On alfalfa the half lives of parent materials are less than a day but that of the heptachlor epoxide is eight days and for dieldrin, it is seven days. Heptachlor fed to animals is concentrated about 20 times and stored as heptachlor epoxide. Thus, these reactions may pose serious problems of environmental contaminations.

The acute oral LD_{50} values to rats range from extremely toxic (teladrin—5 mg/kg) to very low toxicity (β dihydroheptachlor>4000 mg/kg). Most of the cyclodienes are readily absorbed by the mammalian skin and the dermal values in many cases are only twice the oral LD_{50} values. These are primarily nervous stimulants and cause reflux excitability, convulsions, brachycardial and vaso depressions. Phenobarbitol and phentobarbitol are effective antidotes and recovery is usually prompt with no after effects.

Chlorinated Terpenes—Toxaphene

By the chlorination of the naturally occurring terpenes, a group of incompletely characterised insecticidal compounds like, toxaphene have been produced. When camphene is chlorinated to a chlorine content of 67-69%, the mixture of products formerly called chlorinated camphene, Hercules 3956, now called Toxaphene with insecticidal properties, has been produced. Camphene is made by isomerising alpha-pinenen, a major constituent of turpentine. From the structural formula for camphene, it is clear that chlorination of the molecule may take place by addition to unsaturated linkage, by rearrangement or by a combination of two or more of these compounds. Chlorination can also be carried out in the presence of initiators of free radicals such as exo-bis (isobutyronitril).

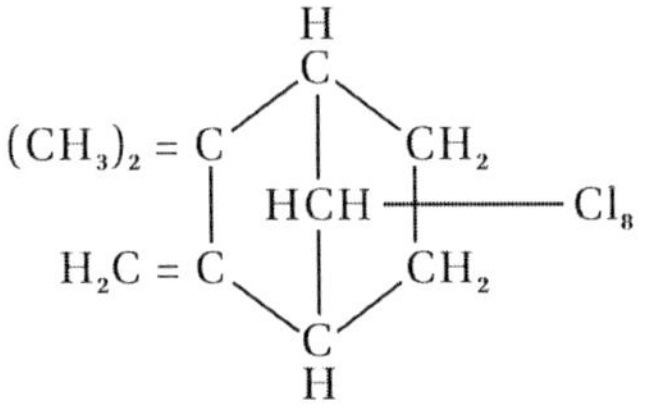

Toxaphene

At first the chlorination of camphene proceeds very vigorously with the evolution of much heat but later as chlorine is introduced the role of chlorination sharply decreases. At this stage, it is necessary to carry the reaction at elevated temperature. At the same time, because of relatively low thermal stability of toxaphene, the reaction temperature should not be high also. Further to decrease the viscosity of the reaction medium, the chlorination process is very often carried out in carbon tetrachloride, which is distilled off after chlorination has ended.

The technical material is a yellow, waxy solid, having a mild piney odour. It is a mixture of isomers and has a melting point range of 65 to 90°C and a density of 1.6. It is insoluble in water but readily soluble in organic solvents, the solubilities in grams/100 ml of solvent being in benzene, carbon tetrachloride, ethylene dichloride, and xylene >450, kerosene 280, mineral oil 55 to 60. Toxaphene is unstable in the presence of heat, UV light, and alkalies as it is dehydrochlorinated yielding HC1 and thus loses its insecticidal activity. Toxaphene has one labile chlorine atom which is attacked by alkaline materials and hence mixtures with the later (lime or sulphur) should be avoided. When moisture is present, the commercial materials attack metal containers. Apparently, dehydrochlorination is catalysed by iron as in the case of DDT.

Toxaphene is a slow acting insecticide but is effective against many species. It is relatively non-phytotoxic though injury on peaches and cucurbits have been reported in a few cases. Chronic toxicity is relatively high. Lehmen considers toxaphene to be primarily a liver poison. Toxaphene is readily decomposed in the soil. Since toxaphene is a mixture of compounds, a specific analytical method is not available though Kenyan has described the infrared spectrum of the materials and also the organochlorine method.

All insecticidal compounds described till now from chlordane have two features in common. They are all polychlor compounds and all contain at least one bridged (bicyclic) ring. Except for toxaphene, the chemical constitution of which is not known, all the other insecticides have two fused rings in the molecule.

Strobane

A compound of composition similar to toxaphene, strobane is prepared by the chlorination of a mixture of camphene and pinone to contain 66% chlorine. The technical material is a straw coloured liquid with a density of 1.60. It is practically insoluble in water and highly soluble in organic solvents. It is used similar to polychloropinene compound and is one and half times weaker than toxaphene in insecticidal activity. However, the polychloroterpenes have some acaricidal effect and also due to low cost and simplicity of production, the polychloroterpenes are being used more and more in agriculture.

Polychloropinene

This is obtained by the photochemical or initiated chlorination of bornyl chloride to chlorine content of 64 to 67%. Polychloropinene $C_{10}H_{10}Cl_8$ is a viscous colourless oil with a density of 1.6. It is practically similar to toxaphene in its solubility, physical and chemical properties. It exceeds toxaphene in phytotoxicity and is also toxic to bees, unlike toxaphene. Most of the polychlor insecticides exhibit a marked specificity against insect pests. Most of them affect cucurbits and similar crops. However, purified formulations appear to cause less injury. Heavy applications in soils may cause a reduction in yield or complete loss of the crop even and those formulations containing BHC may produce off-flavour to food crops when grown on treated soils.

CHAPTER VIII

ORGANOPHOSPHORUS INSECTICIDES

The discovery of the biological activity of organophosphorus esters originated with the chance observation of Lange and Kreuger in 1932 that casual exposure to vapours of diethylphosphorus- fluoridate produced strong cholonergic effects in human beings. The practical development of these compounds as insecticides is mainly due to the original and extensive work of Schrader and his coworkers from 1937. In seeking substitutes for nicotine, Schrader discovered Bladen, which contained tetraethylpyrophosphate as its active principle. A further search for a lime stable product to control the Colorado potato beetle *Leptinotarsa decemlineata* led to the development of parathion. This was followed in succession by Schradan and demeton, the first practical systemic insecticide, Azino- phosmethyl (Guthion)—a broad spectrum insecticide, Trichlorfon (Dipterex), a stomach poison, Coumaphos or Co-Ral for the control of animal parasites and numerous others. As such today, organophosphorus compounds form an important class of pesticides. More than 100,000 different organophosphorus compounds have been synthesised and evaluated as pesticides. Of these more than 80 are widely used in agriculture. The use of organophosphorus compounds as pesticides almost equals that of organochlorine compounds. Substances with a great variety of pesticidal properties like insecticides, acaricides, nematocides, herbicides, defoliants and fumigants, are found among the organophosphorus compounds. Also there are compounds with a very short residual action such as TEPP and Mevinphos, or with prolonged residual action such as diazinon and azinphos. There are also broad spectrum insecticides such as parathion and malathion and materials with very specific action such as Schradan. The unique properties of demeton and dimethoate have resulted in successful plant systemic insecticides and this property has been further refined for seed and soil treatments so that the newly developed seedlings could be protected from insect attack. Also compounds such as trichlorfon having pronounced activity as stomach poison and no contact action have also been developed. Most of the compounds of this group are required only in small quantities for the control of pests. Pesticidal activity is also very rapid. Many of the compounds break down to non-toxic compounds very rapidly in vertebrate animals. Also they do not accumulate in the body of animal. Though many of the early compounds of this group had relatively high toxicity to vertebrates, recently a large number of organophosphorus compounds with moderate to low toxicity to mammals, have been synthesised and have been put in the market. These have helped in the rapid adoption of organophosphorus compounds as pesticides.

In spite of the enormous structural diversity of organophosphorus insecticides, all the compounds can be represented by the classical hypothetical structure as originally proposed by Schrader $\begin{matrix} R \\ R \end{matrix}\!\!>\!P(=O(S))\!-\!X$ where R and R' are short chain alkyl, alkoxy, alkylthio or amide groups and X is a labile leaving group or groups that can be metabolised *in vivo* to a labile entity. The biological activity of the organophosphorus compounds is due to the capacity of the central P-atom to phosphorylate the active site of the enzyme cholinesterase (ChE) which is an essential constituent of the nervous system, not only of insects but also of higher animals. The phosphorylated enzyme is irreversibly inhibited and is therefore no longer able to carry out its normal functions of rapid removal and destruction of neurochrome (Acetylcholine) from the nervous synapse. This results in the accumulation of acetylcholine and consequent disruption of the normal functioning of the nervous system giving rise to typical cholinergic symptoms associated in insects with organophosphorus poisoning, like hypersensitivity, hyperactivity, tremors, convulsions, paralysis and death.

Organophosphorus compounds may be expected to be acutely toxic to many forms of animal life. Parathion and certain related compounds are generally toxic to nearly all types of animals. However by taking advantage of selective activation, detoxication and target sensitivity, it is possible to design such organophosphorus compounds as malathion which has a very low degree of toxicity to higher animals.

The nomenclature of organophosphorus compounds is still not well developed. In many cases they are considered as derivatives of the corresponding acids or hydrogen phosphide (phosphine). Compounds of this group possessing insecticidal action can be classified as follows:

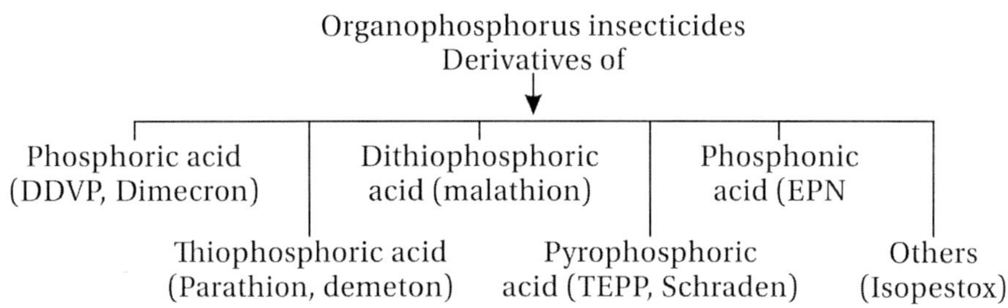

Derivatives of phosphorus acids have the ending "ITE" and those of phosphoric acid the ending "ATE", for example the diethyl ester of phosphorus acid carries the name diethylphosphite and that of phosphoric acid is called diethyl phosphate.

$$(C_2H_5OH)_2P(=O)H \qquad\qquad (C_2H_5OH)_2P(=O)OH$$

Diethylphosphite   Diethylphosphate

Where there is a sulphur atom in the molecule, the name of the compounds will be derived in the following way. The diethyl ester of thiophosphorus acid depending on the position of sulphur is called O,O-diethylthiophosphite or O,S-diethylthiophosphite.

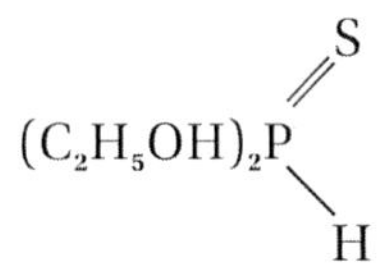

O,O-diethylthiophosphite

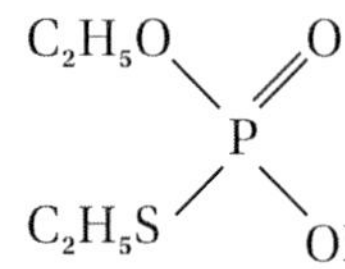

O,S-diethylthiophosphite

The names of the derivatives of thiophosphoric acids are derived in a similar manner:

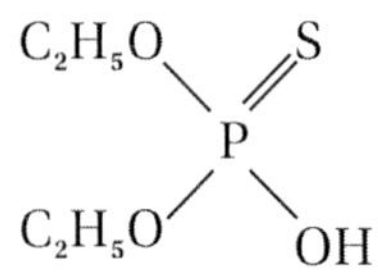

O,O-diethylthiophosphate

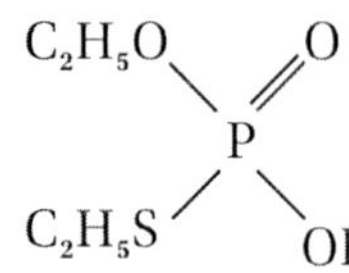

O,S-diethylthiophosphate

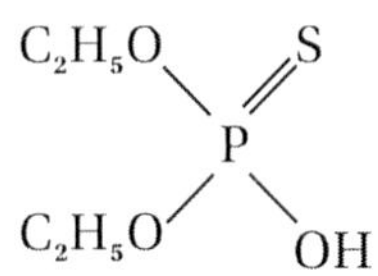

O,S—diethyldithiophosphate

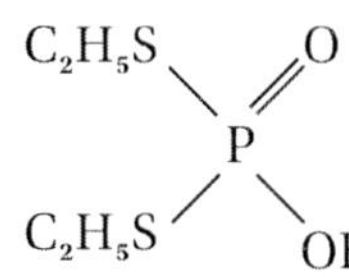

S,S—diethyldithiophosphate

The names of derivatives of other acids and amides of phosphoric acids are also named in the same manner.

DERIVATIVES OF PHOSPHORIC ACIDS

The insecticidal and acaricidal properties increase when we go from phosphites to phosphates. The important derivatives of phosphoric acid having insecticidal activity are dichlorovos, naled, phosphinon, phosdrin, phosphamidon, bidrin, birlane, gardona, dimefox, mipafox, avenin, ruelene, and cyolane.

Dichlorovos ... $(CH_3O)_2\overset{O}{\overset{\|}{P}}OCH_2 = CCl_2$

Naled ... $(CH_3O)_2\overset{O}{\overset{\|}{P}}OCBr.CCl_2Br$

Phosphinon ... $(C_2H_5O)_2\overset{O}{\overset{\|}{P}}O\underset{OCH_2CH_2Cl}{\underset{|}{C}} = CCl_3$

Phosdrin ... $(CH_3O)_2\overset{\overset{O}{\|}}{P}O\overset{\overset{CH_3}{|}}{C}=CHCOOCH_3$

Phosphamidon ... $(CH_3O)_2\overset{\overset{O}{\|}}{P}O\overset{\overset{CH_3}{|}}{C}=CClCON(C_2H_5)_2$

Bidrin ... $(CH_3O)_2\overset{\overset{O}{\|}}{P}O\overset{\overset{CH_3}{|}}{C}=CHCON(CH_3)_2$

Birlane ... 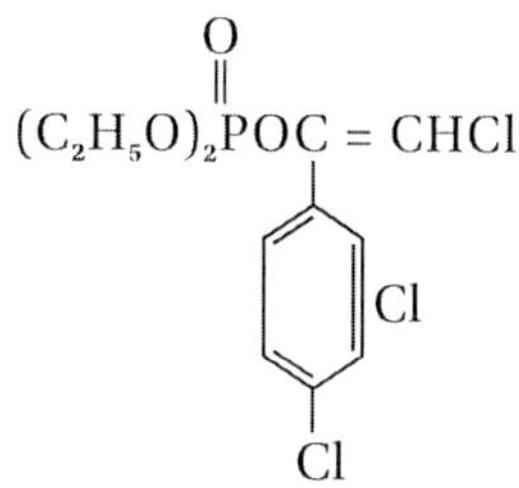

Gardona ... 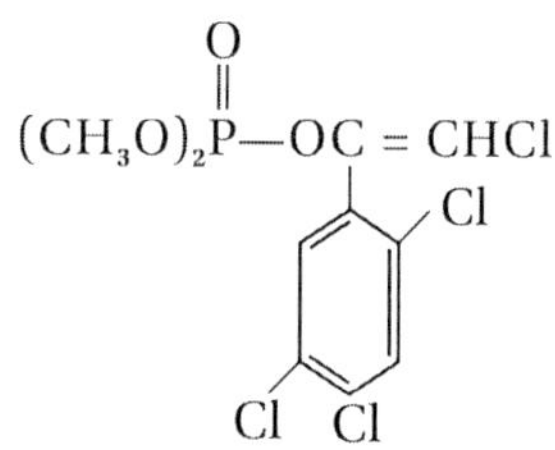

Dimefox ... $\{(CH_3)_2N\}_2\overset{\overset{O}{\|}}{P}F$

Mipafox ... $(C_3H_7)NH)_2\overset{\overset{O}{\|}}{P}F$

Avenin ... $C_3H_7O\overset{\overset{O}{\|}}{C}NH\overset{\overset{O}{\|}}{P}(OCH_3)_2$

Cyolane ... 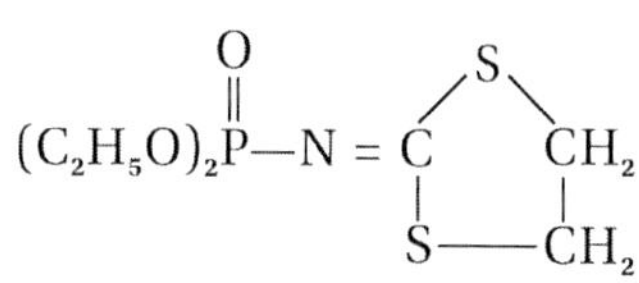

Dichlorovos (DDVP, Nuvan, O,O-dimethyl O-2,2-dichlorovinyl phosphate)

Dichlorovos is the first representative of enolphosphate series. It is produced by the splitting out of HCl from trichlorfon (chlorophos) by caustic alkali in aqueous solutions at 40-50°C.

$$(CH_3O)_2\overset{\overset{O}{\|}}{P}CH(OH)CCl_3 + KOH \rightarrow (CH_3O)_2\overset{\overset{O}{\|}}{P}OCH = CCl_2 + KCl + H_2O$$

It is also obtained by reacting chloral with trimethylphosphite either

$$(CH_3O)_3P + CCl_3CHO \rightarrow CH_3Cl + (CH_3O)_2\overset{\overset{\large O}{\|}}{P}OCH = CCl_2$$

in the presence of solvent like benzene or without a solvent, but with adequate cooling. In both the cases, dichlorovos of 92-93% purity can be produced in both batch type or continuous type processes.

In the presence of traces of moisture, dichlorovos on standing, breaks down with the formation of acidic products which further catalyse the decomposition. To stabilise the technical materials, 2 to 4% epichlorohydrin is added to tie up the acidic substances and improve the conditions for storage. Dichlorovos keeps well in glass containers. The homologs of dichlorovos are considerably less toxic to insects.

Dichlorovos is a colourless liquid, b.p. 74°C. It is highly soluble in most of the organic solvents, and the solubility is about 1% in water. In pure water, 50% dichlorovos is hydrolysed at 20°C in 61.5 days and at 70°C in 25 minutes.

$$(CH_3O)_2\overset{\overset{\large O}{\|}}{P}OCH = CCl_2 + H_2O \rightarrow (CH_3O)_2\overset{\overset{\large O}{\|}}{P}OH + CHCl_2CHO$$

Dimethlphosphoric acid   Dichloroacetal-dehyde

Dimethylphosphoric acid is harmless and the dichloroacetaldehyde rapidly decomposes and evaporates. Therefore when dichlorovos is applied to plants and other materials, no toxic residue is left. The LD_{50} to rats is 80 mg/kg.

Naled

Dichlorovos adds chlorine or bromine at the double bond of dichlorovinyl group and this results in another powerful insecticide called Dibrom (Naled).

$$(CH_3O)_2\overset{\overset{\large O}{\|}}{P}OCH = CCl_2 + Br_2 \rightarrow (CH_3O)_2\overset{\overset{\large O}{\|}}{P}OHBrCCl_2Br$$

O,O-dimethyl-O-2,2-dichloro-1,2-dibromoethylphosphate (Naled)

Naled is a crystalline substance m.p. 25°C, b.p. 110°C. It is insoluble in water, slightly soluble in paraffinic hydrocarbons but highly soluble in most other organic solvents. It is an active insecticide for the control of both sucking and chewing insects. Naled has also slight fungicidal effect. Considering hydrolysis, naled is more stable than dichlorovos. On hydrolysis, dimethylphosphoric acid, hydrogen bromide and dichlorobromoacetic acid are formed. The last one gets further converted to oxalic acid. The LD_{50} of naled for rats is 430 mg/kg.

Phosphinon {Phosthenon, O,O-diethyl-O-(2,2,dichioro-l-chloro-ethoxy vinyl phosphate)}

Phosphinon is the closest analog of dichlorovos. It is produced by reacting triethylphosphite with 2-chloroethyltrichloroacetate. It is

$$(C_2H_5O)_3P + CCl_3\,COOCH_2CH_2Cl \rightarrow (C_2H_5O)_2P(=O)OC(OCH_2CH_2Cl) = CCl_3$$

a mobile liquid, b.p. 124°C. Though it is highly potent insecticide, because of its high toxicity to animals, it has not been used widely. The LD_{50} of phosphinon for rats is 6.8 to 9.7 mg/kg.

Phosdrin {Mevinphos, O,O-dimethyl-O-(1-methyl-2-carbo-methoxy vinyl) phosphate}

Phosdrin is another enol phosphate produced by reacting the sodium enolate of methyl acetoacetate with dimethylchlorophosphate

$$(CH_3O)_2P(=O)Cl + CH_3C(ONa) = CHCOOCH_3 \rightarrow (CH_3O)_2P(=O)OC(CH_3) = CHCOOCH_3 + NaCl$$

or by the Parker reaction from trimethylphosphite and chloroaceto¬acetic ester. The latter method gives better yields of pesticides. The

$$(CH_3O)_2P + CH_3COCHClCOOCH \rightarrow (CH_3O)_2P(=O)OC(CH_3)=CHCOOCH_3 + CH_3Cl$$

corresponding thiomevinphos is several times less toxic for vertebrates than mevinphos.

Phosdrin is colourless liquid, highly soluble in water, b.p. 76°C. It is slightly soluble in petroleum ether but highly soluble in acetone and benzene. In neutral medium, mevinphos is resistant to hydrolysis (50% hydrolysed in 30 to 35 days) but in alkaline medium hydrolysis is appreciably accelerated at pH 11, 50% hydrolysing in 14 hours.

$$(CH_3O)_2P(=O)\text{—}O\text{—}C(CH_3)=CHCOOCH_3 + H_2O \rightarrow (CH_3O)_2\text{—}P(=O)OC(=CH_3)=CHCOOCH$$

Phosdrinic acid

↓

$$CH_3O\text{—}P(=O)(HO)\text{—}OC(=CH_3)CHCOOH$$

Desmethylphosdrinic-acid

The hydrolysis of mevinphos by water proceeds according to the above scheme.

Analogous products are formed when mevinphos breaks down in plant tissues. Usually mevinphos breaks down in plants in 0.8 to 4.2 days. Mevinphos exists in the form of cis and trans isomers of which the trans isomer is more stable. However, the cis isomer exceeds the trans isomer in insecticidal activity, by approximately 100 times. The technical grade product usually contains 60% cis and 40% trans isomer. Mevinphos is toxic to warm-blooded animals. LD_{50} for rats is 3.7 to 6.1 mg/kg. It is a good insecticide of short-term action for sucking and chewing insects. It is marketed in the form of 50% emulsion concentrate.

Phosphamidon {Dimecron, O,O-dimethyl-O (2-chloro-N,N-diethyl- carbamyl)-methylvinylphosphate}

Phosphamidon is produced by reacting equimolecular quantities of trimethylphosphite with the boiling solution of diethyl amide of dichloroaceto-acetic acid in chlorobenzene:

$$(CH_3O)_2P + CH_3COCCl_2CON(C_2H_5)_2 \rightarrow (CH_3O)_2\overset{\overset{O}{\|}}{P}O\overset{\overset{CH_3}{|}}{C}{=}CClCON(C_2H_5)_2 + CH_3Cl$$

The yield of phosphamidon is about 83%.

Pure phosphamidon is a colourless to pale yellow liquid. However the commercial formulation is bright violet in colour due to the presence of a dye. Its b.p. is 70°C. It has a faint but pleasant odour. It is highly soluble in water, alcohol, acetone and slightly soluble in saturated hydrocarbons such as hexane and practically insoluble in kerosene. Compared to other organophosphorus compounds, it has high water solubility and hence the presence of moisture in fillers and emulsifiers do not affect its stability. In many cases there is no necessity of emulsifiers and spreaders. As this does not clog the nozzles, it is well suited for low volume and ultra low volume sprays. Phosphamidon is stable in neutral and weakly acidic aqueous solution. In alkaline medium, it is rapidly hydrolysed. It is stable at least for two years when stored in closed containers. It is compatible with most pesticides except the alkaline ones like Bordeaux mixture, lime sulphur wash, nicotine sulphate and copper oxychloride based fungicides.

When sprayed on plants, the spray marks are not conspicuous and also the residues can be washed away as it is water soluble. It gets absorbed inside the plant tissue within one to three hours and is translocated more towards the top. The active ingredients get degraded into less stable metabolites having low mammalian toxicity than the parent active ingredient. When phosphamidon is added to fungicides like captan, it exhibits synergistic action by not only controlling the pest but also the disease. The insecticidal property is reduced when phosphamidon is mixed with copper oxychloride. In plants, the active ingredient breaks down as follows:

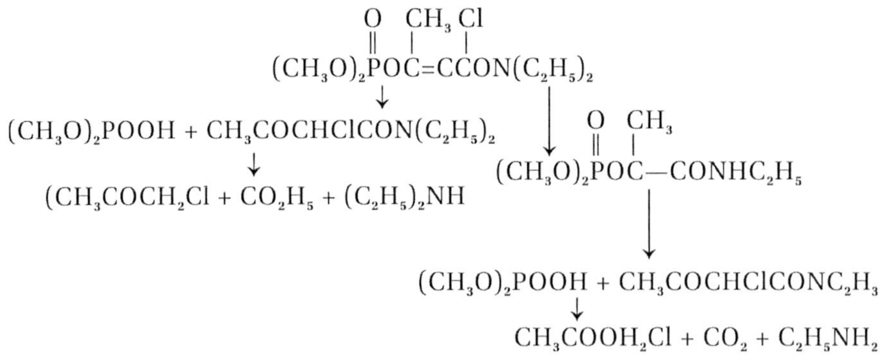

It is practically non-phytotoxic but it is better not to spray just like other systemic organophosphorus insecticides when there is lack of soil moisture, as some sort of marginal burn occurs due to the accumulation of active material along the leaf margin where the stomata are located.

When insects feed on the sap containing active material, the cholinesterase is inhibited and the insects get killed. Foliar applications will not be able to control the soil borne insects like cutworms. In spite of its relatively high toxicity for animals (LD50 for oral administration to rats is 7.5 to 15 mg/kg) it is used as a systemic insecticide for the control of sucking pests on cotton and other crops.

Bidrin {3-(dimethoxyphosphinyloxy)-N-N-dimethyl crotonamide, dimethyl phosphate of 3-hydroxy N-N-dimethyl cis-crotonamide}

Bidrin is a persistent, systemic vinyl phosphate insecticide. It is synthesised by the Parker reaction using dimethylamide of mono-chloroacetoacetic acid and trimethylphosphate:

$$(CH_3O)_3P{=}O + CH_3ClCHClCON(CH_3)_2 + CH_3Cl \rightarrow$$

$$(CH_3O)_2\overset{\overset{\displaystyle O}{\|}}{P}{-}O{-}\overset{\overset{\displaystyle CH_3}{|}}{C}{=}CHCON(CH_3)_2 + CH_3Cl$$

Bidrin is the name applied to the pure cis isomer of the compound. It is a yellow to brown liquid with a mild ester odour. It is miscible with water, ethyl alcohol, acetone, isopropyl alcohol, diacetone alcohol, isopropyl oxitol, hexylene glycol and other organic solvents. It is very sparingly soluble in diesel fuels and kerosene (< 1%). The technical material contains a minimum of 85% by weight of bidrin. It does not deteriorate normally during storage for number of years. At pH 9 and 37°C, 50% of bidrin is hydrolysed in 50 days and at pH 1 in 100 days. When bidrin is hydrolysed by refluxing 10% w/w in distilled water, orthophosphoric acid, methanol, and acetone are formed together with N,N-dimethylacetoacetamide (DMAA). In the alkaline condition, the major products are dimethylphosphate, acetone and dimethylamine. In the presence of 3% water, technical bidrin has been found to suffer a loss of 10% bidrin in several days at 38°C. When applied to air dry sandy soil of moisture content of 1, 4, and 7% by weight, the loss of bidrin after 7 to 8 days increased from 20% at 1%, to 90% at 7% moisture.

Technical bidrin decomposes to a brown viscous liquid at 55°C and the rate of decomposition increases rapidly with temperature. Bidrin is non-corrosive on copper, nickel and aluminium but somewhat corrosive to iron, stainless steel and brass. It does not attack glass, polyethylene and bakelite.

Bidrin is highly toxic to human beings and animals and may be absorbed by inhalation or through unbroken skin. Like other organophosphorus insecticides, it acts by inhibiting the enzyme cholinesterase.

Birlane {Chlorfenvinphos, 2-chloro-l-(2,4-dichlorophenyl) vinyl diethyl phosphate}

Birlane is a moderately persistent contact and stomach insecticide. It is synthesised by the condensation of triethylphosphite with 2,4,1', 1 -tetrachloro-acetophenene.

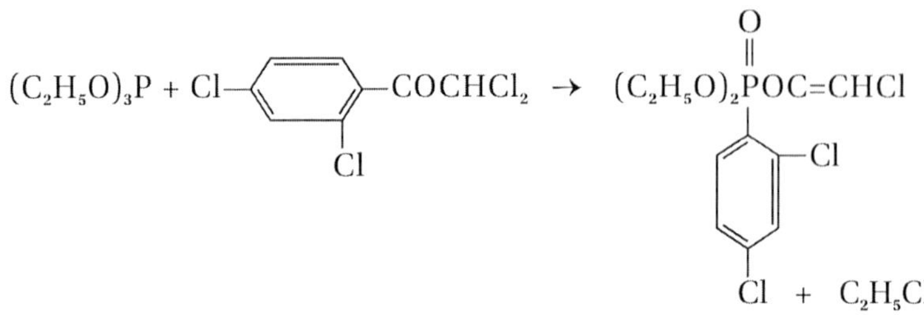

The technical grade product thus obtained is a mixture of cis and trans isomers in the ratio of 8.5 : 1. The trans isomer is insecticidally most active. Technical birlane consists of >92% active material.

Birlane is an only amber coloured liquid, with a weak chemical odour. It is miscible with acetone, xylene, alcohol, kerosene, corn oil, heavy aromatic naphtha and propylene glycol. It is sparingly soluble in water. Birlane decomposes very slowly in the presence of water. In aqueous solution, 50% is hydrolysed in 400 hours at pH 9 and in 700 hours at pH 1.1. Birlane is distinguished by its relatively low vapour pressure in comparison with other common organophosphorus insecticides, i.e. 1/200 that of parathion and malathion, 1/2000th that of diazinon or schradan, 1/200,000th that of dichlorovos at about 25-30°C. Birlane is extremely stable to thermal decomposition and rapid decomposition takes place only at temperature above 150°C. It is relatively stable in the presence of water; corrosive to iron, steel and brass; is stable when stored in glass or suitably lined steel drums. Plastics are unsuitable for storing birlane. Due to relative stability in alkaline and acidic conditions, birlane is compatible with most of the known pesticides in common use.

Birlane has been found to breakdown to 2,4-dichloroacetophenone, 1-(2'-4'-dichlorophenyl) ethan-l-ol, desethyl chlorofenvinphos and salts of conjugates of desethyl chlorfenvinphos.

Possible breakdown path for chlorfenvinphos in soils:

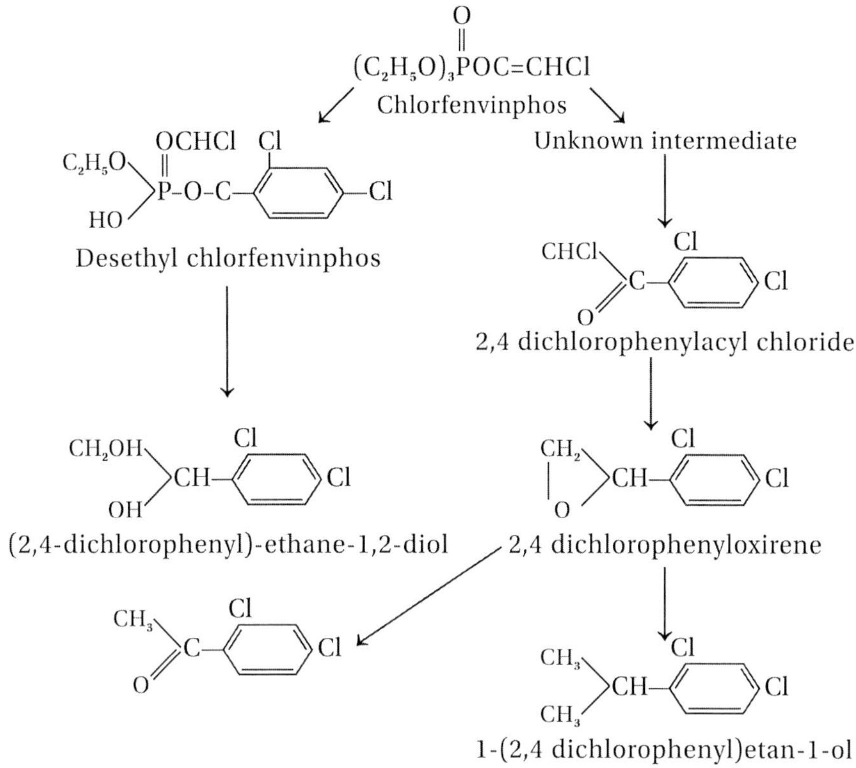

In plants, the major breakdown compounds are similar to that of soils but the breakdown products appear to be more persistent than the parent compounds. Birlane is used in the form of emulsions, wettable powders and granules when it is used to control pests on cabbage, carrot and onion crops at planting time. Residues of this compound in the harvested produce should not exceed 0.05 mg/kg.

Gardona {2-chloro-l-(2,4-5 trichlorophenyl) vinyldimethyl phosphate}

Gardona is a low toxicity organophosphorus insecticide with contact action and has moderate persistence. The technical material is an off-white crystalline solid, sparingly soluble in most common solvents. At 20°C, only 11 ppm of gardona is dissolved in water. It is stable at temperatures up to 100°C. Gardona hydrolyses slowly in acid or neutral pH but rather more rapidly in alkaline conditions. Gardona is compatible with most of the pesticides except those of alkaline nature.

On foliage and soils, gardona breaks down into number of products as shown below:

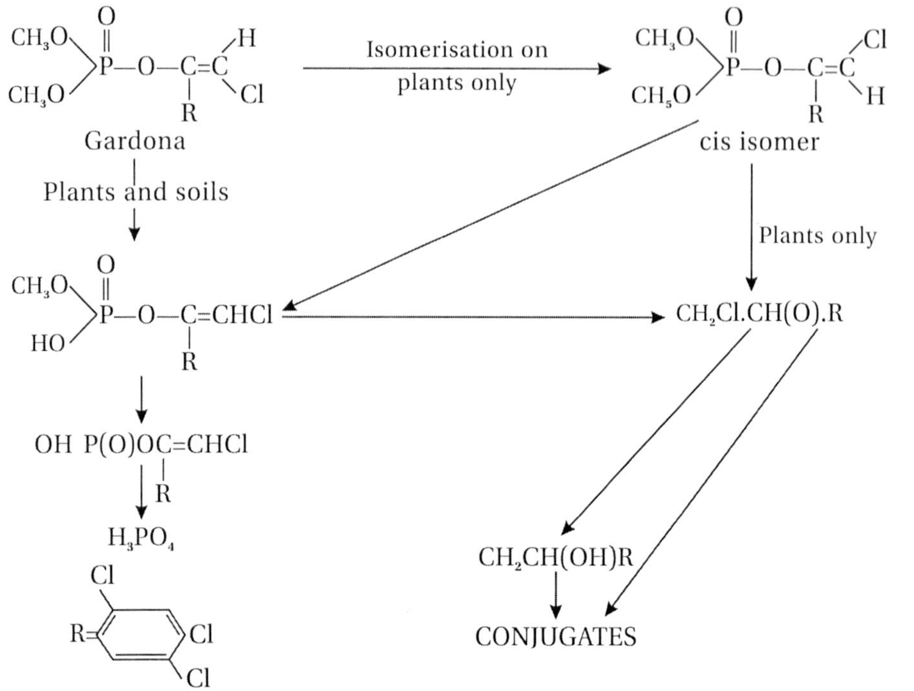

Some of the breakdown products are present in detectable amounts only within a few days of application. On foliage there may be some isomerisation of gardona to its cis isomer but major breakdown product detectable at harvest appears to be 1-(2,'4',5'-trichloro-phenyl) ethan-l-ol. Gardona applied to the soil is lost within 30 days and the loss of it is more rapid in medium loam soil (4 to 5 days) than in clay soils (30 days). Gardona is synthesised by the condensation of trimethyl phosphite with 2,4,5,1, 8-pentachloroacetophenone.

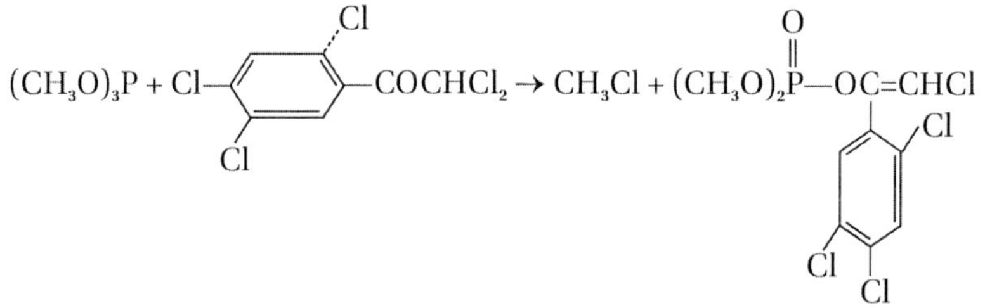

It is used in the form of emulsion, wettable powder and granules for the control of pests. A large number of other similar compounds also have been studied but have not been used in agriculture. In addition to the esters of enolphosphoric acid, amides of fluorophosphoric acid like Dimefox and Mipafox have been used.

Dimefox (Pestox, Hnanne, Terrasystam, Tetramethyl phosphorodi-amidic fluoride, bis dimethyl fluorophosphate)

Dimefox is a powerful systemic insecticide used for the control of sucking pests and vectors of diseases of cacao. This is a colourless mobile liquid with a weak odour, highly soluble in water and in many polar organic solvents, b.p. 80°C. It is stable in aqueous alkaline solution and is rapidly hydrolysed in acid medium. Because of its volatility, dimefox has a very short residual activity and hence is not suited for foliage application. It is highly toxic to mammals, the LD_{50} for rats being 1 to 1.5 mg/kg.

Mipafox (Pestox-15, Isopestox, N, N'-diisopropyl phosphorodiamide fluoride)

Mipafox is a strong systemic and contact insecticide. It is a white crystalline substance with a.m.p. 61-62°C. It is highly soluble in most organic solvents except paraffinic hydrocarbons. Its solubility in water is about 8%. The LD_{50} for rats is 50 mg/kg. Mipafox is stable in aqueous neutral or weakly acid solutions, 50% being hydrolysed in 200 days at pH 6 and in six days at pH 8. The vapours of mipafox when inhaled by animals or man causes irreversible paralysis. Hence, this compound is not in use now.

Avenin {0,0-dimethyl-N(isopropoxy carbamoyl) phosphate}

Avenin is a systemic insecticide of low activity with narrow selective action. It is a light oily liquid, not distillable without decomposition, soluble in most organic solvents, but slightly soluble in water. When it is heated it breaks down with the formation of methyl ester of isocyanotophosphoric acid. The LD_{50} for animals is > 5000 mg/kg.

Cyolane (Dithiolane, Iminophosphate, diethyl-N-1-3-dithiolanyl- 2-iminophosphate)

Cyolane is a stomach and contact insecticide with systemic activity in plants through roots and foliar absorption. It is a colourless to yellow liquid, m.p. 115-118°C, soluble in acetone, benzene, toluene, ethanol, water, cyclohexanone and very slightly soluble in hexane. It is used in the form of emulsion and granules for the control of pests in cotton and on other crops. The LD_{50} for rats is 8 to 9 mg/kg.

Structure-activity Relationship

Among the derivatives of phosphoric acid, the mixed esters having one ester radical of acidic nature are very active as pesticides. However when the length of alkyl radicals in the esters and amides of phosphoric acid is increased, the toxicity of the compound to animals is decreased. The maximum toxicity for many mixed esters of phosphoric acid occurs at the diethyl derivatives. This is probably associated with their high alkylating capability with respect to various nitrogen and sulphur compounds that are present in biological substrates. Not only the nature of the substituent in the aromatic radical, but also its position has great influence on the toxicity of a mixed ester of phosphoric acid for insects and mites. In general the activity of aryl phosphate substituted in the position 4 is higher than in 2 and 3. However, in spite of their high insecticidal activity mixed alkylaryl esters of phosphoric acid have not been practically utilised in agriculture due to their toxicity to animals and man. Certain aliphatic halogen containing esters of phosphoric acid and enol phosphates are used in both agriculture and public health since they are somewhat less toxic.

DERIVATIVES OF THIOPHOSPHORIC ACID

Replacement of one of the oxygen atoms by sulphur in the derivatives of phosphoric acid, generally decreases the toxicity of the compound for mammals without any substantial change in insecticidal or acaricidal activity. Today more than 30 derivatives of thiophosphoric acid are being used as pesticides. These derivatives can be either a thiolo or thiono structure:

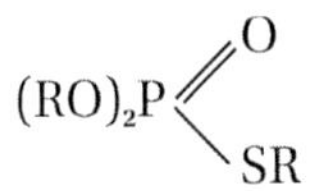

Thiolo isomer

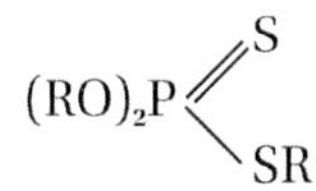

Thiono isomer

Thiolo derivatives are more toxic to mammals than thiono derivatives. When the thionoisomers are heated or acted by the certain reagents, they are converted into thiolo isomers (Pischemeuka rearrangement).

Compounds used as pesticides are mainly mixed esters of thiophosphoric acid. Maximum activity is obtained when the total number of carbon atoms in R and R' does not exceed four. Minimum activity is shown when both R and R' are methyl groups. Among the mixed esters of thiophosphoric acid, maximum insecticidal activity is shown when the ester contains a nitro group in the 4-position than in the 2 or 3-positions. Though introduction of cyano, sulphide, sulphoxide and sulphonyl groups in the 4-position also show increased insecticidal activity, mammalian toxicity is also increased simultaneously.

Introduction of a second group in the aromatic radical has been found to decrease slightly the mammalian toxicity but not the insecticidal activity. If there are more than three substituents in the aromatic radical, insecticidal activity is lowered considerably. Replacement of the aromatic radical by a heterocyclic ones in most cases leads to the production of active insecticides like diazinon, and dursban. Practically all compounds with the structure, shown below, have

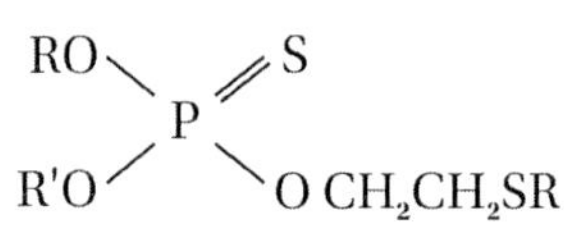

systemic effect. Most of the thiophosphates pesticides are produced by reacting dialkylchlorothiophosphates with phenols or salts of phenol in the presence of HCl acceptors like organic bases (tertiary amines, triethylamine or pyridine), aliphatic ketones (acetone, methyl ethyl ketone) or inorganic bases (caustic alkalies, carbonates of alkali metals).

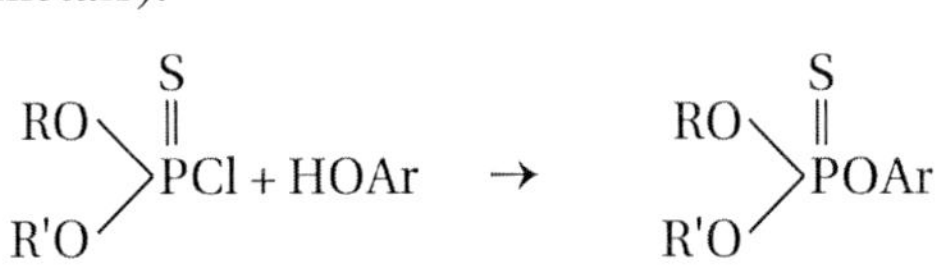

Among the large number of compounds used as pesticides, the following are widely used:

Parathion ... 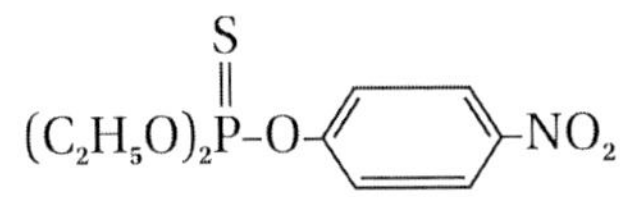

Methyl parathion ... 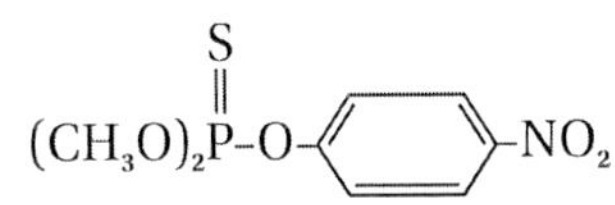

Paraoxon ... 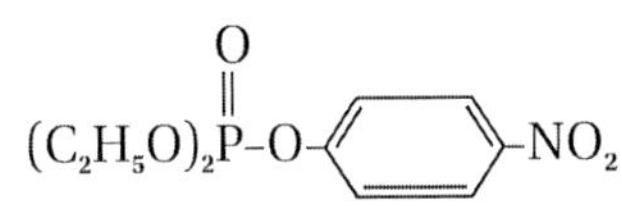

Thiophos ME ... 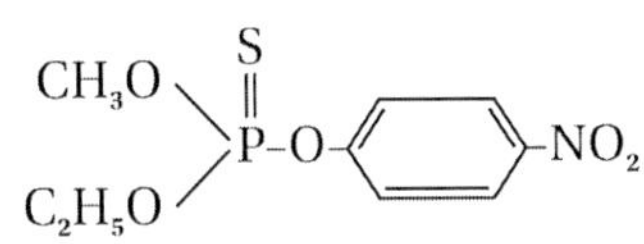

Fenitrothion ... 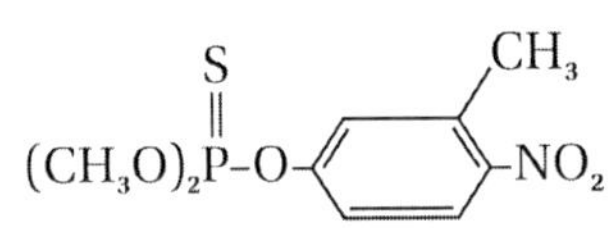

Chlorothion ... 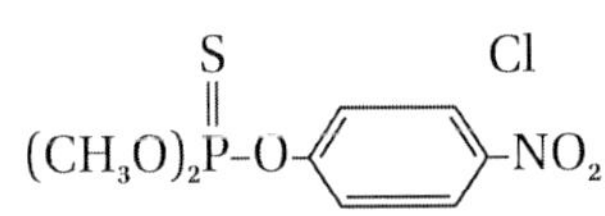

Dicapthon ... 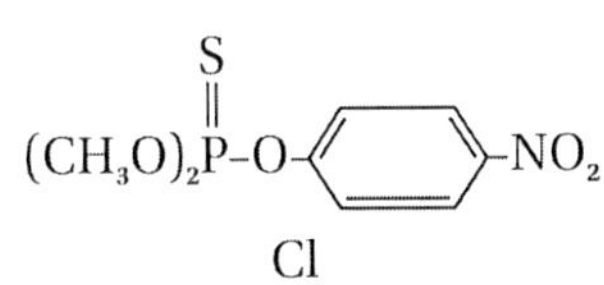

Ronnel ... 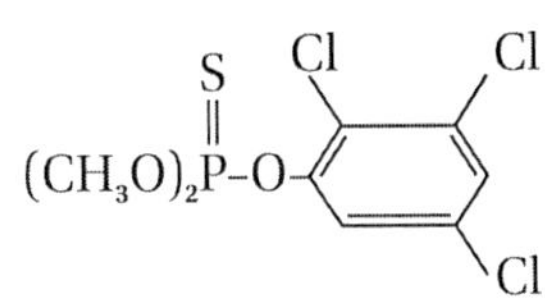

Bromophos ... 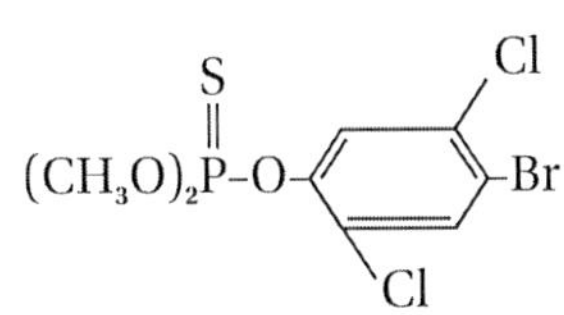

Lebaycid ... 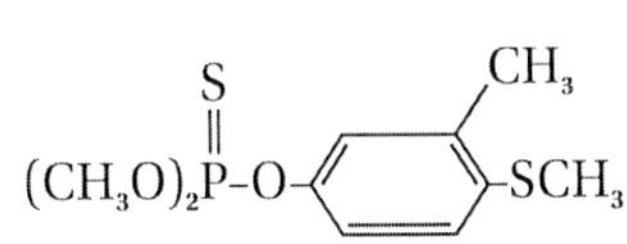

Dasanit ... 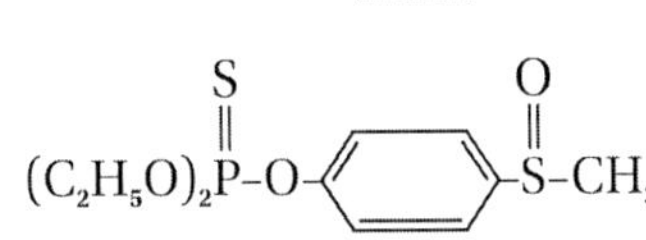

Diazinon ... 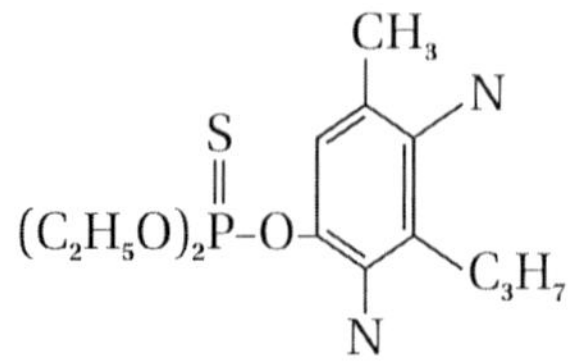

Dursban ... 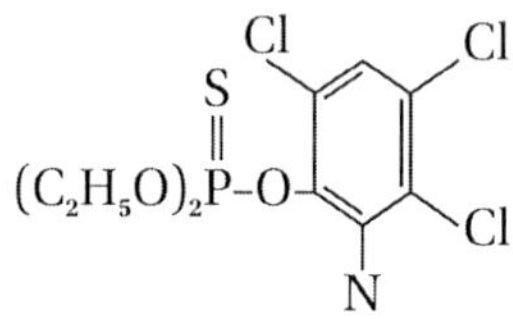

Potasan ... 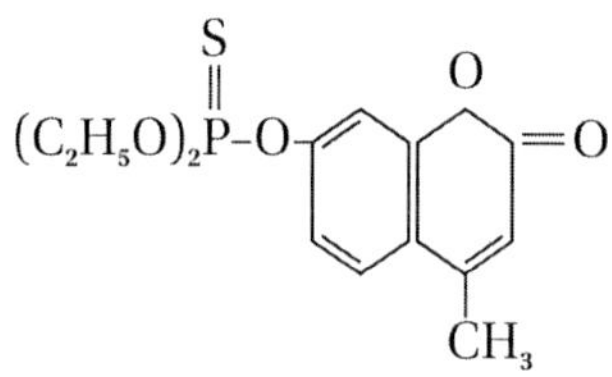

Endothion ... 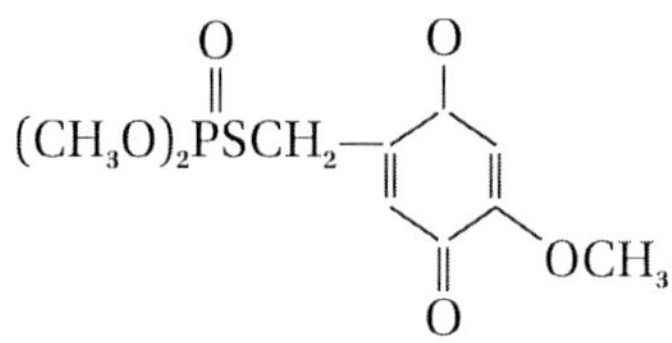

Demeton ... $(CH_5O)_2\overset{S}{\overset{\|}{P}}OCH_2CH_2SC_2H_5$

Methyldemeton ... $(CH_3O)_2\overset{S}{\overset{\|}{P}}OCH_2CH_2SC_2H_5$

Oxydemeton-Methyl ... $(CH_3O)_2\overset{O}{\overset{\|}{P}}OCH_2CH_2SC_2H_5$

Vamidothion ... $(CH_3O)_2\overset{O}{\overset{\|}{P}}SCH_2CH_2S\overset{CH_3}{\overset{\|}{C}}HCONHCH_3$

Acetofos ... $(C_2H_5O)_2\overset{O}{\overset{\|}{P}}SCH_2COOC_2H_5$

Methylacetofos ... $(CH_3O)_2\overset{O}{\overset{\|}{P}}SCH_2COOC_2H_5$

Tinox ... $(CH_3O)_2\overset{S}{\overset{\|}{P}}OCH_2CH_2SCH_3$

Parathion (Thiophos, E-605, 0,0-Diethyl-0-4-nitrophenylthiophosphate, SNP 3422, $C_{10}H_{14}NO_5PS$)

The insecticidal properties of this compound were first discovered by Schrader in 1944. He prepared this compound by reacting diethylchlorothiophosphate with sodium-p-nitrophenolate

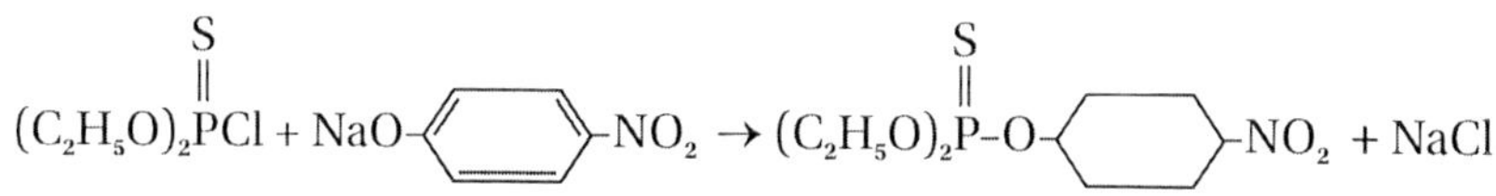

The reaction is carried out in chlorobenzene or in aqueous medium in the presence of emulsifiers. Good yields of this compound are also obtained when diethylchlorothiophosphate reacts with p-nitrophenol in the presence of potassium carbonate (HCl acceptor) in acetone or in other solvents. (Usually the acceptor is a tertiary amine.)

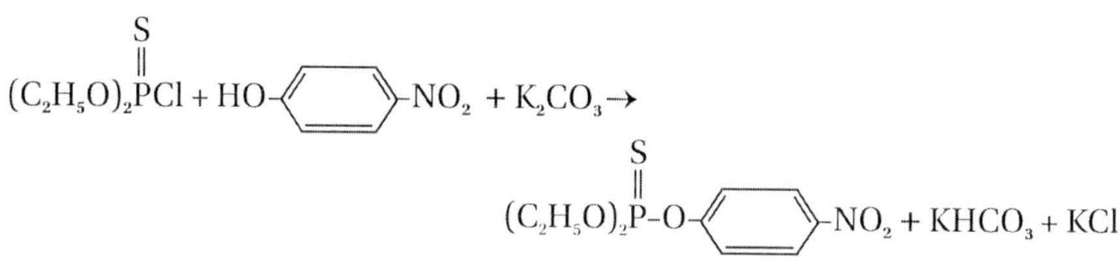

Usually by the aqueous method, technical grade parathion is obtained containing 85-92% of diethyl-0-4-nitrophenylthiophosphate. By the acetone method, it is possible to obtain a compound of 98% purity. The main impurities in the technical grade product are 0-ethyl-0,0-bis (4-nitrophenyl) thiophosphate, nitrophenol, and a small amount of triethylthiophosphate.

It is a clear oily liquid, b.p 113°C. Its solubility in water is 24 mg/l. It is highly soluble in most organic solvents, with the exception of paraffinic hydrocarbons. When heated to 100°C, parathion is gradually converted to the thiolo isomer:

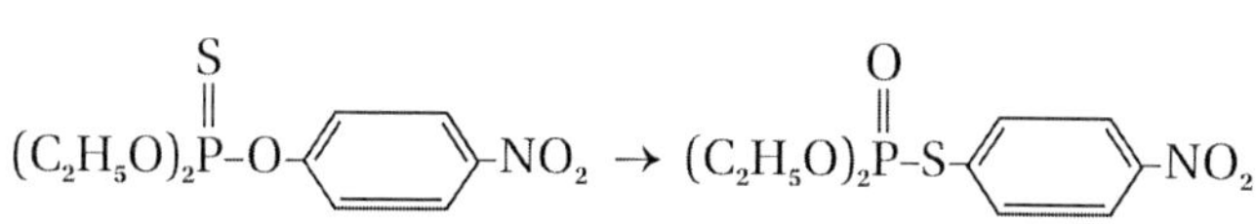

At high temperatures, reaction may take place very violently with explosion. Parathion is slowly hydrolysed by water with the formation of p-nitrophenol and diethylthiophosphoric acid.

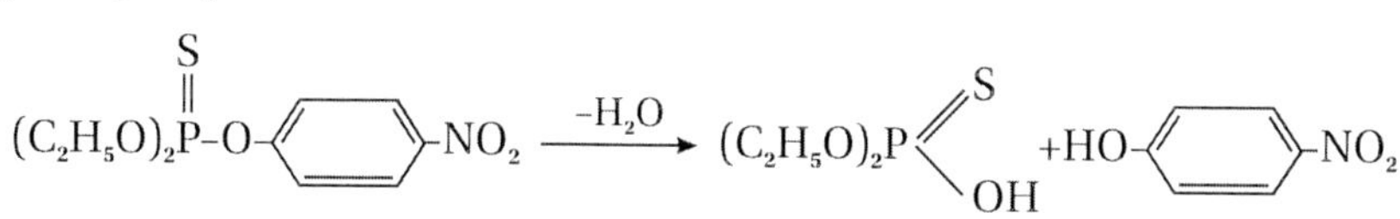

At pH 1 to 5, about 50% is hydrolysed in 690 days at 20°C and in 17 to 20 hours at 70°C. Hydrolysis is more rapid in alkaline medium. This is made use in the quantitative analysis of parathion by the calorimetric method of p-nitrophenol. The rate of hydrolysis in alkaline medium is of the first order with respect to the ester and hydroxyl ion. Below pH 10, 50% hydrolysis took place in 120 days at

25°C whereas in the presence of saturated lime water only eight hours was required.

Oxidising agents convert parathion into 0,0-diethyl-0-nitrophenyl phosphate (paraoxon) which is more toxic for vertebrates than the starting parathion

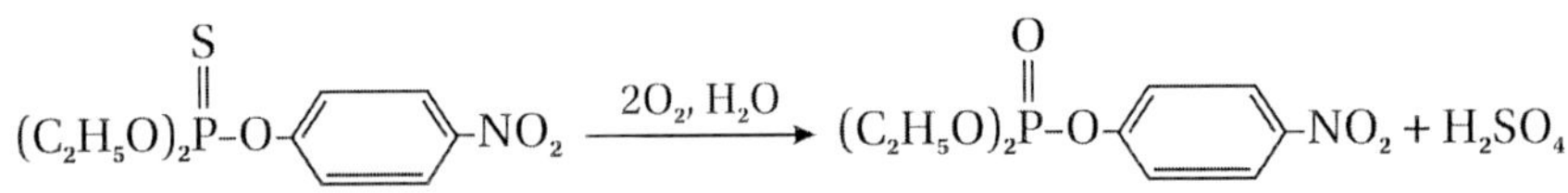

Reducing agents like metals in acid medium convert it to the corresponding amino compound (aminoparathion)

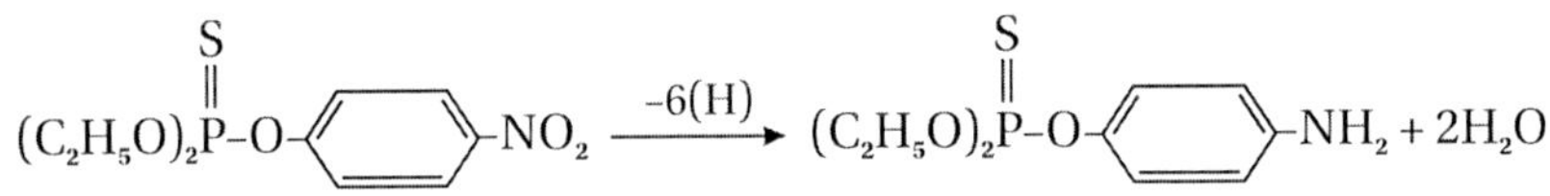

0,0-diethyl-0-4-aminophenylthiophosphate which is non-toxic to animals also does not have any insecticidal effect.

Study of the metabolism of parathion in insects and man has shown that in different species of living organism, the metabolism may proceed in several directions. The principal metabolic products of parathion in man is p-nitrophenol and other hydrolysis products, while in the bovine organism parathion is reduced to aminophenol, which is derivatised in the form of the aminophenyl glucuronide. The pathway of metabolism of parathion can be represented as follows:

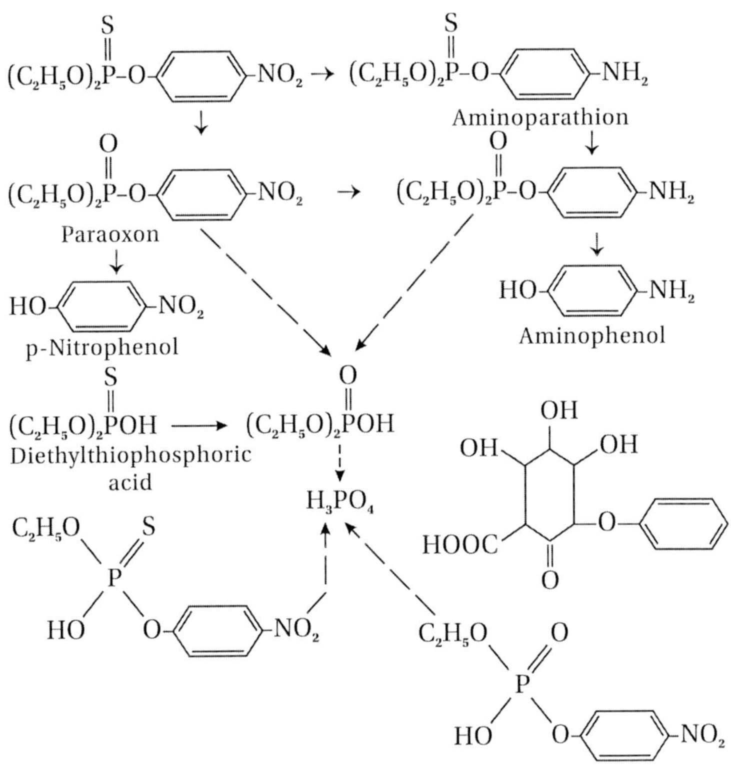

The waiting period (minimum interval after treatment) is determined by the crop that is treated, since the rate of metabolism of parathion depends on the species of the plant. The maximum permissible concentration in air is 0.05 mg/kg.

Chemically, parathion is more stable than TEPP. Formulations include 15, 25, and 50% wettable powders and 25 and 50% emulsifiable concentrates, dusts and aerosols. At the low concentration normally used, parathion is not phytotoxic to crops except certain varieties of apples, pears and ornamental plants. The extent of these injuries apparently is correlated with weather conditions. Parathion is probably translocated through roots in some plants although the systemic action is limited and only small amounts of the compound were recovered from carefully sprayed plants. Parathion has been found in the peel but not in the pulp of harvested citrus fruits. Applications to most crops are usually made one month or more before harvest, so that there will be negligible residue.

Parathion is highly toxic to warm-blooded animals and can be absorbed through the skin, lungs and digestive tract. The LD_{50} for rats is of the order of 2 to 5 mg/kg when parathion is administered orally in propylene glycol solution. Parathion is effective generally against aphids, spider mites and certain other insect pests. It acts mainly as contact insecticide and is rapid in its action although it may be entirely ineffective at temperatures as low as 50 to 60°F. The compound is relatively stable and the spray or dust residues remain somewhat effective for several days, but the residues on plants disappears sufficiently to leave no detectable parathion after about 30 days.

Toxic vapours developed and persisting a week or two in locations such as orchards treated with parathion, have been known to produce serious illness in field workers. Such conditions appear to be associated with hot windless weather. Since parathion inactivates the cholinesterase of the blood and tissue, resulting in symptoms of marked parasympathetic stimulation, atropine is indicated as an antidote for parathion poisoning.

Paraoxon (E-600, oxygen analog of parathion, 0,0-diethyl-0-p-nitrophenylphosphate)

Paraoxon is made by nitrating diethylphenyl phosphate at 0°C or below or from diethylphosphate, carbon tetrachloride, and sodium-p-nitrophenoxide. It was made by Schrader by reacting sodium-p-nitrophenoxide and diethylchlorophosphate using xylene as a solvent.

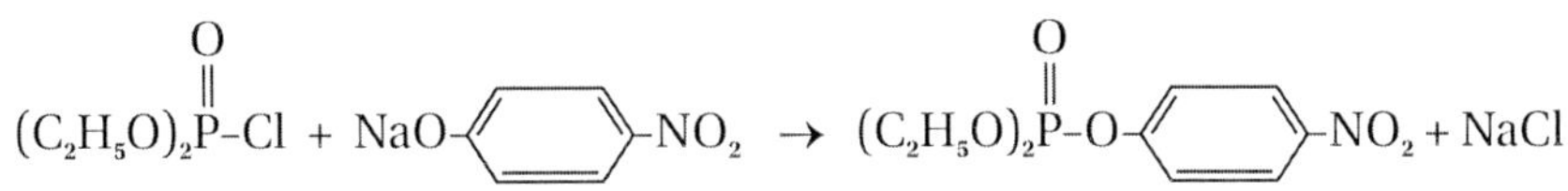

The pure compound is an odourless, reddish yellow oil, b.p 148- 151°C. It is slightly soluble in most organic solvents and of intermediate solubility in oils. Unlike parathion, paraoxon is completely miscible in phosphoric acid. Chemically, paraoxon is relatively stable and is hydrolysed very slowly in water at pH 7 or below. Only 1% hydrolysis takes place in 62 days at pH 5 to 6. This compound is 300 times as stable as TEPP to hydrolysis. In alkaline solutions however hydrolysis is rapid. Little information is available on the phytotoxic properties of paraoxon but insecticidal tests have shown it to be highly toxic to a number of species. It is more toxic than parathion to certain pests like milkweed bugs, german roach and houseflies. To higher animals, paraoxon is also highly toxic. The LD_{50} oral to rats is 3.5 mg/kg.

Methyl Parathion (Metaphos, dimethylparathion, Metacide, Folidoi- 80, Wofatox, 0,0-dimethyl 0-4-nitrophenylthiophosphate)

Methyl parathion is industrially prepared by reacting dimethyl chlorothiophosphate with p-nitrophenol in the presence of HCl acceptors like potassium carbonate in acetone or with sodium p-nitrophenolate. Reaction is carried out in the presence of emulsifiers like ammonium or amine naphthenates,

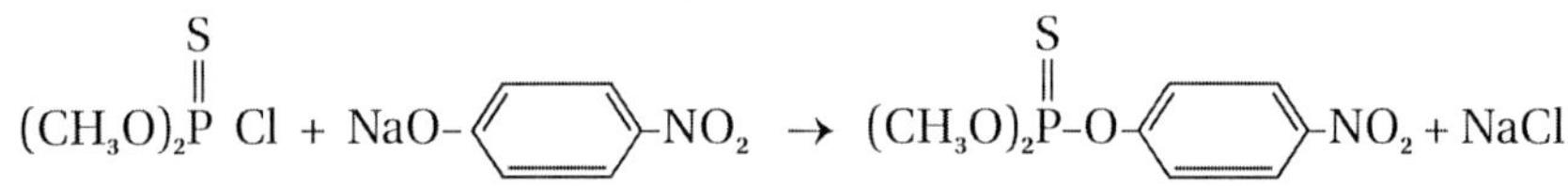

and in water. This is also carried out in organic solvents like chlorobenzene, xylene, or aliphatic ketones (acetone and methyl ethyl ketone). When conducting the reaction in organic solvents, it is necessary to distill off the solvent at the lowest possible temperature in continuously operating film evaporators. Otherwise sometimes, decomposition of the methyl parathion occurs with an explosion. A product of 96 to 98% purity is obtained if the starting materials are pure. This product is diluted with 10 to 15% xylene for the convenience of handling and transportation. However, if different methods are used, compounds of lesser purity are obtained. The yields of methyl parathion by different methods range from 75 to 90%.

Methyl parathion is a white crystalline substance with a m.p. of 35 to 36°C. Its solubility in water is about 55 mg/litre. It is slightly soluble in paraffinic hydrocarbons but highly soluble in aromatic hydrocarbons and most organic solvents. The rate of hydrolysis of methyl parathion is considerably higher than that of parathion, at pH 1 to 5, 50% is hydrolysed in 175 days at 20°C and in 11 hours at 70°C. The rate of hydrolysis in alkaline medium is still greater.

Methyl parathion is relatively unstable thermally and when it is heated to 140 to 160°C, it is completely converted to the corresponding thiolo isomer. This reaction sometimes takes place explosively and a porous mass is formed containing a large amount of carbon.

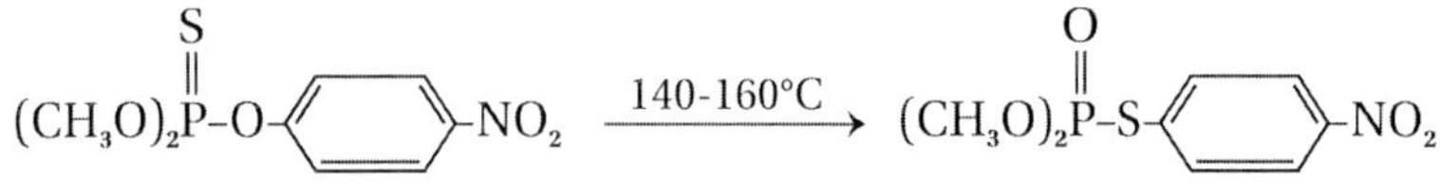

This reaction also occurs on prolonged heating of an alcoholic solution of methyl parathion at 100°C. Among the products of thermal decomposition of methyl parathion are found trimethylsulphonium salts, that are formed as a result of methylation of dimethylsulphide. The formation of this compound also is possible from 0-methyl-S-methyl-4-nitrophenylthiophosphate. Methyl parathion is a powerful alkylating agent and can methylate sulphides, amines, thiourea and many other compounds. This high methylating ability of methyl parathion is the cause of its lower toxicity for mammals because part of the compound is decomposed as a result of demethylation reaction before it reaches the reactive centres of the cholinesterase.

Other chemical properties of methyl parathion are similar to those of parathion. The toxicity of methyl parathion for mammals is considerably lower. It also penetrates the skin with greater difficulty than parathion. The maximum permissible concentration in air is 0.1 mg/m^3.

Methyl parathion is marketed in the form of emulsion, wettable powder and dusts. In selecting diluents for dusts and wettable powders even weak alkaline diluents should not be used as methyl-parathion (as well as parathion) is decomposed. Consequently, it loses its insecticidal activity. Like parathion, methyl parathion is also used for the control of a wide variety of plant pests such as aphids and beetles. Because of its lower toxicity for mammals, methyl parathion is gradually displacing parathion. Now the production of methyl-parathion is higher than that of parathion. A commercial mixture of one part of parathion and four parts of methylparathion is also being manufactured.

Thiophas-ME (Methyl ethyl thiophos, methylethylparathion, 0-methyl -0-ethyl-0-4-nitrophenylthiophosphate)

Thiophos-ME is produced by the condensation of methylethyl chlorothiophosphate with p-nitrophenol and an HCl acceptor in an organic solvent or with sodium p-nitrophenolate in aqueous medium in the presence of emulsifiers (yield is about 80 to 90%). Surface active agents of the type of OP-7 and ammonium or amine naphthalene are used as emulsifiers:

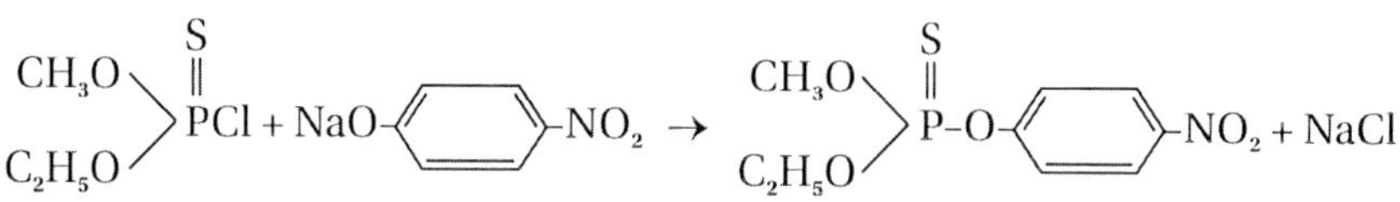

Impurities found in the compound are p-nitrophenol, ethyldimethyl thiophosphate, and some other compounds.

It is a clear oily liquid, b.p. 116°C. Its solubility is very low in paraffinic hydrocarbons but it is soluble in most organic solvents. The resorption toxicity is somewhat lower than that of parathion and the LD_{50} value for rats is 6 to 8 mg/kg. The maximum permissible concentration in air is 0.03 mg/m^3.

The chemical properties of methylethyl parathion are similar to those of parathion. Thus, its rate of hydrolysis occupies an intermediate position between the hydrolysis rates of parathion and methyl parathion. A similar situation is also observed in methylation of organic compounds and in the thermal decomposition. However, because of the presence of methoxyl group in the methylethyl parathion, its chemical properties are closer to those of methyl parathion.. The Pishchemuka isomerisation proceeds mainly with the formation of 0-ethyl-S-methyl-0-4-nitrophenyl thiophosphate. The presence of the isomeric 0-methyl-S-ethyl-0-4-nitrophenyl thiophosphate

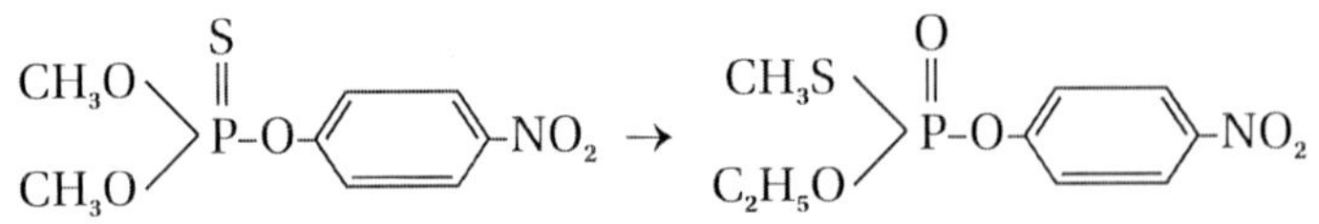

in the reaction mixture is insignificant on plants and in other living organisms. Methylethylparathion breaks down somewhat more quickly than parathion. Their paths of conversion are identical. It is marketed in the form of 20% emulsive concentrate and is used similar to parathion.

Sumithion (Folithion, Metathion, methylnitrophos, fenitrothion, 0,0-dimethyl-0,4-nitro-3-methylphenylthiophosphate)

0,0-Dimethyl-0-4-nitro-3 methylphenylthiophosphate is obtained by the condensation of dimethylchlorothiophosphate with sodium-4-nitro-3-methyl-phenolate in aqueous medium or of free nitrocresol in the presence of anhydrous potassium carbonate in acetone or methylketone. The synthesis of 4-nitro-3-methylphenol present appreciable difficulties since other isomers are also formed and more than 30% of the valuable m-cresol is lost.

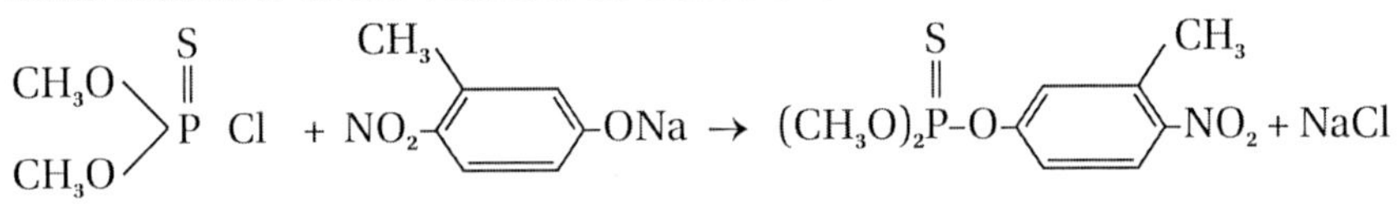

Sumithion is a clear liquid with an unpleasant odour, b.p. is 95°C. Its solubility in water is about 30 mg/litre. It is highly soluble in most organic solvents and is miscible in all proportions with methyl and ethylalcohols, alkylacetates, ketones and aromatic hydrocarbons. Its solubility in kerosene is about 45 and in petroleum ether, about 7%.

The chemical properties of fenitrothion practically do not differ from those of methyl parathion but its rate of hydrolysis by water and alkalies is somewhat lower. Thus, in 0.1N NaOH solution at 30°C, 50% of methylparathion is hydrolysed in 5 minutes and 50% of fenitrothion in 12 minutes. The thermal stablity. of this compound is also lower. When it is heated above 100°C, it undergoes Pishchemuka isomerisation and may decompose explosively. Because of this, overheating of the compound must be avoided both in its production and in storage. The compound must be stored in enameled, aluminium or glass containers. Iron promotes decomposition of the compound like that of most other organophosphorus compounds. Sometimes, a thick jelly-like sediment may occur at the bottom of an emulsion concentrate, possibly due to its coming into contact with iron or its salts. The active material is unstable in alkaline medium. It is compatible with most pesticides except those of alkaline nature such as Bordeaux mixture, nicotine sulphate and lime sulphur.

0,0-dimethyl-0, 6-nitro-3-methylphenylthiophosphate is an active synergist for many organophosphorus insecticides like parathion, methyl parathion, methylethylparathion and fenitrothion. When they are mixed in the ratio of 1 : 1, the mixture is effective at the same concentration as the pure individual compound. For use in agriculture, a mixture of sumithion with 0,0-dimethyl-0-6-

nitro-3-methyl-phenylthiophosphate (called methyl nitrophos) has been recommended. This mixture practically does not differ in insecticidal and acaricidal activity of pure sumithion. The LD_{50} for various experimental animals ranges from 142 to 1000 mg/kg. The nature of action on animals is similar to that of methylparathion.

Chlorothion (0,0-Dimethyl-0,4-nitro-3-chlorophenyl thiophosphate)

Chlorothion is one of the two chlorinated analogs of methyl parathion with greatly reduced mammalian toxicity and is useful as a household insecticide. It is synthesised by reacting dimethylchlorothiophosphate with 3-chloro-4-nitrophenol in methylethylketone in the presence of anhydrous potassium carbonate.

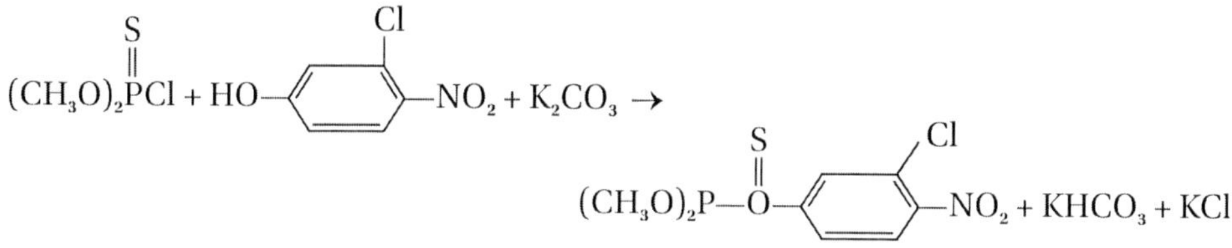

This is a yellow crystalline substance with m.p. 21°C. It is highly soluble in aromatic hydrocarbons and their halogen derivatives but slightly soluble in paraffinic hydrocarbons. Its solubility in water at 20°C is 40 mg/litre. The chemical properties of chlorothion are similar to those of methylparathion. Thus 50% is hydrolysed at 20°C and at pH 1 to 5 in 138 days and at 70°C and pH 5 in 5.3 hours. When acted on by oxidising agents, it is converted to the corresponding ester of phosphoric acid. The LD_{50} for rats is 850 to 980 mg/kg. It is somewhat less active insecticidally than methylparathion and is marketed in the form of 10 and 50% emulsion concentrates and a 20% wettable powder.

Dicapthon (0,0-Dimethyl-2-chloro-4-nitrophenylthiophosphate)

Dicapthon is another chlorinated analog of methylparathion. It is produced by reacting dimethylchlorothiophosphate with 2-chloro-4- nitrophenol in toluene, in the presence of potassium carbonate. Dicapthon is used mainly in the form of emulsion or suspension for the control of flies. Dicapthon is a white crystalline powder m.p. 52 to 53°C and in its chemical properties it is similar to those of

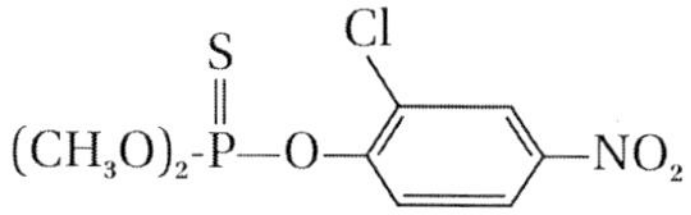

Dicapthon

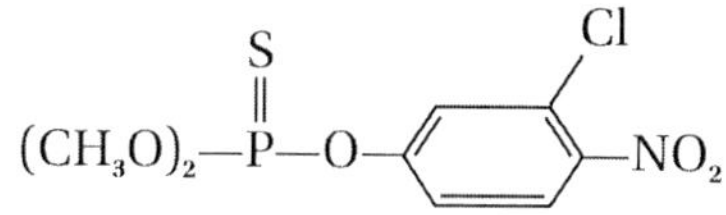

Chlorthion

chlorothion. Its LD_{50} for rats is 330 to 400 mg/kg. However both chlorothion and dicapthon have not been widely used.

Ronnel (Trolene, Etrolene, Dow et-14, Doet 57, Koralan, Nankor, trichlorometaphos, 0,0-Dimethyl-0-2,4,5-trichlorophenylthiophosphate)

Ronnel is a general purpose insecticide which is manufactured by reaction of sodium 2,4,5-trichlorophenolate with dimethylchlorothiophosphate in aqueous medium in the presence of emulsifiers.

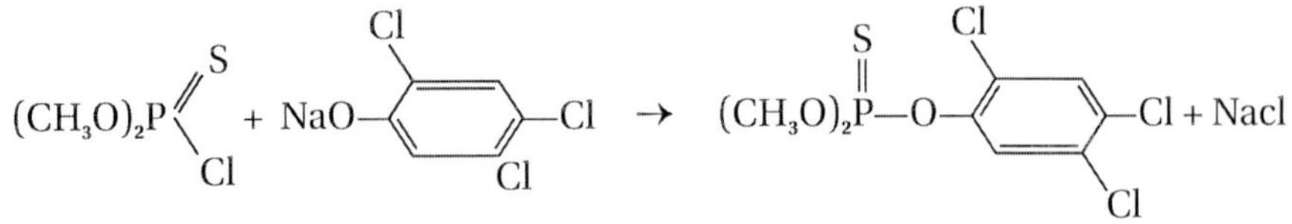

or by methanolysis of 0-2,4,5 trichlorophenyldichlorothiophosphate in the presence of NaOH.

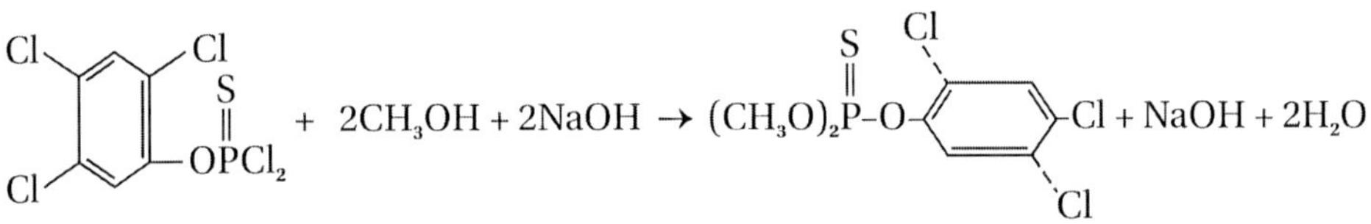

This compound is a white crystalline compound, m.p. 41°C. It is highly soluble in most organic solvents. Its solubility in water is 44 mg/litre and it is stable at temperatures up to 80°C. In weakly alkaline medium, it is hydrolysed with the formation of mainly 0-methyl-0-2,4,5-trichlorophenylthiophosphoric acid while in strongly alkaline medium, the main product formed is 0,0-diniethylthiophos- phoric acid.

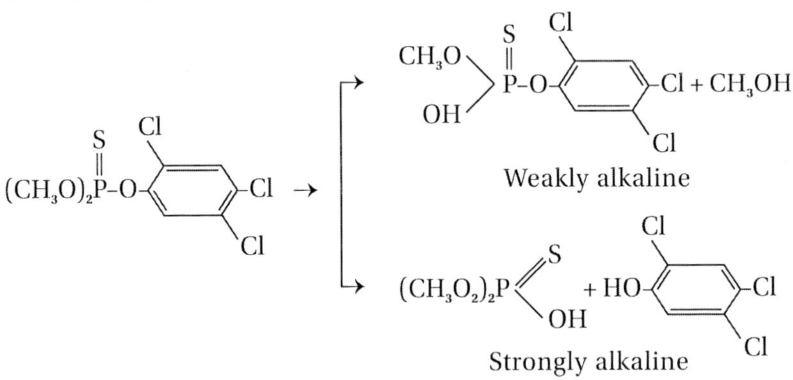

In bovines and in rats about half of the compound introduced is excreted in the urine in the form of 0-methyl-0-2,4,5-trichlorophenyl-thiophosphoric acid while in insects it breaks down predominantly to dimethylthiophosphoric acid and 2,4,5 trichlorophenol.

Ronnel is one of the compounds that is low in toxicity to warm-blooded animals. The LD_{50} for animals is in the range of 400 to 3000 mg/kg. The 4-bromoanalog of ronnel is a persistent general purpose insecticide of low acute toxicity to mammals. It is sold under the name Bromophos.

Bromophos (0,0-Dimethyl-0,2,5-dichloro-4-bromophenylthiophos- phate)

Bromophos is synthesised by all methods that have been used for the production of ronnel. It is a white crystalline substance, m.p. 54°C. Its solubility in water is about 40 mg/litre. It is soluble in various organic solvents like acetone (109 g/1), diesel oil (21 g/1), isopropylalcohol (8 g/1), methylalcohol (10 g/1), methylethylketone (120 g/1), xylene (90 g/1), chlorobenzene (78 g/1) and methylene chloride (112 g/1). In the alkaline medium bromophos is hydrolysed similar to ronnel with the formation of 0,0-dimethylthiophosphoric acid, 2,5-dichloro-4-bromophenol and 0-methyl-0-2,5-dichloro-4-bro-mophenyl phosphoric acid.

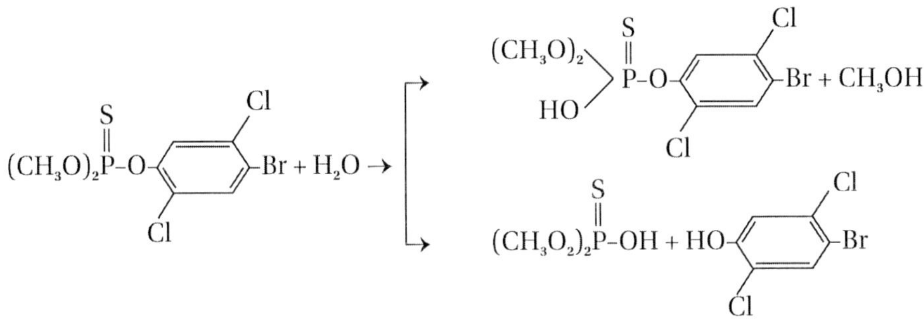

At 22°C, 50% is hydrolysed at pH 13 in 3.5 hours. On plants bromophos is almost completely broken down in 13 to 20 days depending on the species. The LD_{50} value for various species of animals varies from 2000-4000 mg/kg. Bromophos has been suggested for the control of wide variety of pests. It is sold in the form of 20 and 25% emulsion concentrates, 5% granulated preparations and aerosols.

Fenthion {Lebaycid, Baytex, 0,0-Dimethyl-0-(4-methyl mercapto-3-methylphenyl) thiophosphate}

Fenthion is a fast killing insecticide with long residual activity. It is obtained by reacting 4-methylmercapto-3-methylphenol with the dimethylchlorothio-phosphate in methylethylketone at 60°C in the presence of finely ground potassium carbonate.

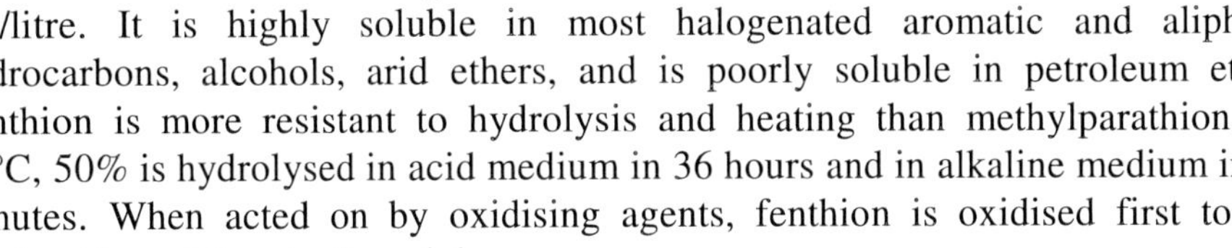

It is a colourless oil, b.p. 87°C. Its solubility in water at 20°C is about 54 mg/litre. It is highly soluble in most halogenated aromatic and aliphatic hydrocarbons, alcohols, arid ethers, and is poorly soluble in petroleum ether. Fenthion is more resistant to hydrolysis and heating than methylparathion. At 80°C, 50% is hydrolysed in acid medium in 36 hours and in alkaline medium in 95 minutes. When acted on by oxidising agents, fenthion is oxidised first to the sulphoxide and then to the sulphone.

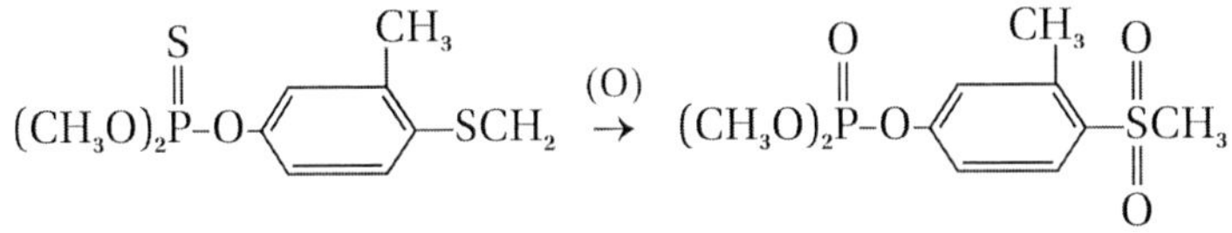

The sulphone under the influence of oxidising agents, further splits off the thionosulphur and goes to the corresponding ester of phosphoric acid.

Under natural conditions fenthion approaches chlorophos but the selectivity of this action is less, since its toxicity for mammals is higher. In animals and plants fenthion is oxidised to the sulphone which then is hydrolysed and breaks down to products that are harmless to animals. It is marketed in the form of a 50% emulsion concentrate, a 25% wettable powder and a 3% dust. Fenthion may be mixed with other insecticides and fungicides. Its high alkaline stability permits it to be combined with Bordeaux mixture also. However mixtures of this kind must be applied immediately after being mixed.

A large number of analogs and homologs of fenthion have been synthesised and studied but most of these compounds have a higher toxicity than fenthion for animals. In many mixed esters of thiophosphoric acid the introduction of a methyl group or halogen in the metaposition to the ester group, lowers the toxicity to vertebrate animals and makes comparatively little change in insecticidal properties.

Dasanit {Fensulphothion, Terracur-P, 0,0-Diethyl-0-p-(methylsul- phinyl) phenylthiophosphate}

Dasanit is produced by the condensation of 0,0-dicthylchlorothio- phosphate with 4-methylsulphinyl-l-hydroxybenzene in the presence of combining reagent:

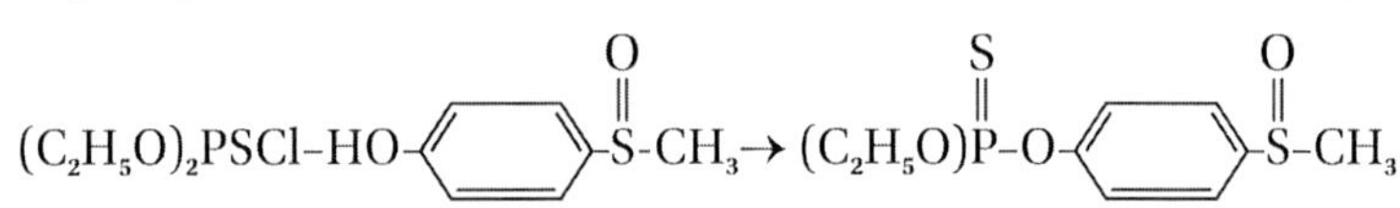

or by reaction of 0,0-diethylchlorothiophosphate with 4-methylthio-hydroxybenzene and oxidation of the resulting ester with hydrogen oxide.

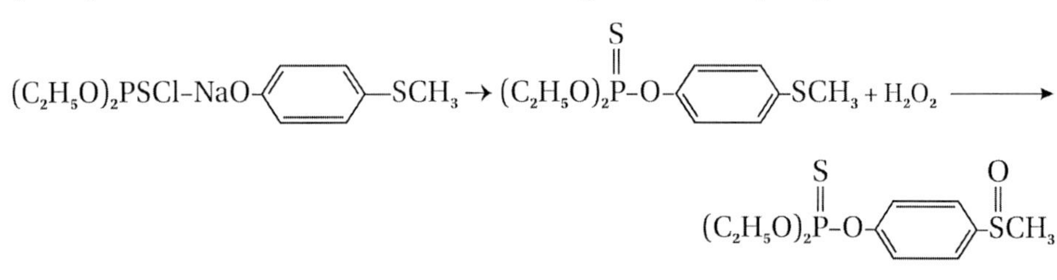

Dasanit is an oily liquid, b.p. 138-141°C. It is soluble in most organic solvents, slightly soluble in water. The LD_{50} for rats is 10.5 mg/kg. It is used in the form of emulsion concentrate, wettable powders, dusts and granules for the control of pests.

Diazinon {0,0-Diethyl-0-(2 isopropyl-4-methylpyrimidyl-6) thiophosphate}

Diazinon is a persistent insecticide for the control of soil insects, pests on various crops, and as a residual spray for dairy and livestock pests. It is produced by reacting 2-isopropyl-4-methyl-6-hydroxy pyrimidine with diethyl chlorothiophosphate in the presence of potassium carbonate.

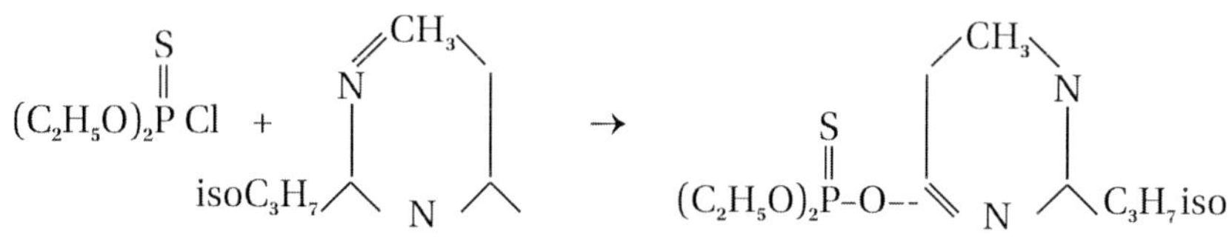

The reaction goes well in dimethylformamide solution at 50-60°C.

It is a colourless oil, boiling point 89°C. Solubility in water is 40 mg/litre at 20°C. It is highly soluble in most organic solvents. Diazinon is not as resistant to hydrolysis as parathion. In acid medium, it is hydrolysed 12 times as rapidly as parathion and in alkaline medium the hydrolysis goes on at practically the same rate. In excess of water the principal products of hydrolysis of diazinon are diethylthiophosphoric acid and 2-isopropyl-4 methyl-6-hydroxy pyrimidine but with insufficient water in acid medium, a small amount of tetraethyl dithio and thiopyrophosphates is formed.

It is marketed in the form of emulsion concentrates, wettable powder, dusts and granulated formulations. In spite of its relatively high toxicity to mammals diazinon is considered in most countries, as a safe compound.

Dursban (0,0-Diethyl-0-3,5,6-trichloropyridylthiophosphate)

Dursban is an active insecticide for the control of sucking and chewing plant pests, soil inhabiting plant pests and household parasites. It is synthesised by reacting the sodium salt of 3,5,6-trichloro- 2-hydroxy-pyridine with diethylchlorothiophosphate in dimethylformamide:

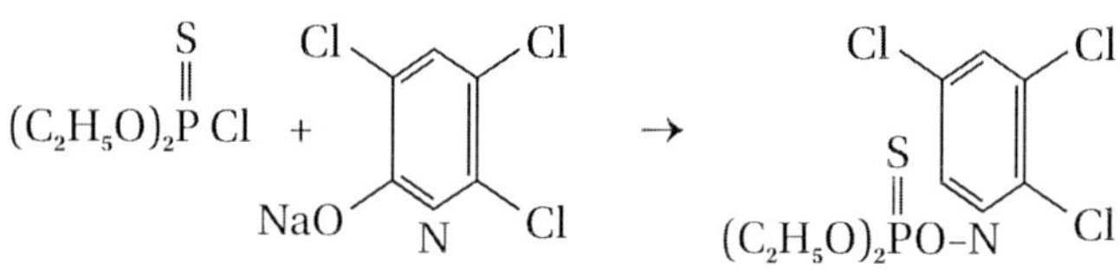

It is a crystalline substance m.p. 41 to 43°C. It is highly soluble in most organic solvents, but almost insoluble in water. In acid or alkaline media, the compound is slowly hydrolysed by water forming diethylthiophosphoric acid and ethylthiophosphoric acids and trichlorohydroxypyridine. By oxidation it may be converted to the corresponding phosphate. The duration of action of the compound on different surfaces is 6 to 11 weeks, but on leaves of plants it is shortacting, apparently due to its rapid hydrolysis under the influence of enzymes. The compound is active on grain plants for several weeks. In addition to the above certain other thiophosphates like Zinophos, Potasan and Endothion are also used in agriculture. The LD_{50} for rats is 150 mg/kg.

Zinophos (Thionazin, 0,0-Diethyl-0-pyrazinylthiophosphate)

Zinophos is produced by reacting diethylchlorothiophosphate with the sodium derivative of 2-hydroxypyrazine. Zinophos is a clear liquid, highly soluble in organic solvents. Its chemical properties are similar to those of other thiophosphates. Zinophos has been used as a systemic nematicide, acaricide and insecticide. The LD_{50} for rats is 10.7 mg/kg.

Potasan {0,0-Diethyl-0-(4-methylcoumarinyl-7) thiophosphate}

Potasan is a selective insecticide and acts as a stomach and contact poison. It is synthesised in greater than 95% yield by reacting at 80 to 90°C, diethyl-chloro thio phosphate with methylcumerleliferone in chlorobenzene and in the presence of potassium carbonate.

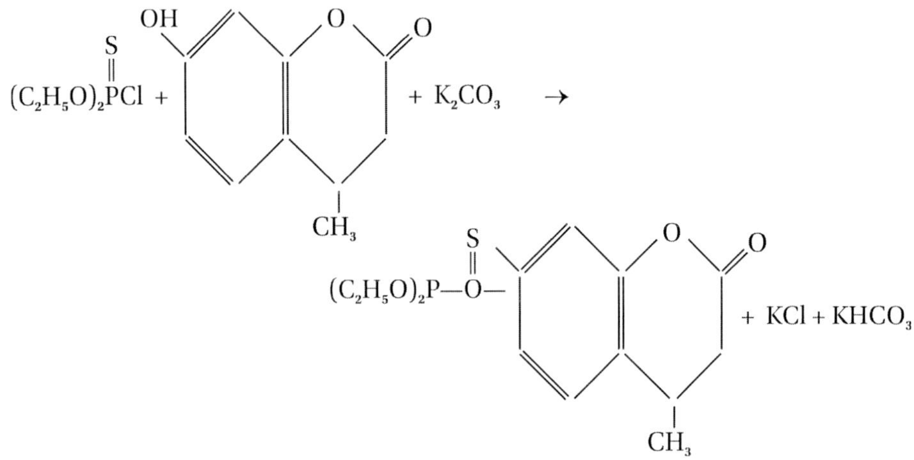

It is a white crystalline substance, m.p. 38°C, practically insoluble in water but highly soluble in most organic solvents. At pH 5-8, it is resistant to the action of water and only when the pH is increased, its hydrolysis is substantially increased. The hydrolysis of potasan takes place at the ethyl group and the first product of hydrolysis that is formed is 0-ethyl-0-(4-methylcoumarinyl-l) thiophosphoric acid, which has low toxicity for both insects and animals. An aqueous solution of Potasan at pH 7-8 exhibits a blue fluorescence. The LD_{50} for rats is 19 to 42 mg/kg.

Endothion {Phosphoryrone, 0,0-Dimethyl-S-(5-methoxy-pyronyl- 2-methyl) thiophosphate}

Endothion is a systemic insecticide produced by reacting sodium 0,0-dimethyl-l-thiophosphate with S-methoxy-2-chloromethyl-pyrone in acetone or other suitable solvent. It is a white crystalline substance, m.p. 90-91 °C, highly

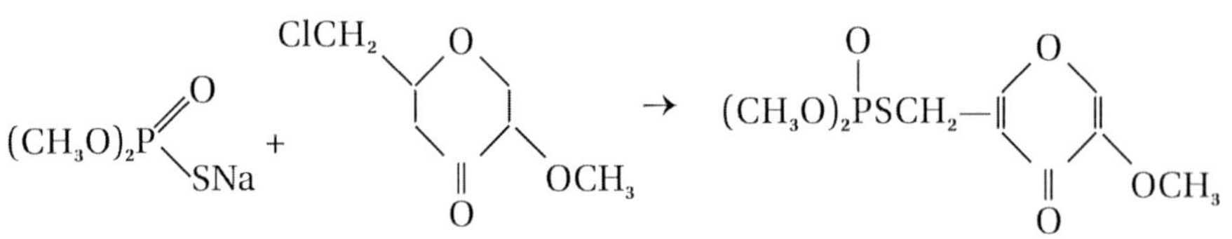

soluble in water (150 g in 100 ml of water) but only slightly soluble in hydrocarbons and oils. When acted on by alkalies endothion is hydrolysed to form as the first product 0-methylthiophosphate which has low toxicity for animals and insects. The effective duration of endothion is 8 to 14 days. The LD_{50} for rats is 30-50 mg/kg. It is marketed in the form of emulsion concentrate and wettable powder.

Demeton (Systox, Mercaptophos, 0,0-Diethyl-2-ethylmercapto-ethylthiophosphate)

Of considerably greater interest as systemic insecticide and acaricide are the mixed esters of thiophosphoric acid containing a sulphide of sulphoxyl group, in one of the ester radicals. At present more than five compounds of this type are used in agriculture.

The technical grade product of demeton is a mixture of thiono (70%) and thiolo (30%) isomers of 0,0-diethyl-2-ethylmercapto- ethylthiophosphate. In industry demeton is produced by the reaction of diethylchlorothiophosphate with a 2-hydroxydiethyl-sulphide in the presence of an HCl acceptor (Na OH or carbonates of the alkali metals).

$$(C_2H_5O)_2\overset{\overset{S}{\|}}{P}Cl + HOCH_2CH_2SC_2H_5 + NaOH \rightarrow (C_2H_5O)_2\overset{\overset{S}{\|}}{P}O(CH_2)_2SC_2H_5 + H_2O + NaCl$$

The compound is partially isomerised directly in the process of preparation and the higher the reaction temperature, the more isomerisation occurs. The pure thiolo isomers of demeton can be obtained in good yield from the salts of diethylthiophosphoric acid and 2-chlorodiethylsulphide.

$$(C_2H_5O)_2P\begin{matrix}\nearrow O \\ \searrow SK\end{matrix} + ClCH_2CH_2SC_2H_5 \rightarrow (C_2H_5)_2\overset{\overset{O}{\|}}{P}SCH_2SC_2H_5 + KCl$$

Thiono isomer is an oily liquid, b.p. 106°C. At 20°C, its solubility in organic solvents including petroleum hydrocarbons is very high. Thiolo isomer is also highly soluble in most of the organic solvents. The boiling point of thiolo isomer is 100°C. The LD_{50} values for rats of the thiono, thiolo isomers and technical grade mixture of isomers are 30, 1.5 and 7 to 10 mg/kg. respectively. The maximum permissible concentration in air is 0.02 mg/m^3.

The thiono isomer of demeton is more resistant to hydrolysis in alkaline medium. The time for 50% hydrolysis of the thiono isomer at 20°C and at pH 13 is 75 minutes and that of the thiolo isomer is 0.85 minutes. In acid medium, the thiolo isomer is hydrolysed half as fast as the thiono isomer. The thiono isomer of demeton rearranges to the thiolo isomer at elevated temperature and in polar solvents probably through the intermediate formation of the sulphonium ions.

$$(C_2H_5O)_2\overset{\overset{O}{\|}}{P}OCH_2CH_2SC_2H_5 \rightarrow (C_2H_5O)_2\overset{\overset{O}{\|}}{P}SCH_2CH_2SC_2H_5$$

This rearrangement probably also takes place in plant tissues. Probably the metabolism of demeton in plants takes place as per the following scheme:

$$\begin{array}{ccc}
(C_2H_5O)_2\overset{S}{\overset{\|}{P}}OCH_2CH_2SC_2H_5 & \rightarrow & (C_2H_5O)_2\overset{O}{\overset{\|}{P}}OCH_2CH_2\overset{O}{\overset{\|}{S}}C_2H_5 \\
\downarrow & & \downarrow \\
(C_2H_5O)_2\overset{O}{\overset{\|}{P}}OCH_2CH_2SC_2H_5 & \rightarrow & (C_2H_5O)_2\overset{S}{\overset{\|}{P}}OCH_2CH_2\overset{O}{\overset{\|}{S}}C_2H_5 \\
\downarrow & & \downarrow \\
(C_2H_5O)_2\overset{O}{\overset{\|}{P}}OCH_2CH_2\overset{O}{\overset{\|}{S}}C_2H_5 & \rightarrow & (C_2H_5O)_2\overset{O}{\overset{\|}{P}}OCH_2CH_2\underset{O}{\underset{\|}{\overset{O}{\overset{\|}{S}}}}C_2H_5
\end{array}$$

Subsequently hydrolysis and further oxidation of the organic part of the demeton molecule takes place with the formation of the simplest decomposition products.

Oxidation of demeton also takes place with such oxidising agents as hydrogen peroxide, bromine water and hypochlorites. Its reaction with bromine is utilised for the quantitative determination of thiolo isomer content in the mixture, since the isomers behave differently with bromine. Demeton yields sulfonium compounds with dimethylsulphate and other alkylating agents. Demeton has powerful systemic effect and is used for the control of many sucking plant pests. It has also some contact and fumigant action and is also used for seed treatments. The duration of action of the compound under field conditions depends on the crop, the period varying from 4 to 6 weeks. It is marketed in the form of emulsion concentrates'. Non-ionic compounds of the type of polyethylene glycol esters are used as emulsifiers with the addition of small amounts of a calcium alkylarylsulphonate. The oral LD_{50} values for rats is approximately 15 mg/kg.

Methyldemeton (Metasystox, methylmercaptophos, 0,0-Dimethyl-2-ethylmercapto ethylthiophosphate)

Methyldemeton is the methyl analog of demeton. Technical grade methyldemeton is a mixture of thiono (70%) and thiolo (30%) isomer of 0,0-dimethyl-2-ethylmercaptoethylthiophosphate. Methyldemeton is produced by the reaction of 2-hydroxy diethylsulphide with dimethylchlorothiophosphate in the presence of an HCl acceptor.

$$(CH_3O)_2\overset{S}{\overset{\|}{P}}Cl+HOCH_2CH_2SC_2H_5+NaOH \rightarrow (CH_3O)_2\overset{S}{\overset{\|}{P}}OCH_2CH_2SC_2H_5+NaCl+H_2O$$

The technical grade product contains a small amount of trimethylthiophosphate, 2-hydroxydiethylsulphide and some other esters of thiophosphoric acid as impurities. The total content of thiono and thiolo isomers usually is less than 90%.

Thiono isomer is a liquid with a characteristic, unpleasant odour, b.p. 93°C. It is slightly soluble in water (330 mg/litre). The thiolo isomer is a liquid boiling at 102°C. It is more soluble in water than the thiono isomer (3300 mg/litre). Both are highly soluble in organic solvents.

The chemical stability of methyldemeton is lower than that of demeton. For example 50% of the methyldemeton is hydrolysed at pH 3 and 70°C in 4.9 hours while it is 10 hours in the case of demeton. In alkaline medium the difference is still appreciable. Rearrangement of the thiono to thiolo isomer also takes place more rapidly. While the thiono isomer of demeton at 40°C is converted to a mixture of 90% thiono and 10% thiolo isomers in 91 days, only 10% of the thiono isomer of methyl demeton under the same conditions is isomerised in eight days. In the presence of traces of water, methyldemeton can enter into a intermolecular methylation reaction which in many instances is the cause of spoilage of the product on prolonged storage. The stability of methyldemeton is also affected negatively by metallic iron and some iron compounds. So it should not be kept in iron containers but it can be kept in enameled or pure aluminium containers.

The metabolism of methyldemeton in plants and insects is similar to that of demeton except that when methyldemeton enters the plant, it is readily isomerised to the thiolo isomer which is further metabolised by the same route as demeton. Oxidation products of methyldemeton have not been detected but they have been obtained synthetically. The conversion and decomposition of methyldemeton in plants takes place relatively quickly, as the duration of insecticidal activity does not exceed three weeks. Methyldemeton is less toxic, than demeton. It is used in the form of aqueous emulsion and is marketed as concentrate containing 30 and 50% of the mixture of isomers. Methyldemeton is a very active systemic insecticide particularly suitable for granular application to the soil. The LD_{50} of the technical product, thiono isomer and thiolo isomer are 80-100, 180, and 40 mg/kg respectively. The maximum concentration permissible in air is 0.1 mg/m^3.

Tinox (0,0-Dimethylmercaptoethylthiophosphate)

Tinox is similar to methyldemeton. The technical grade product is a mixture of thiolo isomer (60 to 70%) and thiono isomer (30 to 40%). Tinox is manufactured by methods similar to those of demeton and methyldemeton. Both the isomers are liquids. The solubility of Tinox in water is about 500 mg/litre. It is highly soluble in most organic solvents. The chemical properties, insecticidal activity and duration of action of Tinox are similar to that of methyldemeton. It is marketed in the form of 50% emulsion concentrates.

Metasystox-R {Oxydemeton-methyl, 0,0-dimethyl-S-(2-ethyl-sulphinyl) ethylthiophosphate}

Metasystox-R is another compound similar to that of methyldemeton. It is produced by the oxidation of thiolo isomer of methyldemeton with various oxidising agents like hydrogen peroxide, hypochlorite and bromine.

$$(CH_3O)_2\overset{\overset{\displaystyle O}{\|}}{P}SCH_2CH_2SC_2H_5 \rightarrow (CH_3O)_2\overset{\overset{\displaystyle O}{\|}}{P}SCH_2CH_2SOC_2H_5$$

This is a yellow liquid with a weak odour, b.p. 106°C. It is highly soluble in water but slightly soluble in hydrocarbons of the aliphatic series, highly soluble in methylene chloride and halogenated aromatic hydrocarbons; In acid medium, it is more resistant to hydrolysis than methyldcmeton but in alkaline medium, it is hydrolysed more rapidly than the latter. Oxydemetonmethyl is close to methyldemeton in insecticidal activity. It is marketed in the form of a 25% emulsion concentrate. The LD_{50} for rats is 65 to 75 mg/kg.

Metasystox-S {0-Dimethyl-S-(2 ethylsulphinyl) isopropylthiophosphate}

Metasystox-S is an oily liquid, boiling at 115°C. It is rather highly soluble in water, alcohol, ketones and chlorinated derivatives of hydrocarbons, slightly soluble in hydrocarbons of the aliphatic series. The compound is resistant to hydrolysis in acid medium but is hydrolysed several times faster in alkaline medium. Its insecticidal activity is almost the same as of oxydemeton methyl. It is marketed in the form of 50% emulsion concentrates. The hydrolytic stability of sulphoxide is almost 100 times greater than that of sulphides. The LD_{50} for rats is 105 mg/kg.

Another compound Metasystox-I (0,0-Dimethyl-S-ethylmercaptoethyl-thiophosphate) is a rapidly acting systemic and contact insecticide. It is the thiolo isomer of methyldemeton. It is marketed in the form of 50% emulsion concentrate.

Acetofos (0,0-Diethyl-S-carboethoxymethylthiophosphate)

Acetofos is a contact insecticide but its activity approaches that of trichlorofon. It is produced by the reaction of the ethyl ester of monochloroacetic acid with ammonium salt of dicthylthiophosphoric acid in a suitable organic solvent like benzene.

$$(C_2H_5O)_2\overset{\overset{\displaystyle O}{\|}}{P}SNH_4 + ClCH_2COOC_2H_5 \rightarrow (C_2H_5O)_2\overset{\overset{\displaystyle O}{\|}}{P}SCH_2COOC_2H_5 + NH_4Cl$$

This is colourless liquid with a specific unpleasant odour, boiling at 120°C. It is highly soluble in water and in most organic solvents. When acetofos reacts with water in acid or alkaline medium, hydrolysis of the compound occurs, with splitting off of the ester groups. Break down of compound proceeds differently in warmblooded animals and in insects. Consequently, acetofos has little toxicity for mammals but is of highly insecticidal value. Hydrolysis in warm-blooded animals possibly takes place by the following path:

$$(C_2H_5O)_2\overset{\overset{\displaystyle O}{\|}}{P}SCH_2COOC_2H_5 + H_2O \rightarrow (C_2H_5O)_2\overset{\overset{\displaystyle O}{\|}}{P}SCH_2COOH + C_2H_5OH$$

In insects splitting off of the acetic acid group and inhibition of vitally important enzymes takes place.

Methylacetofos (0,0-Dimethyl-S-carboethoxymethylthiophosphate)

Methylacetofos is the methyl analog of acetofos, and is synthesised similarly to acetofos. However because of side reactions the yield does not exceed 70%. The amine impurity in the technical grade product is trimethylthiophosphate which greatly increases the toxicity of methylacetofos to vertebrate animals.

Methylacetofos is a clear liquid with an unpleasant odour boiling between 116-120°C. It is highly soluble in water and most of the organic solvents. It is similar to acetofos in chemical properties but is less stable and in storage may undergo conversions as a result of its high alkylating ability. The main decomposition products are compounds of the sulphonium type. Methylacetofos is similar to acetofos in insecticidal activity. The LD_{50} of pure methyl acetofos for rats is 1000 mg/kg.

Vamidothion {0,0-Dimethyl-S-2-(1-N-methylcarbamoyl ethylmercapto) ethyl-thiophosphate}

Another important insecticide is vamidothion. It is produced by reacting ammonium dimethylthiophosphate or an alkali dimethylthiophosphate with an a-(2-hacethylmercapto) propiomethylamide.

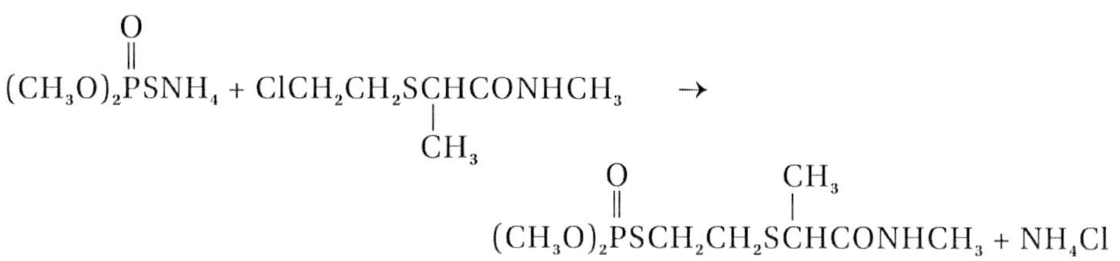

Vamidothion is a white crystalline substance, m.p. 46 to 48°C. The technical grade product is highly soluble in water, acetone, methylethylketone, ethylacetate, and acetonitrile, not easily soluble in hexane, cyclohexane and other paraffinic and cycloparaffinic hydrocarbons. The insecticidal activity of vamidothion approaches that of methyldemeton but considerably exceeds it in duration of effect. It is somewhat more resistant to hydrolysis than methyldemeton and demeton. Oxidising agents convert it first to sulfphoxide and then further to sulphone. The first product of metabolism in plants is also the suiphoxide. Simultaneously with

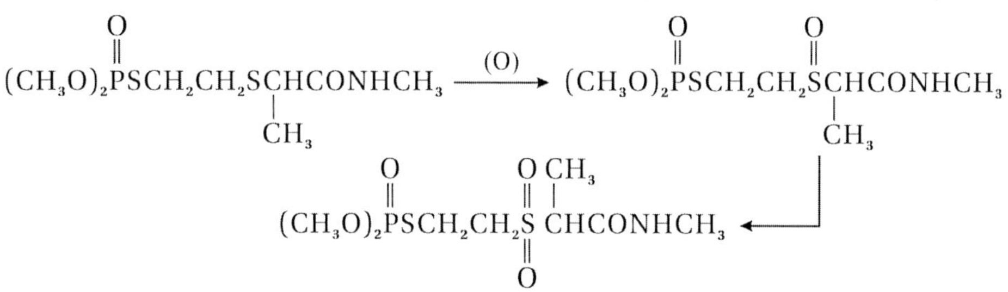

oxidation in plants, hydrolysis of the compound also occurs with the formation of dimethylphosphoric acid and phosphoric acid. The LD_{50} for rats is 64 to 100 mg/kg.

DERIVATIVES OF DITHIOPHOSPHORIC ACID

Today more than 25 different derivatives of dithiophosphoric acid are being used as insecticides in agriculture. In most cases, the toxicity of the derivatives of dithiophosphoric acid is less than the corresponding compound of thiophosphoric acid. But the chemical stability of the compound is increased in dithiophosphoric acid derivatives. The mixed esters of dithiophosphoric acid derivatives are prepared by either reacting the salts of dialkyl dithiophosphoric acid with the appropriate halogen derivatives:

$$(RO)_2\overset{\overset{\displaystyle S}{\|}}{P}SNa + ClCH_2R'' \rightarrow (RO)_2\overset{\overset{\displaystyle S}{\|}}{P}SCH_2R'' + NaCl$$

or by the addition of dialkylthiophosphoric acids at double bond of the unsaturated compounds:

$$(RO)_2\overset{\overset{\displaystyle S}{\|}}{P}SH + \overset{\overset{\displaystyle CHR}{\|}}{CHR} \rightarrow (RO)_2\overset{\overset{\displaystyle S}{\|}}{P}S\underset{\underset{\displaystyle CHR}{|}}{CHR}$$

or by reacting dialkylchlorothiophosphates with the appropriate thiols in the presence of HC1, caustic alkali or various organic bases:

$$(RO)_2\overset{\overset{\displaystyle S}{\|}}{P}Cl + NaSR' \rightarrow (RO)_2\overset{\overset{\displaystyle S}{\|}}{P}SR' + NaCl$$

Some of the important derivatives of dithiphosphoric acid are:

Malathion ... $(CH_3O)_2\overset{\overset{\displaystyle S}{\|}}{P}S\underset{\underset{\displaystyle CH_2COOC_2H_5}{|}}{CH}COOC_2H_5$

Dimethoate ... $(CH_3O)_2\overset{\overset{\displaystyle S}{\|}}{P}SCH_2CONHCH_3$

Thiocron ... $(CH_3O)_2\overset{\overset{\displaystyle S}{\|}}{P}SCH_2CONHCH_2CH_2OCH_3$

Morphothion ... 

Formothion ... $(CH_3O)_2\overset{\overset{\displaystyle S}{\|}}{P}SCH_2CO\overset{\overset{\displaystyle CH_3}{|}}{N}CHO$

Mecarbam ... $(C_2H_5O)_2\overset{\overset{\displaystyle S}{\|}}{P}SCH_2CO\overset{\overset{\displaystyle CH_3}{|}}{N}\,COOC_2H_5$

Thimet ... $(C_2H_5O)_2\overset{S}{\overset{\parallel}{P}}SCH_2SC_2H_5$

Carbophenothion ... 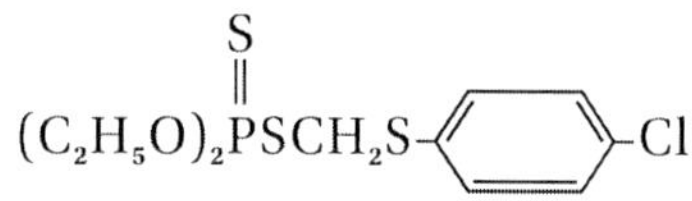

Methyl trithion ... 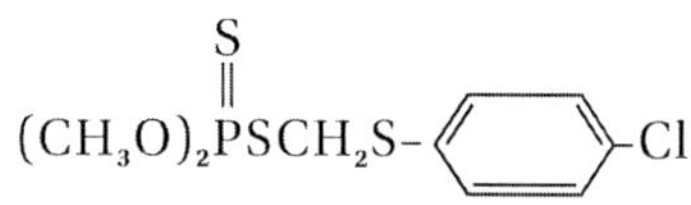

Phenkapton ... 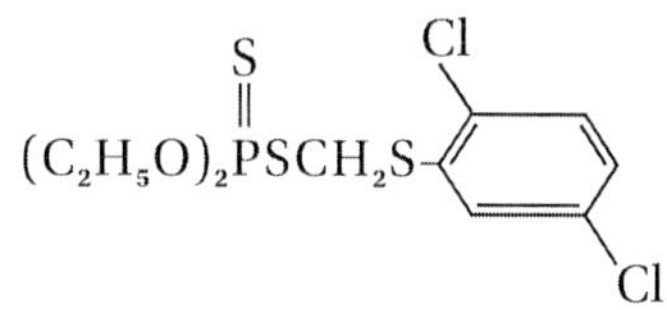

Ethion ... $(C_2H_5O)_2\overset{S}{\overset{\parallel}{P}}SCH_2S\overset{S}{\overset{\parallel}{P}}(O\ C_2H_5)_2$

Thiometon ... $(CH_3O)_2\overset{S}{\overset{\parallel}{P}}SCH_2CH_2SC_2H_5$

Disyston ... $(C_2H_5O)_2\overset{S}{\overset{\parallel}{P}}SCH_2CH_2SCH_5$

Tetrathion ... $(CH_3O)(C_2H_5O)\overset{S}{\overset{\parallel}{P}}\text{-}SCH_2CH_2SC_2H_5$

Imidan ... 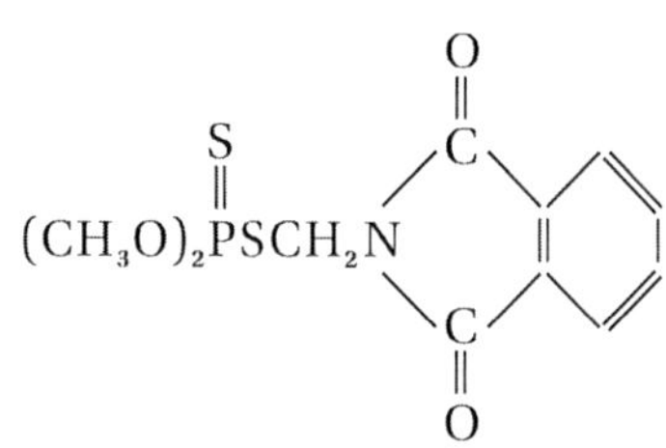

Phosalone ... 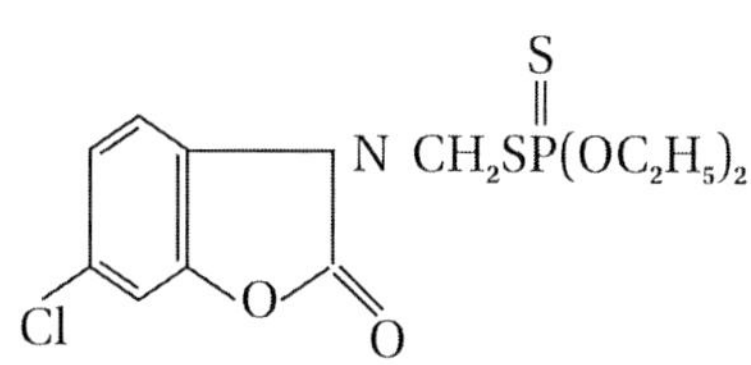

Guthion ... 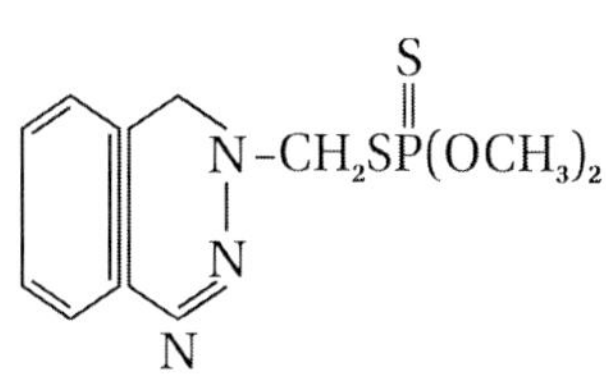

Menazon ... 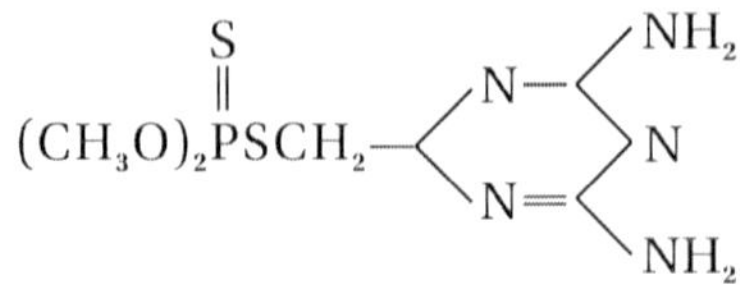

Delnav ... 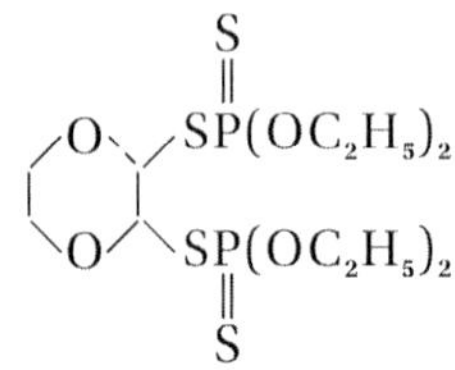

Cidial ... 

Malathion (Carbophos, 0,0-Dimethyl-S-1,2-dicarboethoxyethyl-dithio phosphate)

Malathion is one of the safest organophosphorus compounds now marketed in the form of emulsion concentrate. It is also possible to use malathion in the form of dusts or suspensions prepared from a wettable powder. It is produced by the addition of dimethyldithiophosphoric acid to maleic acid ester in the presence of basic catalysts in various organic solvents or without solvents. The technical

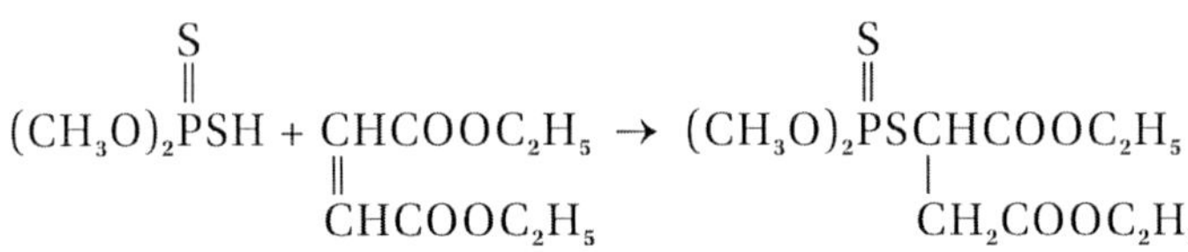

grade compound obtained by this method contains small amounts of impurities like trimethyldithiophosphate, dimethylmaleate and solvent which decompose malathion very rapidly. Xylene is most often used as a solvent for malathion synthesis. Decomposition takes place even at very low temperatures.

Pure malathion is a colourless liquid boiling at 120°C. At 20°C, it is sparingly soluble in water but highly soluble in most organic solvents with the exception of saturated hydrocarbons. Malathion on prolonged heating at 150°C is isomerised and goes over to the thiolo isomers.

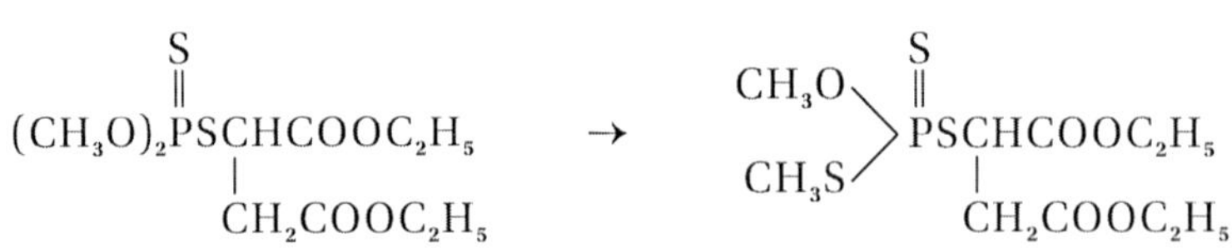

At higher temperatures the reaction proceeds violently and a considerable part of the product decomposes, sometimes even explosively.

Hydrolysis of malathion follows different paths in acid and alkaline media. In acid medium, the main products of hydrolysis are dimethyldithiophosphoric acid and the ester of mercaptosuccinic acid while in alkaline medium the salt of dimethyldithiophosphoric acid and ester of fumaric acid are formed.

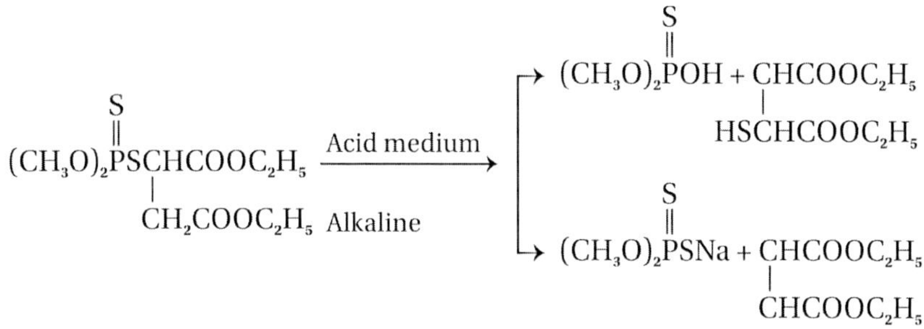

When malathion is oxidised by nitric acid or by other strong oxidising agents, the thiono sulphur atom is split off and the corresponding ester of thiolo phosphoric acid is formed. Malathion on prolonged contact with iron or materials

$$(CH_3O)_2\overset{\overset{\displaystyle S}{\|}}{P}S\underset{\underset{\displaystyle CH_2COOC_2H_5}{|}}{C}HCOOC_2H_5 + 8HNO_3 \rightarrow (CH_3O)_2\overset{\overset{\displaystyle O}{\|}}{P}S\underset{\underset{\displaystyle CH_2COOC_2H_5}{|}}{C}HCOOC_2H_5 + 8NO_2 + H_2SO_4 + 3\,H_2O$$

containing iron breaks down and completely loses its insecticidal properties. Therefore it cannot be kept in iron containers. It is best kept in glass containers.

Malathion is a persistent general purpose insecticide particularly suitable for household, homegarden, vegetable and fruit pests control, It is highly toxic to mites and a number of insect species while being generally non-phytotoxic. Malathion has relatively low toxicity to higher animals. Its acute oral toxicity is 1/5 to 1/10th that of parathion, to laboratory animals. Chronic toxicity and inhalation effects are said to be low. A large number of analogs and homologs of malathion have been synthesised and studied but most of them have higher toxicity to animals and therefore are not used. Of interest for the control of flies is 0,0-diethyl-S-carbothioxymethyl-dithiophosphate which is known as acethion. This has low toxicity for animals.

Dimethoate {Rogor, phosphamide, BI-58, 0,0-Dimethyl-S-(N-methyl- carbamoyl methyl) dithiophosphate}

Dimethoate is a systemic insecticide produced by reacting salts of dimethyl-dithiophosphoric acid with N-methylchloroacetamide, in aqueous medium in the presence of some organic solvents. It is also produced by reacting

$$(CH_3O)_2\overset{\overset{\displaystyle S}{\|}}{P}SK + ClCH_2CONHCH_3 \rightarrow (CH_3O)_2\overset{\overset{\displaystyle S}{\|}}{P}SCH_2CONHCH_3 + KCl$$

0,0-dimethyl-S-(carboxymethyl) or S-(carbo-alkoxy-methyl) dithiophosphate at low temperature with aqueous methylamine. Both the phenyl and ethyl esters are used. However the dimethoate obtained by this method is less stable due to the

$$(CH_3O)_2\overset{\overset{\displaystyle S}{\|}}{P}SCH_2COOC_6H_5 + CH_3NH_2 \rightarrow (CH_3O)_2\overset{\overset{\displaystyle S}{\|}}{P}SCH_2CONHCH_3 + C_6H_5OH$$

presence of impurities, like methylamine, traces of which can decompose dimethoate easily.

Pure dimethoate is a white crystalline substance with camphor-like odour. The technical material is a yellowish brown (amber) coloured oily liquid with sulphurous acid smell. The emulsion concentrate is a pale, straw coloured liquid with a characteristic pungent odour. Pure dimethoate melts at 51-52°C. It is highly soluble in water (39 g/litre) and most organic solvents but is slightly soluble in paraffinic hydrocarbons. Dimethoate is thermally unstable and on heating it decomposes, first undergoing the Pischemuka isomerisation.

$$(CH_3O)_2\overset{\overset{\displaystyle S}{\|}}{P}SCH_2CONHCH \rightarrow \begin{matrix} CH_3O \\ CH_3S \end{matrix} > \overset{\overset{\displaystyle O}{\|}}{P}SCH_2CONHCH_3$$

The 0,S-dimethyl-S-(N-methylcarbamoylmethyl) dithiophosphate obtained is more toxic to mammals than the starting dimethoate (LD_{50} for rats being 100 mg/kg).

When dimethoate is oxidised by various oxidising agents or by the oxygen of the air (oxidation by the oxygen of the air proceeds especially well on green leaves of the plants) it is converted to 0,0-dimethyl-S-(N-methylcarbamoylmethyl) thiophosphate.

$$(CH_3O)_2\overset{\overset{\displaystyle S}{\|}}{P}SCH_2CONHCH_3 + (O) \rightarrow (CH_3O)_2\overset{\overset{\displaystyle O}{\|}}{P}SCH_2CONHCH_3 + H_2SO_4$$

The hydrolysis of dimethoate takes place most rapidly in alkaline medium. In acid medium it is more stable (50% is hydrolysed in 0.8 hours at pH 9 and at 70°C but at pH 5 it requires 21 hours). Depending on the dosage of dimethoate applied, it is usually broken down in plants in 15 to 20 days. While applying dimethoate, care should be taken to see that there is no water stress, as it will cause injury to the plants.

In storage, dimethoate is relatively unstable and rapidly breaks down particularly at an elevated temperature. Impurities present in dimethoate catalyse its decomposition. It decomposes very quickly when it contains traces of organic bases or comes in contact with iron or salts of iron. Even stainless steel on standing, promotes a rapid decomposition of the compound. Hence, iron containers should not be used for storing dimethoate. Solutions of dimethoate in organic solvents are more stable. Usually dimethoate is marketed in the form of 40% emulsion concentrate, containing an emulsifier and an organic solvent in addition to the active ingredient. It is compatible with all pesticides of neutral reactions. The LD_{50} of pure- compound for rats is 250 to 265 mg/kg while that of the technical material is 150 to 165 mg/kg. Impurities in the technical material increase the toxicity of the compound, as the most important impurity 0,0-dimethyl-S-(N-methylcarbamoyl) methylthiophosphate has a LD_{50} value of 55 mg/kg.

A number of compounds similar to that of dimethoate have been synthesised. One of them is Thiocron {Amidithion, 0,0-diraethyl-S- (N-2-methoxyethyl-carbamoyl methyl) dithiophosphate}. It is a good contact insecticide with prolonged systemic action for 15 to 20 days. It is synthesised by reacting alkali or ammonium dimethyl dithiophosphate with the methoxyethylamide of monochloroacetic acid.

$$(CH_3O)_2\overset{\overset{S}{\|}}{P}SK + ClCH_2CONHCH_2CH_2OCH_3 \rightarrow (CH_3O)_2\overset{\overset{S}{\|}}{P}SCH_2CONHCH_2CH_2OCH_3 + KCl$$

It is a crystalline substance, m.p. 46°C. The technical grade product is an unpleasant smelling liquid and is moderately soluble in water, alcohols and esters but poorly soluble in aliphatic hydrocarbons. The chemical properties of amidithion are similar to those of dimethoate. Its metabolism in plants is apparently analogous to that of dimethoate. The LD_{50} value for rats is 600 to 660 mg/kg. However, this compound is yet to be used on a large scale.

Ekatin-M {Ekatin F, Morphothion, 0,0-Dimethyl-S-(morpholino-carbamoyl-methyl) dithiophosphate}

Ekatin-M is another organophosphorus compound similar to dimethoate. It is produced by reacting salts of dimethyldithio-phosphoric acid with monochloro-

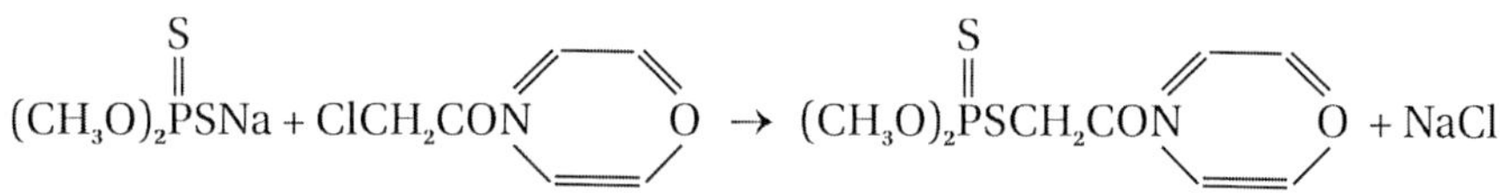

acetomorpholide. This is a white crystalline solid m.p. 63 to 64°C. It is poorly soluble in water, highly soluble in alcohols, ketones and esters of acetic acid and only slightly soluble in aliphatic saturated hydrocarbons. The chemical properties of Ekatin-M are similar to that of other organophosphorus compounds of this group. It is marketed as a emulsion concentrate. The LD_{50} value for rats is 190 mg/kg.

Formothion {Anthio, 0,0-Dimethyl-S-(N-methyl-N-formylcarbamoyl- methyl) dithiophosphate}

Formothion is another organophosphorus compound similar to dimethoate. Formothion is produced by reacting salts of dimethyldithiophosphoric acid with N-formyl-N-methylchloroacetamide.

$$(CH_3O)_2\overset{\overset{S}{\|}}{P}SK + ClCH_2CO\overset{\overset{CH_3}{|}}{N}CHO \rightarrow (CH_3O)_2\overset{\overset{S}{\|}}{P}SCH_2CO\overset{\overset{CH_3}{|}}{N}HCHO + KCl$$

Formothion is a clear liquid, b.p. 25-26°C, poorly soluble in water but highly soluble in organic solvents. Its chemical properties are similar to those of dimethoate, but it is more stable in storage and on heating, one of its principle metabolites in animal is 0,0-dimethyl-S-(carboxy- methyl) dithiophosphate which is further hydrolysed to dimethyl-thiophosphoric acid and phosphoric acid. It is marketed as 25% emulsion concentrate for the control of various sucking and some chewing pests. Its LD_{50} for rats is 330 mg/kg.

Mecarbam {0,0-Diethyl-S-(N-methyl-N-carboethoxycarbamoyl-methyl) dithiophosphate}

Mecarbam is another insecticide similar in action to dimethoate. It is produced by reacting sodium (or potassium) diethyldithiophosphate with N-carboethoxy-N-methylchloroacetamide. It is a clear liquid, b.p. 144°C. It is highly soluble in acetone, acetonitrile and many other organic solvents. However, only 0.1% of the compound dissolves in water, 4% in hexane and 2% in kerosene. It is rather stable under normal conditions of storage. At pH 3 it is hydrolysed. The metabolism of mecarbam in plants is similar to that of dimethoate. It is marketed as 40% emulsion concentrate for the control of sucking and chewing plant pests. The LD_{50} of this compound is 36 to 39 mg/kg for rats.

Thimet {Phorate, 0,0-Diethyl-S-(ethylthiomcthyl) dithiophosphate}

Thimet is a systemic insecticide of low water solubility applied to the soil for the protection of plants. It is produced by reacting diethyldithiophosphoric acid with formaldehyde and ethyl mercaptan at room temperature and then extracting the product with hydrocarbons or by reacting salts of diethyldithiophosphoric acid

$$(C_2H_5O)_2\overset{\overset{S}{\|}}{P}SH + CH_2O + C_2H_5SH \rightarrow (C_2H_5O)_2\overset{\overset{S}{\|}}{P}SCH_2SC_2H_5 + H_2O$$

with chloromethyl ethylsulphide in acetonitrile or other suitable solvents. The second method is somewhat complicated than the former one as it requires

$$(C_2H_5O)_2\overset{\overset{S}{\|}}{P}SNa + ClCH_2SC_2H_5 \rightarrow (C_2H_5O)_2\overset{\overset{S}{\|}}{P}SCH_2SC_2H_5 + NaCl$$

preliminary preparations. The yield of the compound ranges from 70 to 80%.

Phorate is a clear liquid, b.p. 100°C. It is highly soluble in most organic solvents, its solubility in water is about 70mg/litre. Phorate is relatively unstable to hydrolysis and at 70°C and pH 8, 50% of it hydrolysed in two hours. In acid medium it is more stable. It is easily oxidised to the corresponding sulphoxide which is more resistant to hydrolysis than the starting compound. This explains the comparatively long insecticidal action of phorate in plants. The metabolism of phorate in plants is probably as follows:

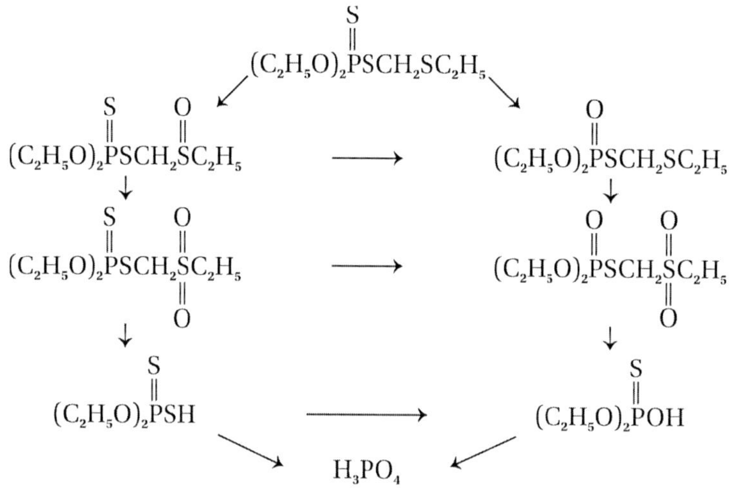

The metabolites of phorate in most cases are as highly toxic to animals as the starting compound. The final products of metabolism are particularly inactive biologically. Phorate is marketed as dust or granules containing 25 or 44% active ingredient. In many cases it provides a complete protection to plants for at least 20-25 days. It belongs to the category of very toxic compounds. The LD_{50} of this compound to rats is 1.1 to 2.3 mg/kg.

Trithion {Carbophenothion, 0,0-Diethyl-S-(4-chlorophenylthio-mcthyl) dithiophosphate}

Trithion is rather powerful contact acaricide and insecticide. It is used for the control of various sucking plant pests. It is produced by reacting salts of dialkyldithiophosphoric acid with p-chlorome-thylsulphide. Trithion is a light yellow oil, b.p. 130°C. It is almost insoluble in water but highly soluble in organic solvents. Its LD_{50} for rats is 10-30 mg/kg.

Methyltrithion

Methyltrithion is the methylhomolog of carbophenothion. It is considerably less toxic to mammals and to insect plant pests. It is similar to trithion. Its LD_{50} for rats is 180 mg/kg.

Phenkapton {0,0-Diethyl-S-(2,5-dichlorophenylthiomethyl) dithio- phosphate}

Phenkapton is a selective acaricide of prolonged action. It is produced in the same way as carbophenothion. Phenkapton is a clear oil, b.p. at 120°C. It is practically insoluble in water but highly soluble in most organic solvents. It is marketed in the form of 25% wettable powder and as an emulsion concentrate. The LD_{50} is 200-250 mg/kg.

Ethion {0,0,0,0-Tetraethyl-S, S-methylene bis (dithiophosphate)}

Ethion is used for the control of plant feeding mites and aphids. It is produced by the condensation of salts of diethyldithiophosphoric acid with chlorobromo-methane or dibromomethane through heating in an organic solvent.

$$2(C_2H_5O)_2\underset{\underset{S}{\|}}{P}SNH_4 + CH_2Br_2 \rightarrow (C_2H_5O)_2\underset{\underset{S}{\|}}{P}SCH_2S\underset{\underset{S}{\|}}{P}(OC_2H_5)_2 + 2NH_4Br$$

When dibromomethane is used, the yield does not exceed 50% while with chlorobromomethane it may reach 80%. It is a yellow oil, b.p. 164-165°C. It is poorly soluble in aliphatic hydrocarbons, highly soluble in aromatic hydrocarbons and their halogen derivatives and practically insoluble in water. It is marketed in the form of wettable powder (25%) and as emulsion concentrate. The LD_{50} for rats is 55 mg/kg. A large number of different analogs and homologs of ethion have been studied and among them a few have been found to be active as insecticide and acaricide but they have not been put to use till now.

Ekatin {Thiometon, Intrathion, 0,0-Dimethyl-S-(2-ethylmer-captoethyl) dithiophosphate}

Ekatin is a good systemic insecticide with about the same duration of action as methyldemeton; but its contact insecticidal properties are somewhat less. It is produced by reacting salts of dimethyl-dithiophosphoric acid with 2-chlorodiethyl-

$$(CH_3O)_2\overset{\overset{S}{\|}}{P}SNa + ClCH_2CH_2SC_2H_5 \rightarrow (CH_3O)_2\overset{\overset{S}{\|}}{P}SCH_2CH_2SC_2H_5 + NaCl$$

sulphide in either aqueous medium with good stirring or in organic solvents. It is also synthesised by reacting 2-hydroxydiethylsulphide with dimethyl dithiophosphoric acid in the presence of sodium hydroxide and p-toluene sulphonyl chloride. However the yield is less than the first method.

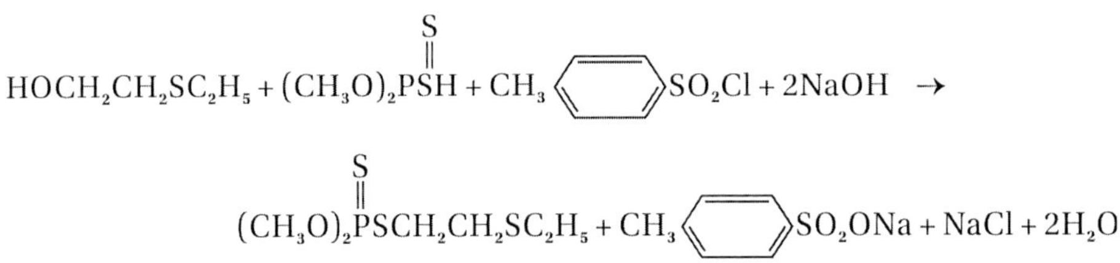

Thiometon is a colourless oil with a strong unpleasant odour, b.p. 104°C. It is sparingly soluble in water (200 mg/litre), highly soluble in most organic solvents with the exception of aliphatic and some alicyclic hydrocarbons. At normal temperatures, thiometon is stable but it may breakdown when heated. The first step in its thermal conversion is a Pishchemuka rearrangement. It is hydrolysed by water in the same manner as methyldemeton.

The metabolic pathway of thiometon in plants can be represented as follows:

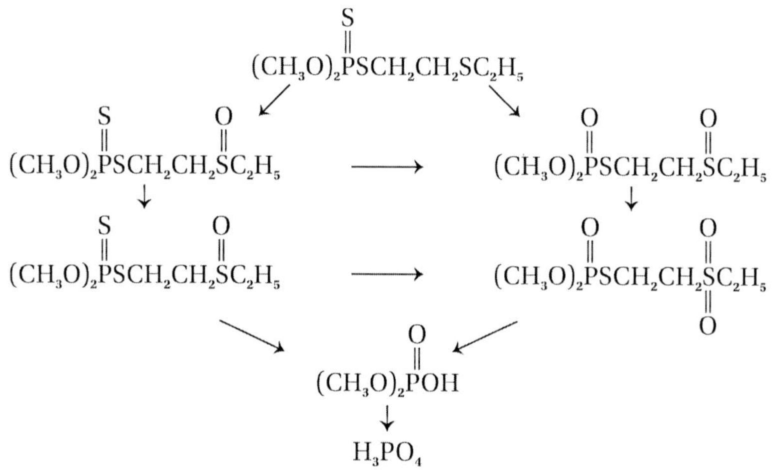

It is assumed that its contact insecticidal properties develop after conversion of the compound to the thiolo isomer of methyldemeton or its metabolites. Thiometon is marketed in the form of emulsion concentrate containing from 20 to 50% active ingredients and also in granulated form. The LD_{50} of thiometon for different animals varies from 70-120 mg/kg. The maximum permissible concentration in air is 0.1 mg/m^3.

Disyston {Dithiosystox, Disulphoton, Compound M. 74, 0,0- Diethyl-S-(2-ethylmercaptoethyl) dithiophosphate}

Disyston is widely used for the control of sucking pests of cotton and other crops through soil application. In certain cases, it is used for seed treatment. Disulphotan is prepared by reacting salts of diethyldithiophosphoric acid with 2-chlorodiethylsulphide:

$$(C_2H_5O)_2\overset{\overset{\Large S}{\|}}{P}SK + ClCH_2CH_2SC_2H_5 \rightarrow (C_2H_5O)_2\overset{\overset{\Large S}{\|}}{P}SCH_2CH_2SC_2H_5 + KCl$$

or by heating diethyldithiophosphoric acid with 2-chlorodiethylsulphide at about 100°C. The product obtained by the second method is slightly impure. Further the

$$(C_2H_5O)_2\overset{\overset{\Large S}{\|}}{P}SH + ClCH_2CH_2SC_2H_5 \rightarrow (C_2H_5O)_2\overset{\overset{\Large S}{\|}}{P}SCH_2CH_2SC_2H_5 + HCl$$

yield of the products is also low (< 80%) in the second method as compared to the first, where it is greater than 90%.

Disulphotan is a colourless liquid with an unpleasant odour, b.p. 113°C. Its solubility in water is very low (25 mg/litre). It is highly soluble in most organic solvents. In acid medium, disulphotan is resistant to hydrolysis but in alkaline solution, it is hydrolysed more rapidly. At pH 5 and 70°C, 50% of it is hydrolysed in 60 hours while at pH 9 it takes only 7.2 hours. The LD_{50} value for rats is 2.5 to 12.5 mg/kg. The chemical properties of disulphotan are similar to those of thiometon but disulphotan is considerably more stable. Its metabolism in plants follows an identical path. Like thiometon it is capable of adding various alkyl halides and dimethylsulphate at the sulphur of the ester group yielding sulphonium compounds. These sulphonium compounds are toxic to mammals but do not show contact insecticidal properties. This apparently is due to the low solubility of sulphonium compounds in lipoids and low penetrative ability of this group of compounds through the calcareous covering of insects. The anticholinesterase activity of this group of compounds is very great.

Tetrathion {0-Methyl-0-ethyl-S-(2-ethylthioethyl) dithiophosphate}

Tetrathion is similar to those of thiometon and disulphotan in chemical properties. It is produced in the same manner as the earlier compounds except that the starting material is 0-methyl-0- ethyldithiophosphoric acid. The final product contains thiometon and disulphotan as impurities. Tetrathion is a thick, oily liquid with an unpleasant odour. It is marketed in the form of a 50% emulsion concentrate and is intended for application to the soil by irrigation, for the control of plant pests. Its LD_{50} for experimental animals is 1.2 to 22.3 mg/kg.

Imidan {Phthalophos, 0,0-Dimethyl-S-(naphthalimidomethyl) dithiophosphate}

Imidan is used for the control of various pests on cotton, fruits and other crops. Its use in the place of DDT, to control the codling moth is of interest since it breaks down relatively easily and does not leave toxic residues on fruits. It is produced by reacting chloromethylphthalimide with salts of dimethyldithio-phosphoric acid.

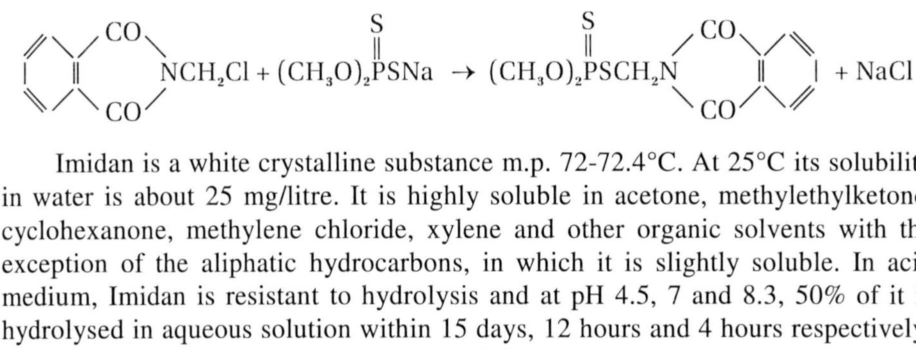

Imidan is a white crystalline substance m.p. 72-72.4°C. At 25°C its solubility in water is about 25 mg/litre. It is highly soluble in acetone, methylethylketone, cyclohexanone, methylene chloride, xylene and other organic solvents with the exception of the aliphatic hydrocarbons, in which it is slightly soluble. In acid medium, Imidan is resistant to hydrolysis and at pH 4.5, 7 and 8.3, 50% of it is hydrolysed in aqueous solution within 15 days, 12 hours and 4 hours respectively. The final products of hydrolysis are phthalimide dimethyl thiophosphoric acid and formaldehyde. However, these compounds are capable of further hydrolysis to

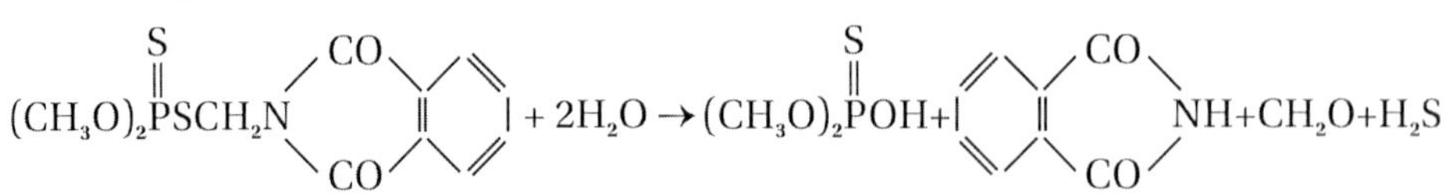

phosphoric and phthalic acids. Oxidising agents also breakdown Imidan and oxidation is directed at the sulphur atom. Imidan is marketed in the form of 20% emulsion concentrate, 50% wettable powder and a 10% granulated formulation. The LD_{50} for experimental animals is 40 to 200 mg/kg.

Phosalone {0,0-Diethyl-S-(6-chlorobenzolinyl-3-methyl) dithiophosphate}

Phosalone has been used for the control of pests of various crops and it is very promising substitute for DDT and other persistent chlorine containing insecticides. It is also used for testing seeds to protect the seedlings from insect damage. Phosalone is produced by condensation of sodium or ammonium diethyldithiophosphate with 6-chloro-3-chloromethylbenzoxazolone. It is practically insoluble in water but highly soluble in many organic solvents.

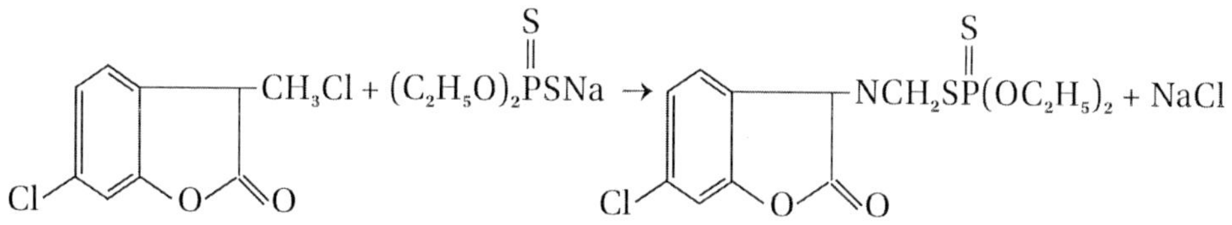

This is a white crystalline substance with a garlic odour, m.p. 45-47°C. It is relatively stable in acid medium but in alkaline medium it is rapidly hydrolysed. The principle products of the hydrolysis are 6-chloro-benzencxazolone, diethylthiophosphoric acid and formaldehyde. Its LD_{50} value for rats is 135 mg/kg.

Guthion {Gusathion, Azinophos methyl, 0,0-Dimethyl-S-(3,4 dichloro-4-keto-1,2,3-benzotriazinyl-3-methyl) dithiophosphate}

Guthion is a broad spectrum insecticide. It is produced by reacting N-halomethyl derivative of azimidobenzoyl with salts of dimethyldithiophosphoric acid. Guthion is a white crystalline compound, m.p. 73-74°C. When heated in high

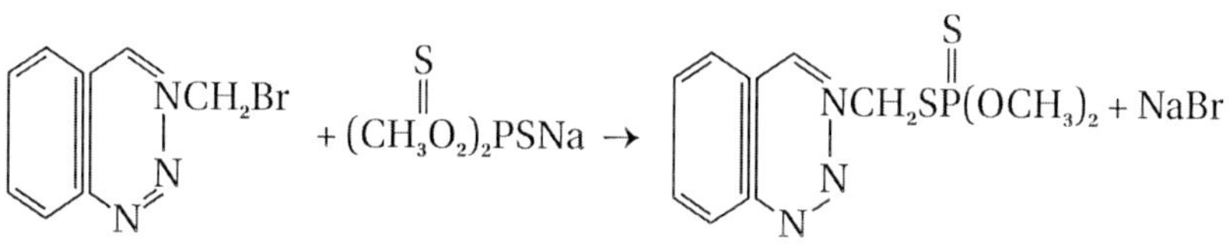

vacuum it breaks down with the evolution of gaseous decomposition products. At 20°C, only about 30 mg of guthion dissolves in one litre of water. It is chemically stable and can be stored under ordinary conditions for an unlimited length of time. At pH 5, and 70°C, 50% of it is hydrolysed in 8-9 hours and at 20°C in 240 days. In alkaline medium, azinphos methyl breaks down several times faster. The action of oxidising agents on guthion causes splitting off of thionosulphur and formation of p-0-azinphosmethyl. Guthion is marketed in the form of emulsion concentrate and wettable powders. Its LD_{50} value for rats is 15-17 mg/kg. The ethylhomolog of guthion (azinphos ethyl, Gusathion-A) has similar properties.

Menazon {Sayfos, 0,0-Dimethyl-S-(4,6-diamino-1,3,5-triazinyl-2-methyl) dithiophosphate}

Menazon has given good control of vectors of viruses that cause potato diseases. It is produced by the reaction of alkali or ammonium dimethyl dithiophosphate with 4,6-diamino-2-chloromethyl- 1,3,5 triazine.

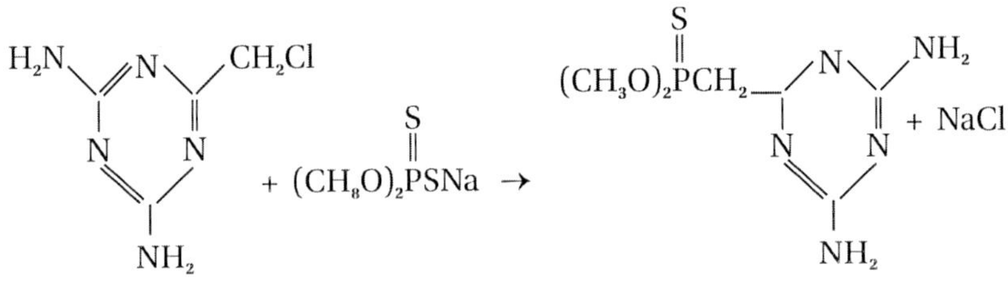

This is a white crystalline substance m.p. 160-162°C. Its solubility in water is only about 0.1%. It is slightly soluble in most organic solvents. It is a systemic insecticide with low toxicity to mammals. Menazon is capable of penetrating plants through the root system and imparting an insecticidal effect of long duration. Its LD_{50} value is 900 mg/kg. Its chemical properties are similar to those of other dithiophosphate of this type. It is marketed as wettable powder and also as granules.

Delnav {Dioxathion, 1,4-Dioxane-2,3-S,S'-bis(0,0-diethyl-dithiophosphate}

Delnav is a persistent, general purpose insecticide. It is synthesised by the addition of bis (diethoxythiophosphone) disulphide to the 1,4-dioxane. Delnav is a yellow oil with an unpleasant odour which does not distill without decomposition

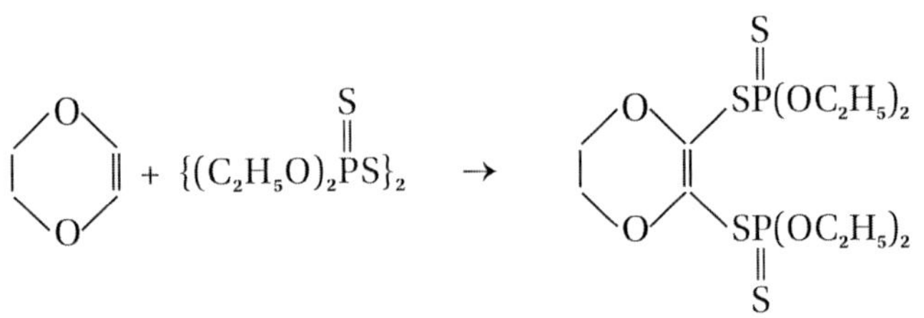

in vacuum. The technical grade product is a mixture of 70% cis and trans isomers of 1,4-dioxane-2,3-S,S'-bis (diethyldithiophosphate) with 30% tetraethyl-trithiophosphate and 0,0-diethyl S-(1,4-dioxoyl) dithiophosphate. The pure compound (trans isomer) is a crystalline substance m. p. 80-81°C while the cis isomer is a liquid. The cis isomer is an effective insecticide but the toxicity of the trans isomer to insects is only about one-third as great. Dioxathion is thermally unstable and even at 130-140°C, it decomposes giving off one molecule of diethyl-dithiophosphoric acid. The trans isomer breaks down very easily.

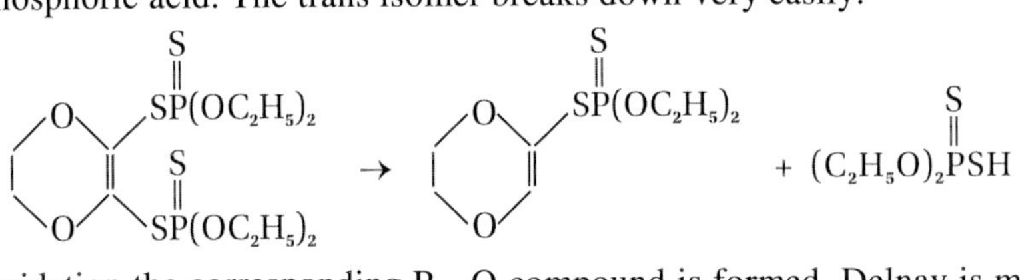

Upon oxidation the corresponding P—O compound is formed, Delnav is marketed as 25% wettable powder and also emulsion concentrate.

Cidial {0,0-Dimethyl-S-(l-carboethoxy benzyl) dithiophosphate}

Cidial is an insecticide with a broad spectrum of action. It is produced by reacting alkalimethyldithiophosphate with the esters of halophenyl acetic acids:

$$C_6H_5CHXCOOC_2H_5 + (CH_3O)_2\overset{\overset{S}{\|}}{P}SNa \rightarrow (CH_3O)_2\overset{\overset{S}{\|}}{P}S\overset{\overset{C_6H_5}{|}}{C}HCOOC_2H_5 + NaX$$

Cidial is an oily liquid with a characteristic unpleasant odour, not distillable in vacuum without decomposition, highly soluble in most organic solvents and insoluble in water. It ignites at temperatures above 168-172°C. It is thermally unstable and when heated to 120°C for 110 hours, almost 50% decomposes. It withstands 30 days heating at 50°C without noticeable decomposition. Cidial may show insecticidal effects for 12 to 25 days in plants. The LD_{50} for rats is 250-300 mg/kg.

Recently number of new organophosphorus insecticides like amophos {0,0-dimethyl-S-2 (acethylamide) ethyldithiophosphate} have been synthesised. Amophos is highly soluble in most organic solvents. It has a systemic effect and is used in the form of aqueous emulsion. Its LD_{50} for experimental animals is 200-400 mg/kg.

DERIVATIVES OF PYROPHOSPHORIC ACID

In 1943, the first organophosphorus compound called tetraethylpyrophosphate was manufactured in Germany and was sold as Bladen. When the insecticidal properties of TEPP and the related compounds were known, a systematic investigation into the pesticidal properties of organophosphorus compounds was started. The relationship of insecticidal activity of the compounds to the structure of esters of pyrophosphoric acid was determined later. The insecticidal activity of tetraalkylpyrophosphate was found to decrease with increase in size of ester radical. Among the compounds, the most active compound is the TEPP. The tetraalkylmonothiophosphates have greater insecticidal activity than the tetraalkylpyrophosphate but their activity on mammals was also high. Introduction of a second sulphur atom in phosphorus, lowers the toxicity of the compound to mammals without appreciable lowering of the insecticidal activity. Introduction of a third sulphur on the phosphorus, decreases the insecticidal activity of the compound.

All the pyrophosphates had strong contact insecticidal effect and practically no systemic effect. Octamethylpyrophosphoric amide possesses systemic action but has no contact insecticidal activity. The tetraalkylpyrophosphates are prepared by reacting trialkyl phosphate with phosphorus oxychloride or other acid chloride at a comparatively high temperature.

$$5\,(RO)_3PO + POCl_2 \rightarrow 3\,(RO)_2\overset{\overset{O}{\|}}{P}—O—\overset{\overset{O}{\|}}{P}\,(OR)_2 + 3\,RCl$$

The technical grade products are usually contaminated with other derivatives of phosphoric acid. These can also be prepared by reacting dialkylchlorophosphate with water in the presence of bases like organic tertiary amines or alkali carbonates.

$$2(RO\overset{\overset{S}{\|}}{P}Cl)_2 + H_2O \rightarrow (RO)_2\overset{\overset{O}{\|}}{P}—O—\overset{\overset{O}{\|}}{P}\,(OR)_2$$

Few of the important pyrophosphates that are manufactured are:

Tetraethylpyro phosphate	...	$(C_2H_5O)_2\overset{\overset{O}{\parallel}}{P}—O—\overset{\overset{O}{\parallel}}{P}(OC_2H_5)_2$
Sulfotepp	...	$(C_2H_5O)_2\overset{\overset{S}{\parallel}}{P}—O—\overset{\overset{S}{\parallel}}{P}(OC_2H_5)_2$
Schradan	...	$\{(CH_3)_2N\}_2\overset{\overset{S}{\parallel}}{P}—O—\overset{\overset{S}{\parallel}}{P}\{N(CH_3)_2\}_2$

Tetraethylpyrophosphate (TEPP, tepp, Bladen)

This was first synthesised by Clement in 1854 but its insecticidal properties were discovered by Schrader in 1943. The first of the organophosphorus insecticides to find commercial application was the so-called "hexaethyltetraphosphate" tested by Schrader. This material was made by reacting one mole of phosphorus oxychloride with 3 moles of triethylphosphate. Hall and Jacobson found that the product of this reaction was a mixture of ethylpolyphosphate which was actually the major insecticidally active ingredient. By increasing the ratio of the triethylphosphate to phosphorus oxychloride to 5.1, the final product containing up to 40% tetraethylpyro-phosphate can he obtained.

$$5(C_2H_5O)_3\overset{\overset{O}{\parallel}}{P} + Cl_3\overset{\overset{O}{\parallel}}{P} \rightarrow 3(C_2H_5O)_2\overset{\overset{O}{\parallel}}{P}—O—\overset{\overset{O}{\parallel}}{P}(C_2H_5O)_2 + 3C_2H_5Cl$$

In another method, phosphoric anhydride is used instead of phosphoric oxychloride:

$$4(C_2H_5)_3PO_4 + P_2O_5 \rightarrow 3(C_2H_5)_4P_2O_7$$

It can also be prepared by reacting five moles of ethylalcohol and two moles of phosphorus oxychloride under reduced pressure.

Commercial TEPP contains approximately 40% of pure tetraethylpyrophosphate. The pure compound is a colourless liquid with an unpleasant odour, b.p. 82°C. It is miscible with water in all proportions and highly soluble in alcohol, acetone, aromatic hydrocarbons and carbon tetrachloride, but poorly soluble in petroleum ether and ligorin. When TEPP is heated to 170°C, it breaks down with the formation of ethylene. At 208-213°C, the decomposition takes place violently.

TEPP is rapidly hydrolysed by water, even at room temperature, forming diethylphosphoric acid which is non-toxic for both insects and animals. Hence the problem of toxic residues does not arise.

$$(C_2H_5O)_2\overset{\overset{O}{\parallel}}{P}OH \rightarrow H_3PO_4 + 2CH_3CHO \rightarrow H_3PO_4 + 2CH_4 + 2CO$$

Under the influence of sunlight, the diethylphosphoric acid breaks down completely to phosphoric acid, methane and carbon monoxide. The formation of the last two products goes through a stage of acetaldehyde formation.

TEPP easily reacts with alcohols and amines. In the first case, the trialkylphosphate and diethylphosphoric acid are obtained and in the second case the amide and salt of diethylphosphoric acid are obtained. The reaction of TEPP with anhydrous HF or potassium bifluoride yields diethylfluorophosphate which is toxic to mammals.

Commercial TEPP is miscible with most organic solvents but not with petroleum oils. The compound is corrosive to iron and other metals. TEPP is formulated with a surface active agent as 20 and 40% anhydrous concentrate or in aerosols with a propellent gas. TEPP is highly toxic to many species of insects and arachinids. It is most widely used against aphids and some scale insects and mites. It is extremely toxic to all warm-blooded animals. Its LD_{50} for rats is 1.12 mg/kg. Cutaneous toxicity is also very high. Repeated exposure to sublethal amounts of TEPP may predispose animals to acute toxic symptoms later, perhaps by decreasing the cholinesterase activity of the tissues. Like other organophosphorus insecticides, TEPP acts on the enzyme and death is apparently due to the reduction of cholinesterase activity below a critical level.

Workers exposed to TEPP should be protected from constant exposure. Symptoms normally noticed in persons handling and spraying TEPP are shortness of breath with a sense of constriction on the chest, due to the destruction of cholinesterase. Atropine is an antidote for the organophosphorus insecticide. The insecticidal activity of TEPP is only about 1/3 that of parathion and its use is continuously decreasing. Other compounds related to TEPP are tetraisopropylpyrophosphate which hydrolyses much more slowly than TEPP and is considerably less toxic to animals.

Sulfotepp (Pirofos, Bladafum, Tetraethyldithiopyrophosphate)

Sulfotepp is closely related to TEPP and can be made by reacting the latter compound with sulphur or by the interaction of diethylchlorothiophosphate with water in the presence of pyridine. Pure tetraethyldithiopyrophosphate is a colourless or slightly yellow oily liquid, b.p. 110-118°C. It is slightly soluble in water and soluble in most organic solvents and in methylchloride, but not in aliphatic hydrocarbons. Sulfotepp is similar to parathion in insecticidal activity. Unlike TEPP this compound is relatively stable in water solutions and is hydrolysed very slowly. Sulfotepp can be considered to be relatively stable in the presence of lime and other alkalies. This compound is highly effective against a number of insects both as a spray and as an aerosol. Like other phosphorus insecticides, the material is a powerful anticholinesterase and is highly toxic to all forms of animal life. Its LD_{50} for rats is about 5 mg/kg. The dithiopyrophosphates and monothiopyrophosphates have been found to be less toxic to houseflies than parathion.

Schradan (Octamethyl, CMPA, Pestox III, System, Octamethyl pyrophosphoramide, Octamethyltetramide of pyrophosphoric acid)

Schrader first noted that certain toxic compounds could be absorbed through the leaves or roots of plants and could thus protect the plant from insect attack. Octamethylpyrophosphoramide was one such compound which was made by reacting tetramethyl-diamidochlorophosphate with water in the presence of potassium carbonate, pyridine or triethylamine. Equimolecular quantities of the reactants are used and condensation is carried out in boiling xylene

$$\{(CH_3)_2N\}_2\overset{\overset{\displaystyle O}{\|}}{P}{-}Cl + \{(CH_3)_2N\}_2P{-}OC_2H_5 \rightarrow \{(CH_3)_2N\}_2\overset{\overset{\displaystyle O}{\|}}{P}{-}O{-}\overset{\overset{\displaystyle O}{\|}}{P}\{N(CH_3)\}_2 + C_2H_5Cl$$

or by reacting tetramethylchlorophosphoramide which partially hydrolyses. The hydrolytic products then condense with unhydrolysed compound in the presence of pyridine or methyldibutylamine.

$$\{(CH_3)_2N\}_2\overset{\overset{\displaystyle O}{\|}}{P}Cl + \{(CH_3)_2N\}_2\overset{\overset{\displaystyle O}{\|}}{P}{-}OH \rightarrow \{(CH_3)_2N\}_2\overset{\overset{\displaystyle O}{\|}}{P}{-}O{-}\overset{\overset{\displaystyle O}{\|}}{P}\{N(CH_3)\}_2 + HCl$$

The commercial product contains dimethylamide of the higher phosphoric acids, of which the triphosphoric acid derivatives predominate. Pure OMPA is a colourless oily liquid, b.p. 118-122°C. It is miscible with water in all proportions. It is highly soluble in alcohols, ketones and halogen derivatives of hydrocarbons, but not easily soluble in petroleum ether and ligrorin.

An aqueous solution of the pure compound is neutral and can be stored for an unlimited time. OCMA is relatively stable in neutral and alkaline aqueous media but is rapidly hydrolysed under acid conditions to orthophosphoric acid and dimethylamine. Thus, 50% of the compound is hydrolysed in 100 years in neutral aqueous solutions, in 200 minutes in 1N HCl and in 70 days in 1N NaOH. It is commonly formulated as 30 or 60% aqueous solution and is also available as a commercial concentrate, which is a dark brown viscous liquid containing 75 to 80% of the pure compound. It is compatible with alkaline materials but combinations with acidic pesticides should be avoided.

Schradan can be oxidised to the hydroxy methyl compound by potassium permanganate or other oxidising agents. When this compound is chlorinated it forms as the first step, the corresponding chloromethylderivative. In total, four atoms of chlorine can be introduced.

OCMA is not phytotoxic to most plants. It is not highly toxic to insects as a contact insecticide when applied directly, but it is readily absorbed by the roots and leaves of plants and translocated, so that the plant becomes toxic to many insects and mites which feed on plant sap. Some of the metabolites formed are more toxic than the original compound. Using radioactive phosphorus, it was found that OCMA was absorbed more rapidly than the higher homolog, but both were lost from the treated plants at similar rates. The rate of disappearance of the undecomposed insecticide was independent of the species but depended upon the season, being most rapid in the early growing season. In the plant 90% of the material disappeared in 40 days in midsummer, probably by decomposition by the plant system. Both the major constituents of the commercial preparation are toxic to insects.

The OCMA is readily translocated upward in plants but moves laterally with less facility. The surface active agents alter the rate of absorption of the compounds by plants. This compound is highly toxic to animals, the LD_{50} to rats being 18 mg/kg.

MISCELLANEOUS ORGANOPHOSPHATIC COMPOUNDS

Number of derivatives of phosphoric acid have been synthesised but only a few like Trichlorfon, EPN have been used in agriculture. These compounds are produced by methods similar to those of the corresponding derivatives of phosphoric acid and thiophosphoric acid. These compounds are less toxic than similar derivatives of thiophosphoric acid.

Trichlorfon {Diptrex, Dylos, Chlorophos, 0,0-Dimethyl-(1-hydroxy-2,2,2-trichloroethyl) phosphonate}

Trichlorfon is a systemic poison used for the control of chewing insects. It is produced on an industrial scale by the condensation of dimethylphosphite with

$$(CH_3O)_2PHO + CCl_3CHO \rightarrow (CH_3O)_2\overset{O}{\overset{\|}{P}}\text{-}\overset{OH}{\overset{|}{C}}H\text{-}CCl_3$$

chloral at room temperature, with the evolution of heat. Greater yields are obtained if pure dimethylphosphite is used. When technical grade chloral and dimethylphosphite react, the yield is only 90% and it contains number of impurities.

In the other method the production of dimethylphosphite and its reaction with chloral are combined in one step and the process is carried out in an organic solvent with the removal of heat. The yield of trichlorfon by this method is > 80%.

$$3CH_3OH + PCl_3 + CCl_3CHO \rightarrow (CH_3O)_2\overset{O}{\overset{\|}{P}}—\overset{OH}{\overset{|}{C}}HCCl_3 + CH_3Cl + 2HCl$$

To purify the trichlorfon it is recrystallised from water or from organic solvent. The mother liquors from the recrystallisation of trichlorfon are usually used for the production of dichlorovos. The solvents generally used are carbon tetrachloride, methylchloride, and chlorobenzene.

Trichlorfon is a white crystalline powder, m.p. 73 to 74°C. It is slightly soluble in paraffinic hydrocarbons and soluble in water, benzene and chloroform. It is stable in acid medium and is rapidly hydrolysed in alkaline medium. Its hydrolysis under different conditions follow two different paths. In acid medium, the first product of hydrolysis is 0-methyl-(1-hydroxy-2,2,2-trichloroethyl) phosphoric acid.

$$(CH_3O)_2\overset{O}{\overset{\|}{P}}—CHCCl_3 + H_3O \rightarrow CH_3OH + \begin{matrix} OH \\ CH_3O \end{matrix}\!\!>\overset{O}{\overset{\|}{P}}—\overset{OH}{\overset{|}{C}}HCCl_3$$

In dilute solutions, trichlorfon breaks down rapidly in the presence of light. In alkaline medium, trichlorfon is dehydrochlorinated and simultaneously gets rearranged. The principal product of the reaction is 0,0-dimethyl-0-(2,2-dichlorovinyl) phosphate (dichlorovos) Dichloroacetaldehyde, dimethylphosphoric

$$(CH_3O)_2\overset{\overset{O}{\|}}{P}—CH_2CCl_3 + KOH \rightarrow (CH_3O)_2\overset{\overset{O}{\|}}{P}—CH{=}CCl_2 + KCl + H_2O$$

acid and some other compounds are formed as products of the side reactions. Trichlorfon is a good methylating agent.

Upon prolonged storage of aqueous solutions of trichlorfon, partial hydrolysis of the compound occurs, as a result of which the solution becomes acid and contains phosphoric acid, dimethylphosphoric acid and HCl. Trichlorfon should not be stored in containers of unprotected iron. Reducing agents also decompose trichlorfon. Above pH 6, trichlorfon is rapidly converted to DDVP or dichlorovos, dimethyldichlorovinylphosphate, a reaction which accounts for its toxic action. The LD_{50} of trichlorfon is 630 mg/kg. for rats. The maximum permissible concentration in air is 0.5 mg/km^3.

EPN {0-Ethyl-0-(4-nitrophenyl) benzene thiophosphate}

EPN is a powerful acaricide. Pure EPN is produced by reacting PCl_3 with benzene in the presence of $AlCl_3$ as per following series of reactions. Pure EPN is a white crystalline substance with m.p. 36°C. It is practically insoluble in water,

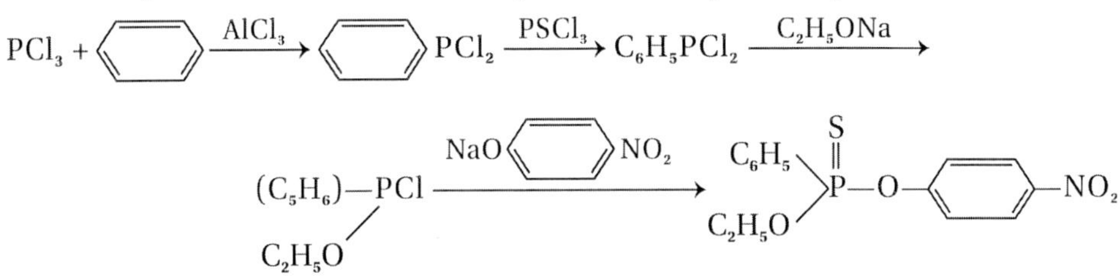

highly soluble in most organic solvents. In alkaline medium, it is relatively hydrolysed with the formation of benzene thiophosphoric acid, alcohol and p-nitrophenol. In acid and neutral media, EPN is more stable. The commercial material is a dark amber coloured liquid, having a specific gravity of 1.27 at 25°C. The impurities include the bis (p-nitrophenol-thiobenzene phosphate) and p-nitrophenol. Formulations include dusts and wettable powders. As these decompose in alkaline media, the formulations should not be mixed with lime, lime sulphur, calcium arsenate and similar materials. These chemical formulations are highly toxic to insects and many species of mites. They are relatively non-phytotoxic except for a few varieties of apples. EPN is highly toxic to warmblooded animals, the acute oral toxicity (LD_{50}) of the technical product being 40 mg/kg for rats. The LD_{50} of pure EPN is 8 to 36 mg/kg. Vapours and dusts from the formulations are also highly toxic.

In addition to the above phosphates, two fluorophosphates, bis (dimethylamide)-fluorophosphate and bis (isopropylamide) fluorophosphate (Isopestox) have also been found to possess systemic insecticidal properties.

$$\{(CH_3)_2N\}_2P(=O)\text{–}F$$

bis (diethylamide) fluorophosphate

$$\{(CH_3)_2CHN\}_2P(=O)\text{–}F$$

bis (isopropylamide) fluorophosphate

The methyl derivative is a colourless liquid, boiling at 67°C. It is completely miscible with water and is stable in neutral solutions. Isopropylamide fluorophosphate exists as odourless white crystals melting at 60°C. Its solubility in water is 8% at room temperature and it is soluble in polar organic solvents but only slightly soluble in oils. Chemically, the isopropyl derivatives are hydrolysed slowly in water. In acid solutions or in plants, decomposition is more rapid. When used as a systemic poison, the compound is highly toxic to insects and appears to be less toxic to warm-blooded animals than OMPA.

CHAPTER IX

CARBAMATE INSECTICIDES

The N-methyl and N,N-dimethylcarbamic esters of phenols and heterocyclic enols possess useful insecticidal properties. Heterocyclic N-N-dimethyl carbamate compounds were discovered by Gysin. The two important insecticides that were isolated are Isolan and Dimetelan. Aromatic N-methylcarbamates are derivatives of phenyl-N-methylcarbamates with a great variety of groups like chloride, alkyl, alkylthio, alkoxy and dialkylamino in the side chain. The carbamate insecticides are generally crystalline materials of low water solubility but soluble in organic solvents. Among the derivatives of carbamic acid, the aryl esters of N-methyl carbamic acid are used for control of insect pests. Maximum insecticidal effect is shown by the aryl esters of N-methylcarbamic acid. Replacement of methyl groups by radicals with a greater number of carbon atoms or by carbonyl or ester oxygen by sulphur also result in decreased activity.

All the insecticidal esters of N-alkylcarbamic acids cause inhibition of cholinesterase. The monosubstituted phenyl N-methyl carbamates depress the cholinesterase activity more strongly as the resistance of the compound to hydrolysis increases. The carbamates behave as competitive inhibitors of the cholinesterase. The relationship of insecticidal and anticholinesterase activity of carbamates to their structure is a more complex one. For high cholinesterase activity the carbamates should fit the esterase molecule like a "Key in the lock". The greatest activity is shown by 1-napthyl- N-methyl carbamate which is used in agriculture under the names of Sevin, carbaryl and naphthylcarbamate. The corresponding 2-isomer has little activity. Esters of dihydro and tetrahydro-1-naphthalo show high insecticidal activity. Many of the esters of methylcarbamic acid are highly toxic to mammals.

Carbaryl ...

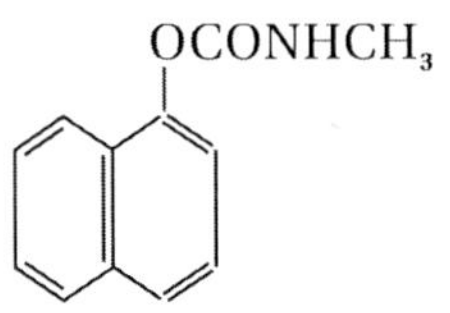

Baygon ...

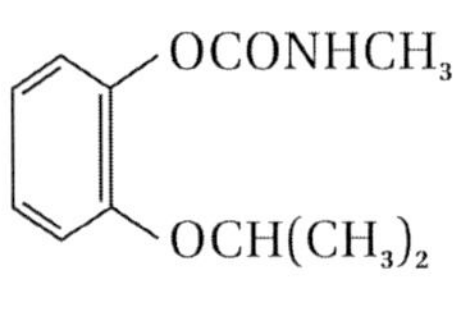

Mesurol ... 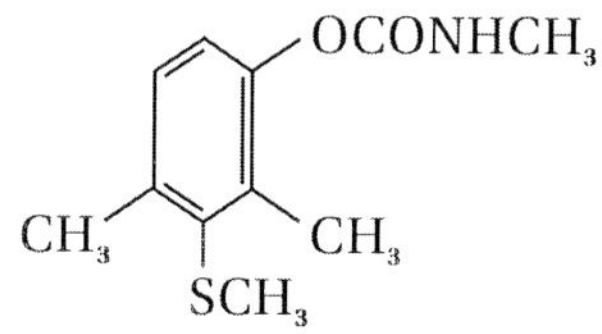

Zectran ... 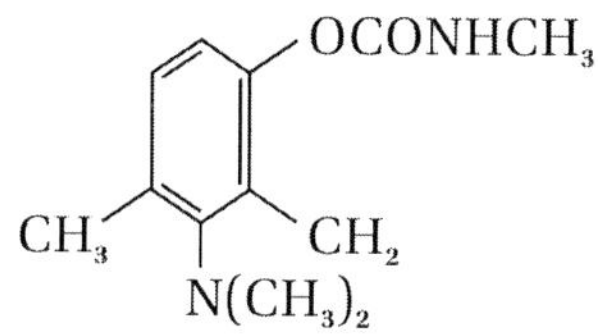

Dimetan ... 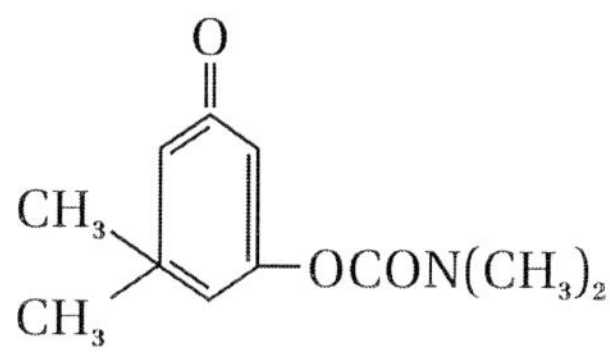

Isolan ... 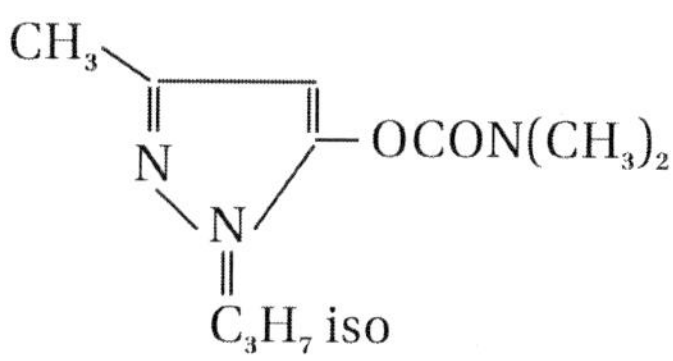

Pyrolan ... 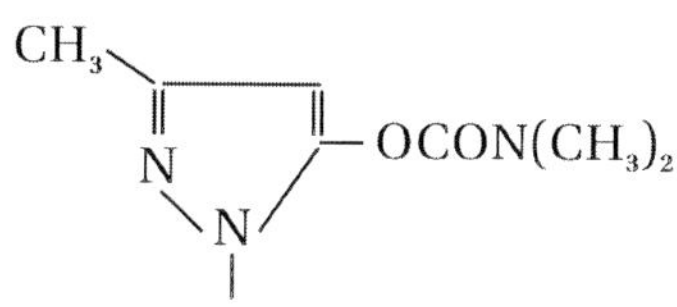

Temik ... 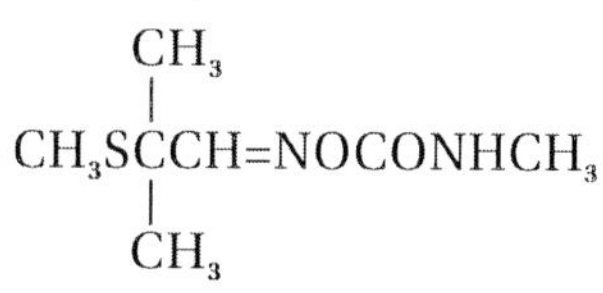

Furadan ... 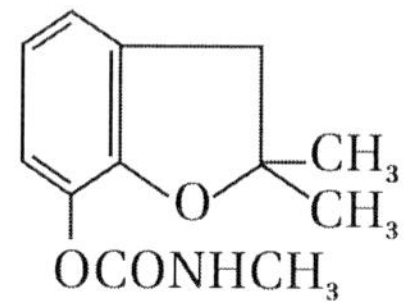

The aryl esters of alkylcarbamic acids are produced by either reacting alkyl isocyanates or alkycarbamoyl chloride with phenols or aromatic alcohols at comparatively low temperatures:

$$ArOH + RNCO \rightarrow ArOCONHR$$

$$ArOH + RNHCOCl \rightarrow ArOCONHR + HCl$$

or by reacting arylcarbonate with an amine in the presence of HC1 acceptors

$$ArOCOCl + RNH_2 + NaOH \rightarrow ArOCONHR + NaCl + H_2O$$

or by reacting methylamine on the diarylcarbonates with heating

$$(ArO)_2CO + CH_3NH_2 \rightarrow ArOCONHCH_3 + ArOH$$

Carbaryl (Sevin, Naphthylcarbamate, 1-Naphthyl-N-methyl carbamate)

Carbaryl is a good contact insecticide with occasional systemic action in certain cases. It is produced by any of the general methods of preparation of carbamates, like reacting 1-naphthol with methylcarbamoyl chloride at room temperature for about 10 to 20 hours.

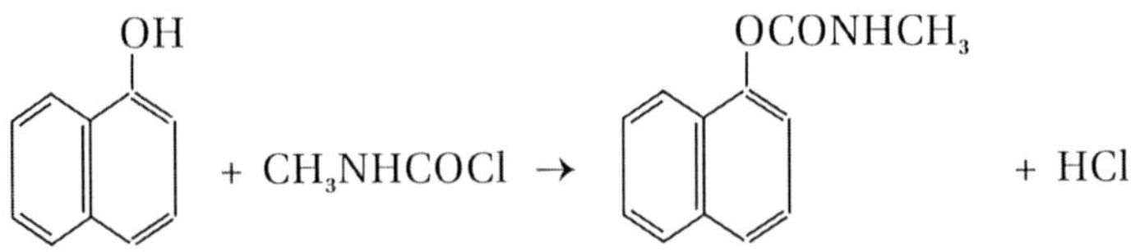

The rate of reaction is increased by removing the HCl that is formed with a stream of air or nitrogen.

Purer compound of carbaryl is obtained by reacting 1-naphthol with methylisocyanate:

$$CH_3NHCOCl + NR_3 \rightarrow CN_3NCO + R_3NHCl$$

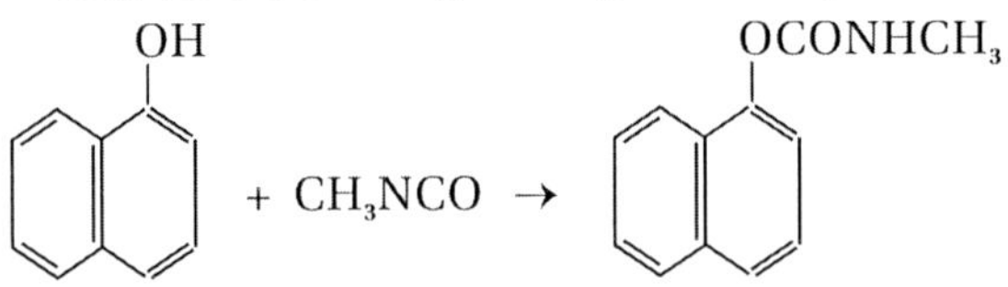

Carbaryl is also synthesised by reacting 1-naphthylchlorocarbonate with methylamine in the presence of HCl acceptors:

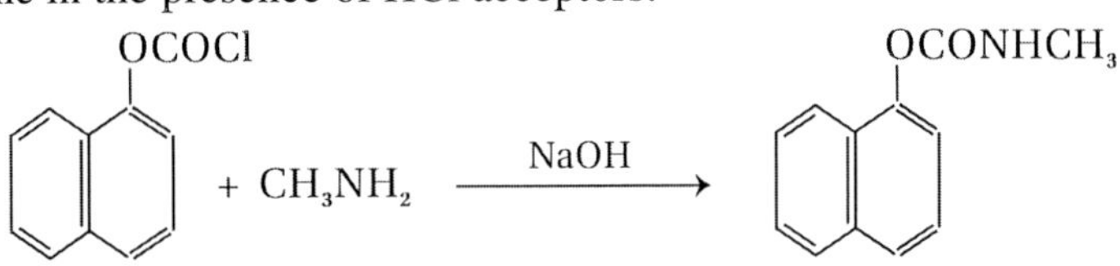

Naphthyl-N-methylcarbamate is a white crystalline compound, m.p. 142°C. It is soluble to less than 0.1% in water but highly soluble in organic solvents. At room temperatures, it is resistant to the action of water, light and oxygen of the air. In alkaline medium, it is rapidly hydrolysed and therefore it is not compatible with compounds of alkaline nature like Bordeaux mixture, calcium and barium polysulphides.

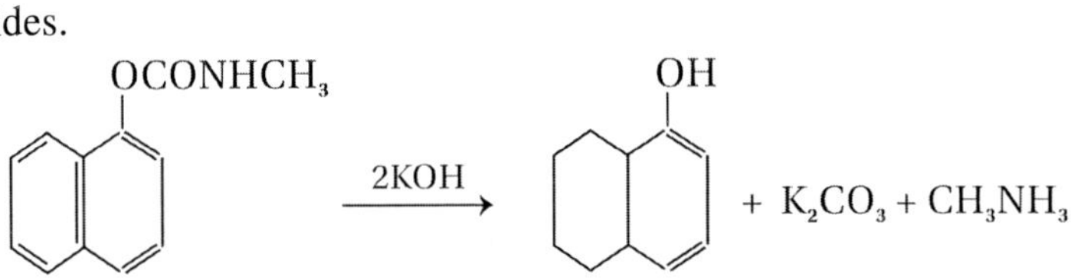

The technical material is tinted but still crystalline and is unstable. It is only weakly phytotoxic. It has been established that the principal metabolites of decomposition of carbaryl in plants and animals are the products of oxidation and hydrolysis. Oxidation is directed primarily at the methyl group attached to the nitrogen atom. The other metabolites 4-hydroxy and 5-hydroxynaphthyl-N-methyl-carbamates that are formed are further broken down with rupture of the aromatic ring.

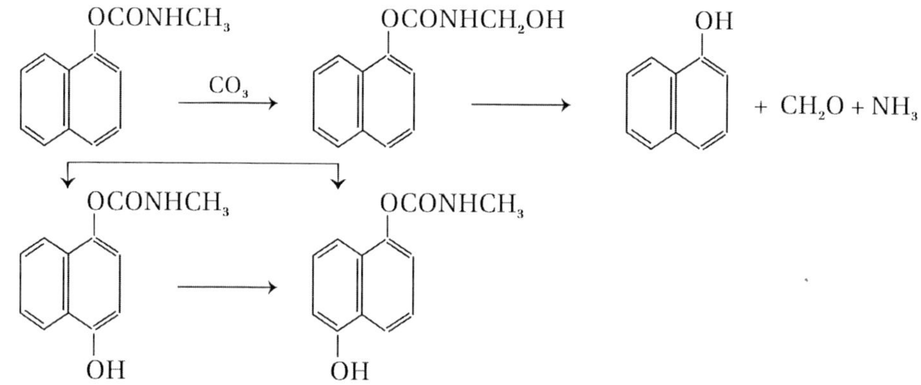

It is marketed in the form of dusts and wettable powders. The LD_{50} for rats is 500 mg/kg. When rats were fed on a diet containing 200 mg of carbaryl per kg of body weight for two years, no harmful effects were observed. The compound does not irritate human skin. The maximum permissible concentration of carbaryl in air is 1 g/m^3.

Baygon {Propoxur, PHC (Japan), Bayer 39007, 2-Isopropoxy phenyl-N methylcarbamate}

Baygon is a broad spectrum insecticide, produced by reacting methylcarbamoyl chloride or methyl isocyanate with 2-isopropoxy phenol.

Baygon is a white to cream coloured crystalline compound with mild phenolic odour. It is highly soluble in organic solvents and slightly soluble in water. Its m.p. is 91.5°C. It is formulated as emulsion concentrate (20%), wettable powder (50%) and dusts. The active ingredient has essentially a contact action. It produces rapid knockdown effect and has an extremely long residual action. Its LD_{50} for rats is 100 mg/kg.

Mesurol (Metmercapturon, 3,5-Dimethyl-4-methyl-mercaptophenyl-N-methylcarbamate)

Mesurol is an active insecticide and acaricide with a wide spectrum of action and relatively long residual activity. Mesurol is synthesised by reacting 4-methylmercapto-3,5-dimethylphenol with methyl isocyanate in the presence of triethylamine or in the presence of an organic solvent like benzene.

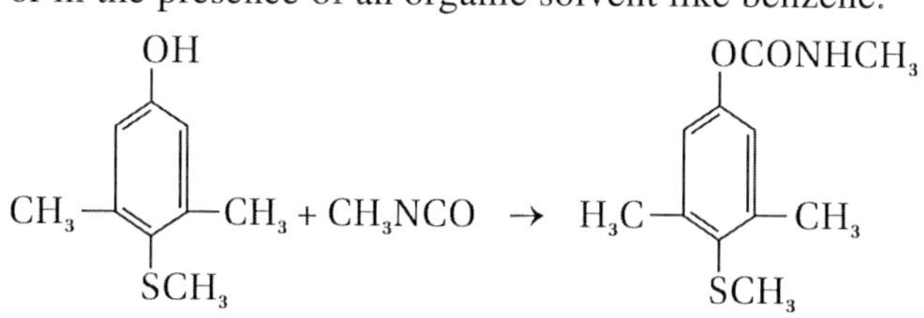

Mesurol is a white crystalline compound, practically odourless, highly soluble in most organic solvents but almost insoluble in water. Its m.p. is 121.5°C. It is resistant to hydrolysis in neutral and weakly acid media but is rapidly hydrolysed by the alkalies. It acts as a stomach and contact poison with contact action predominating. As mesurol does not penetrate into the plant, it is essential that all plant parts are copiously wetted. Though no phytocidal effect has been observed, still like all insecticidal carbamates, it has a thinning effect on fruit trees. It is sold as wettable powder (50%). It is compatible with all conventional insecticides and fungicides. However mixtures of compounds having an alkaline reaction must be sprayed immediately after mixing. Its LD_{50} for rats is 100 mg/kg.

Zectran {4-(N,N-Dimethylamino)-3,5-dimethylphenyl-N-methyl- carbamate}

Zectran is a wide spectrum insecticide. It is prepared by reacting methyl isocyanate with 4-dimethylamine-3,5-dimethylphenol in the presence of small quantities of tertiary amines in organic solvents. Zectran is a white crystalline substance, m.p. 85°C, soluble in organic solvents but poorly soluble in water. With mineral acids, it forms salts that are highly soluble in water but poor in contact insecticidal activity. This compound is resistant to hydrolysis in neutral medium but is rapidly hydrolysed in alkaline medium. Its LD_{50} value is 160-163 mg/kg.

Dimetan (5,5-Dimethyl (dihydoresorcinol) N,N-dimethylcarbamate)

Dimetan is prepared by reacting monosodium salt of dimedon with dimethylcarbamoylchloride.

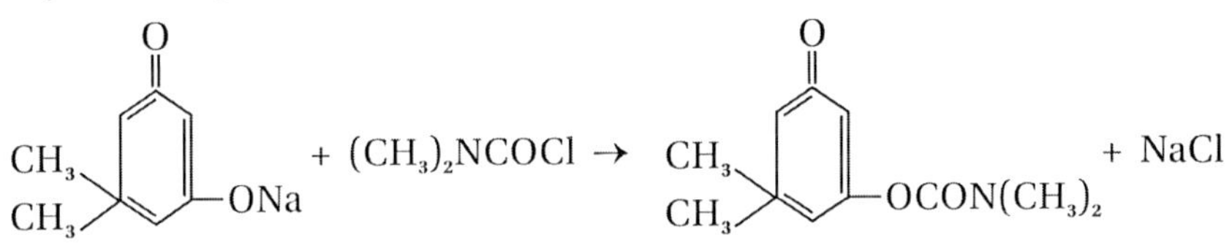

It is a yellow crystalline substance, highly soluble in most organic solvents, soluble in water only up to *3%*. The compound is easily hydrolysed by alkalies with the formation of dimethylamine. The LD_{50} for rats is about 150 mg/kg.

Isolan (1 -Isopropyl-2-methylpyrazolyl-5-dimethylcarbamate)

Isolan is an active aphicide with systemic action. It is produced by the action of dimethylcarbamoyl chloride on the potassium derivative of the enol form of 1-isopropyl-3 methylpyrazolone. Isolan is a colourless liquid that steam distills, b.p. 105 to 107°C. It is miscible with water, alcohol, acetone and xylene in all proportions and soluble in paraffinic hydrocarbons up to 3%. Its LD_{50} value is 11 to 23 mg/kg.

Pyrolan (1-Phenyl-3-methylpyrazolyl-5-dimcthylcarbamate)

Pyrolan is a systemic insecticide. It is produced by the condensation of the potassium derivative of l-phenyl-3-methylpyrozolone with 5 dimethylcarbamoyl chloride with heating in an organic solvent. Pyrolan is also a colourless crystal, m.p. 50°C, soluble in alcohol, acetone and aromatic hydrocarbons, poorly soluble in petroleum oils and water. Its LD_{50} for rats is 62 mg/kg.

Temik (Aldicarb, 2-Methyl-2-methylthiopropyonal-oxime-0,N-methyl carbamate)

Temik is a systemic nemoticide. It is synthesised by reacting the corresponding oxime with methylisocyanate

$$CH_3S\overset{CH_3}{\underset{CH_3}{C}}CH{=}NOH + CH_3NCO \rightarrow CH_3S\overset{CH_3}{\underset{CH_3}{C}}CH{=}NOCONHCH_3$$

It is a white crystalline substance with a m.p. 100°C. It is insoluble in paraffinic hydrocarbons, sparingly soluble in water, soluble in acetone, ethylalcohol, chloroform, toluene, and chlorobenzene. Its LD_{50} for rats is 0.93 mg/kg.

Carbofuran (Furadan, 2,3-Dihydro-2,2-dimethyl-7-benzofuranyl methylcarbamate)

Carbofuran is a systemic carbamate insecticide of broad spectrum utility. It is white crystalline substance, odourless when pure, m.p. 153-154°C. It is stable in acid and neutral media, but unstable in alkaline medium. It is sparingly soluble in water (250-700 ppm) slightly soluble in organic solvents, methylene chloride. It is compatible with all non-alkaline pesticides and fertilisers. It is not phytotoxic to rice. Carbofuran applied to the soil in which plants are growing, is rapidly absorbed into the roots of the plants and distributed to the stems and leaves where it is slowly metabolised to non-toxic compounds over a period of 30 days. That portion that is not picked up by roots but removed from the soil is degraded through hydrolysis, with the half life depending on the pH and clay content of soil involved. Toxic residues do not remain in the soil from one season to the next. Radioactive carbon studies have shown that carbofuran is very rapidly taken into roots and distributed into the aerial parts of plants. The oral LD_{50} for rats is 11 mg/kg.

Mode of Action

The carbamate insecticides are highly cholinergic and poisoned insects and other animals exhibit violent convulsions and neuromuscular disturbances. They are strong inhibitors of cholinesterase and may have a direct effect on acetylcholine receptors because of their pronounced structural resemblance to acetylcholine. These insecticides penetrate rapidly into the insects within 4 to 8 hours. From available information, it is found that two basic but different processes are operating either separately or together in the carbamate metabolism in biological systems. In one of them, the carbamate ester linkage is broken with the consequent liberation of the phenolic or enolic moiety either directly by esterase activity or indirectly through metabolic conversion of the carbamyl moiety into unstable intermediates. Direct split of the carbamate linkage may also occur by the non-enzymatic base-acid catalysed hydrolysis, depending on the intrinsic stability of the carbamates. There is evidence supporting the hydrolysis after prior oxidation of the carbamate moiety.

In mammals, the phenols are rapidly eliminated as conjugates in the urine. The carbamates are not accumulated in the body or excreted in milk. They are rapidly metabolised and excreted in the urine. They are reversible cholinesterase inhibitors and recovery from toxic dosage which produces characteristic symptoms is unusually rapid.

CHAPTER X

MISCELLANEOUS SYNTHETIC ORGANIC INSECTICIDES

In addition to various synthetic organic compounds described earlier, number of other synthetic organic compounds like nitrophenols (Dinitrocresol, Dinex, Dinoseb, Karathane), phenothiazone, thiocyanates and others have been used at one time or other. However, due to the availability of highly potent insecticides with low phytotoxicity and mammalian toxicity, the use of these insecticides have declined considerably now.

NITROPHENOLS

Nitrophenols have been used as contact and stomach poisons for number of years. However, due to their high phytotoxicity, their use has been limited. Further, many of these dinitro compounds are dangerous to handle as they are absorbed much more than other insecticides through the skin. Dinitro compounds are incompatible with oils, aramite, captan, quinones, lime sulphur, and lime. They should be used with caution with calcium arsenate, cryolite, 40% nicotine, dormant nitros, EPN, metacid, organo mercurial, Bordeaux mixture, fixed coppers, wettable sulphur and zinc sulphate plus lime.

Dinitro-O-Cresol (DNOC 8 DINOC, 2,4-Dinitro-6 methylphenol, 4,6-dinitro-O-cresol)

Dinitro-O-cresol was the earliest synthetic organic compound sold under the name of Antinonin in 1892 in Germany. It is produced by the direct nitration of O-cresol with nitrating mixture at low temperature. In some cases the O-cresol is first sulphonated with 70-93% sulphuric acid arid then nitrated.

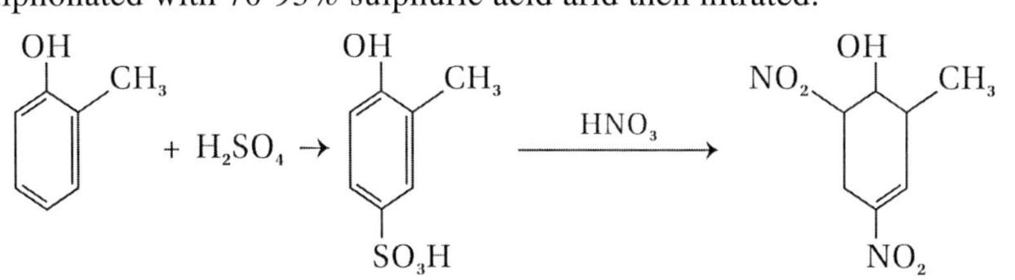

It is obtained in small quantities from the oxidative nitration of toluene in the presence of mercuric nitrate. 2,4-Dinitro-6-methyl- phenol is a yellow crystalline substance, m.p. 86.4°C. The solubility of the compound in g/100 g of solvent is : acetone 100.6, benzene 37.15, methanol 7.33, and petroleum ether 0.31.

Dinitro-O-cresol forms highly water soluble salts with caustic alkalies, ammonia and organic amines. These salts explode readily either from shock or detonation in a dry state. Hence, these salts are prepared either with inorganic diluents like sodium sulphate, ammonium sulphate, or urea or in aqueous solutions. Reducing agents convert it to amines (primarily one nitro group) particularly in warm-blooded animals. The product is 2-amino-4-nitro-6-methylphenol, a yellow substance with m.p. 173-174°C. In the soil, dinitro-O-cresol is decomposed fairly rapidly probably both as a result of reduction and splitting off a nitro group and subsequent complete destruction of the molecule.

DNOC stains the skin yellow. The maximum permissible concentration in air is 0.05 mg/m^3. The DNOC is used to control pests and diseases before opening of the buds either in the form of emulsion with oils or more often in the form of aqueous solution of its salts. The effectiveness of the compound increases when surface active agents are present. DNOC causes severe foliage injury. Replacement of phenolic hydrogen in the molecule with an acetyl or methyl group gave high toxicity. However, no relationship between chemical constitution and toxicity has been observed.

Dinex (2,4-Dinitro-6-cyclohexylphenol, DNOCHP)

Dinex has found wide use as insecticide since it is better than 2,4-dinitro-6-methylphenol. It is produced by the nitration of 2- cyclohexylphenol with a mixture of sulphuric and nitric acids at 60-90°C. The 2-cyclohexylphenol can be prepared by the condensation of phenol with cyclohexane in the presence of aluminium phenolate or other catalysts at 200-250°C. Dinex is a yellow crystalline substance, m.p. 106°C. Its solubility in water changes with pH of the medium. Only 1.8 and 15 mg of dinex dissolves at pH 1 and 6 respectively. With caustic alkalies and ammonia, it forms salts that are highly soluble in water. The biological properties of dinex are similar to those of DNOC. Though it is an excellent insecticide and ovicide, its strong phytotoxic properties limit its use in agriculture. The LD_{50} of this compound is 50-125 mg/kg. The residues do not present a serious health hazard. The dicyclohexamine salt of dinex called DN III is practically insoluble in water, sparingly soluble in most organic solvents, freely soluble in polar organic solvents. The pure compound occurs in orange crystals and this salt may be used for the treatment of green plants for controlling mites, aphids, scale insects and thrips. The increase in molecular size due to the introduction of the dicyclohexylamine moiety reduced the phytotoxicity of the parent phenol, thus making it safe for use on growing plants.

Dinoseb (DNBP, DNOSBP, DNSBP, 2,4-Dinitro-6-sec-butylphenol)

Dinoseb surpasses DNOC almost three times in insecticidal and herbicidal effect. As very little of the compound remains on the plant, the residue hazard is less. It is prepared by direct nitration of 2-sec-butylphenol with nitric acid in the presence of nitric acid. The 2-sec-butylphenol is first treated with 80-90% sulphuric acid and the sulphomixture that is obtained is further treated with 2-4 moles of nitric acid per mole of phenol, depending on the conditions.

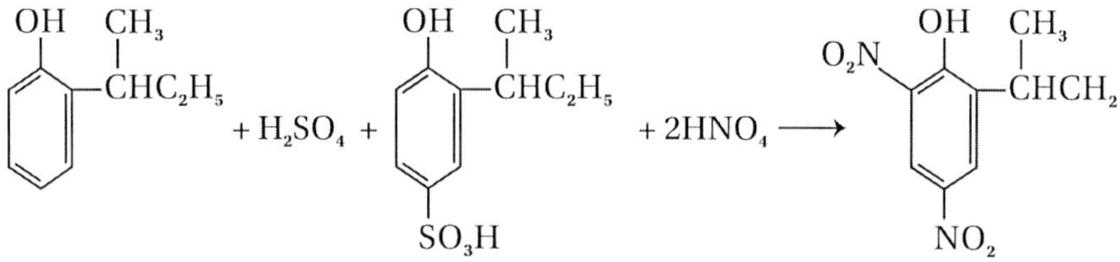

It is also formed by alkylation of p-phenolsulphonic acid with butylene or butylalcohol and subsequent nitration of the 2-butylphenol- 4-sulphonic acid without separating it from the reaction mixture.

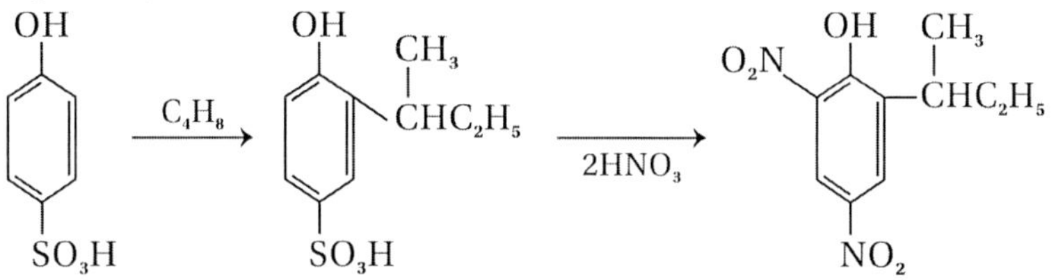

The second step is same in both the methods. The insecticidal properties of dinoseb are similar to those of other nitrophenols. The amino salt is less phytotoxic than DNOC because of its larger molecular size. Its use is limited to dormant application of fruit trees. This compound is relatively toxic to warm-blooded animals.

Karathane (Dinocap, Arathane, Isothan, 2,4-Dinitro-6-sec-octylphenyl cretonate)

Karathane is used as an acaricide and also for the control of powdery mildew on various crops. It is produced by nitrating 2-secoctylphenol and dinitrophenol thus produced is treated with cretonylchloride in the presence of HCl acceptors like tertiary amines in hydrophobic solvent.

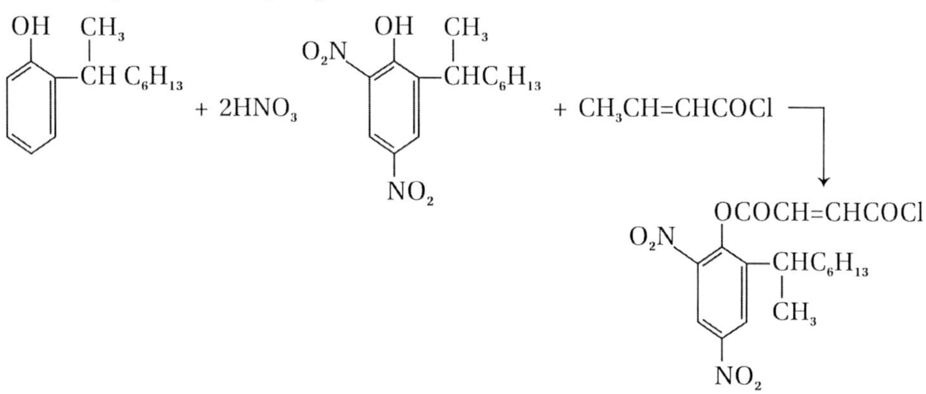

Karathane is a brown liquid which does not distill without decomposition at pressures greater than 0.1 mm of Hg. Its boiling point is 38 to 40°C. It is practically insoluble in water but highly soluble in organic solvents. The LD_{50} for rats is 980-1100 mg/kg.

ORGANIC SULPHUR COMPOUNDS

Phenothiazine ($C_{12}H_9NS$)

In 1934, Campbell Sullivan, Smith and Haller tested a number of sulphur containing organic compounds against mosquito larvae, the most toxic of them being the thiodiphenylamine (phenothiazine). This was more toxic than rotenone and was effective against the larvae at dilutions as low as 1 : 1,000,000. Kipling found phenothiazine to be non-toxic to higher animals. It is prepared by fusing of sulphur and diphenylamine. Presence of a catalyst improves the rate of reaction and also yield.

Phenothiazine is a light yellow crystalline substance, insoluble in chloroform and water. It crystallises from alcohol as colourless diamond shaped plates, with a m.p. 185°C. Phenothiazine is readily oxidised to phenothiazone on exposure to sunlight or air. In the presence of a finely divided inert carrier it may be oxidised to thionol. Lime and bentonite accelerate the oxidation process.

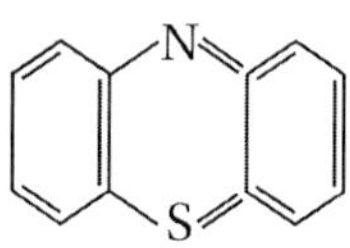

Phenothiazine

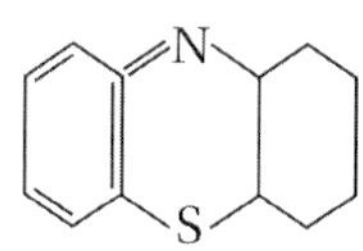

Phenothiazone

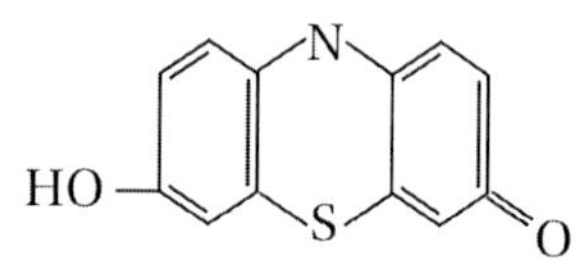

Thionol

Phenothiazone prepared by the oxidation of phenothiazine occurs in reddish brown leaflets, soluble in alcohol and slightly soluble in water, m.p. 162°C. Thionol (hydroxy phenoxy thiazone) may result from the spontaneous oxidation of phenothiazine or may be readily prepared by treatment of phenothiazine with sulphuric acid or hydrogen peroxide. It is a greenish black substance with an indefinite melting point, not easily soluble in water but soluble in alkalies and alcohol.

Both phenothiazone and thionol show the properties of reversible oxidation-reduction systems in the presence of their respective leuco-bases. Phenothiazine is toxic to number of insects and also has fungicidal action. Because of its variable action against some insects, possibly caused due to the oxidation process, this has not been widely used. Phenothiazine is compatible with most pesticides except Bordeaux mixture. Some foliage burn results when it is applied with oil. Bentonite should not be used as a wetting agent or dust with phenothiazine sprays containing phenothiazone, as it has a tendency to photosensitise the skin of persons coming in contact with them. Hence operators sunburn easily. Later sprays sometimes interefere with the proper colouring of fruits. Samples of phenothiazine can be divided by extraction with ether into two fractions, one—the ether soluble portion which is deep red in colour and is toxic to insects and the other insoluble fraction, which is non-toxic. Phenothiazone is highly potent as a fungicide while the unoxidised phenothiazine is not.

Thiocyanates

Murphy and Peet suggested the organic thiocyanogen compounds as contact insecticides. Neifert and his associates in 1925 reported that methyl and ethyl thiocyanates were more toxic to granary weevil and other insects than carbon disulphide. Thiocyanates frequently known as sulfocyanides contain the radical-SCH. Among the number of aliphatic thiocyanates synthesised, laurylthiocyanate $CH_3(CH_3)_{10}CH_2SCN$ which is produced by the reactions of lauryl chloride and sodium thiocyanate was found to be most toxic to aphids. Certain aliphatic thiocyanates containing ether linkage like β-butoxy-β' thiocyanodiethylether ($C_4H_9OCH_2CH_2OCH_2CH_2SCN$) have also been found to be toxic to insect species. Another commercially available organic thiocyanate called Lethane 60, contains the β-thiocyanoethylesters of fatty acids with 10 to 18 carbon atoms. Several alicyclic and aromatic thiocyanates like thanite, isocornyl thiocyanoacetate are also commercially available.

The organic thiocyanates have low toxicity to warm-blooded animals and are rarely phytotoxic. They are compatible with many other pesticides but being esters they are hydrolysed by strong alkalies and acids. Their chief use at present is in kerosene based sprays and aerosals. The thiocyanates are rather specific in their contact action but show little promise as stomach poison. Certain organic thiocyanates exhibit certain amount of fungicidal action.

DERIVATIVES OF DITHIOCARBONIC ACIDS

Morestan (Chinomethionat, Oxythioquinox, Quinomethionate, 6- methyl-2,3-quinoxaline dithiocarbonate: 6-methyl quinoxoline-2,3- dithio-cyclocarbonate)

Morestan has acaricidal action and also fungicidal action. It is synthesised by reacting phosgene with 2,3-dimercaptoquinoxaline in the presence of bases to bind the HCl.

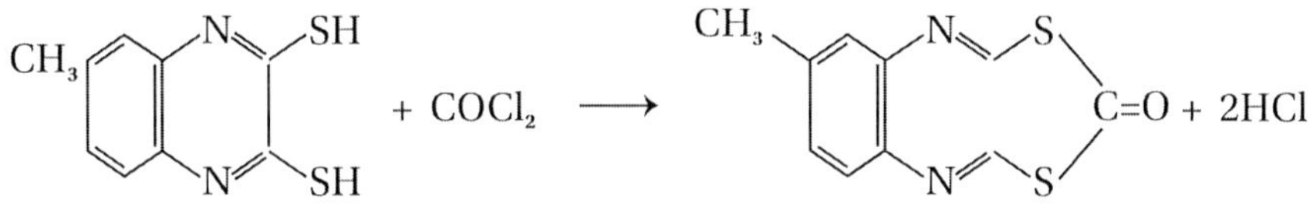

Morestan is a yellow crystalline substance, m.p. 172°C. It is sparingly soluble in water, moderately soluble in most organic solvents, and soluble in hot aromatic hydrocarbons, dioxane and dimethylformamide. It is a odourless compound and is relatively unstable in alkaline medium. It is sold in the form of wettable powders, and dusts. It can be mixed with the conventional insecticides and fungicides. However it is recommended that morestan be mixed with wettable powders and not with emulsion concentrates. Morestan should not be mixed with products of alkaline nature such as Bordeaux mixture, lime sulphur, solbar, TMTD, oils and oil-based products. The LD_{50} for this compound is 2500 mg/kg.

Eradex (Quinoxaline, Bayer 30686, Quinoxaline--2,3-trithio- carbonate)

Eradex is an active acaricide, produced by reacting 2,3-dimercaptoquinoxaline with thiophosgene.

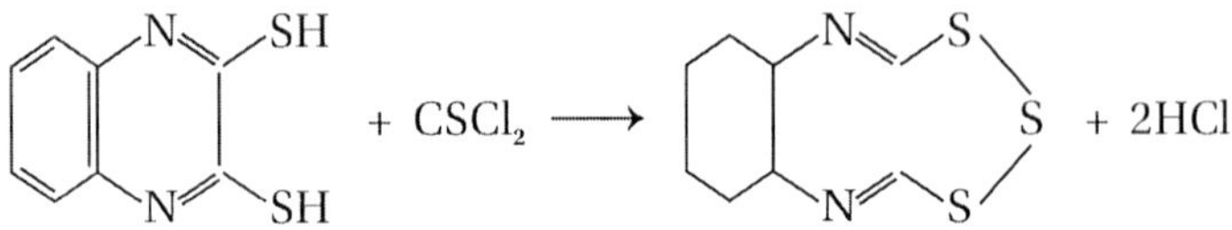

It is a brown compound, m.p. 180°C, practically insoluble in water and most soluble in solvents. At room temperature, eradex is resistant to hydrolysis by water but is relatively easily oxidised by oxygen of the air with the formation of S-oxides, without loss of biological activity. The LD_{50} of the compound is 3400 mg/kg.

OTHER ORGANIC INSECTICIDES

A number of other organic compounds like arsenomethane, sodium fluroacetate, naphthalene, O-dichlorobenzene, p-dichloro-benzene, phenol, pentachlorophenol, O-phenylphenol, cresol, nitrobenzene and cyclohexylamine have been tried for their insecticidal activity. Since most of them are either phytotoxic to some plants or highly toxic to warm-blooded animals or highly specific or costly when compared to other available insecticides, or possess very low insecticidal activity, they have not been widely used now. Possibly with further research some of them could become valuable insecticides at a later date

FUNGICIDES

CHAPTER XI

SULPHUR FUNGICIDES

Considering the present-day need for increasing food production control of plant diseases ranks only next to the control of insects. Even as early as 1000 B.C., Homer mentioned the use of sulphur as a fungicide. Seed treatment has been practiced even in the 17th century. Schulthess in 1761 suggested for the first time the use of copper sulphate as seed dresser. The discovery of Bordeaux mixture in 1882 has been a big step in the control of diseases since this mixture has withstood the test of time and even today continues to hold the field.

In India, control of diseases has been practiced even from a ancient times. In earlier days they have been using certain cultural practices and some natural compounds for the control of diseases. Ozanna in 1888 appears to be the first person in India to advocate the use of chemical compounds like copper sulphate for the control of sorghum smut. In India, Lawrence in 1904 used the Bordeaux mixture for the first time. Subsequently, a number of synthetic fungicides were developed and have been widely used in India.

The various fungicides now in use can be classified as follows:

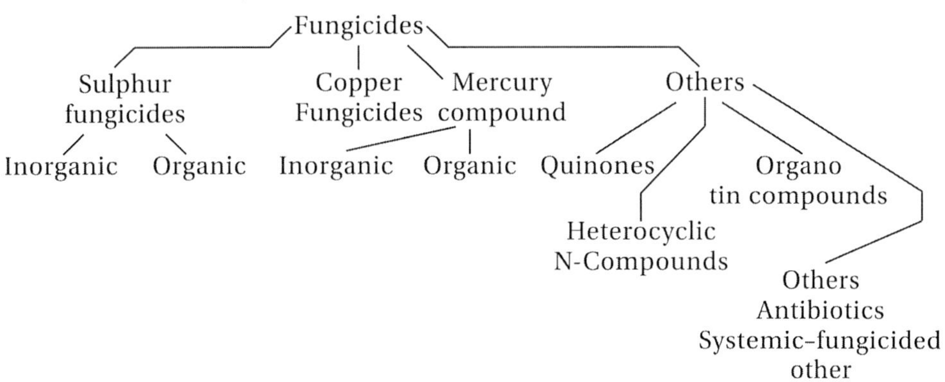

SULPHER COMPOUNDS

Sulphur is one of the most important fungicides used for the control of various diseases, particularly powdery mildews. Elemental sulphur occurs in both free state and also in combination. Free sulphur is found in deposits of volcanic origin. Sulphur occurs in many allotrophic forms, of which α or the rhombic form, is the stable one at ordinary temperatures. The other form β or monoclinic, passes into a form on standing. Sulphur is insoluble in water, slightly soluble in alcohol and ether and freely soluble in carbon disulphide, sulphurchloride and hot benzene. The specific gravity of the rhombic form is 2.07 and the m.p. is 112.8°C.

Every year large quantities of elemental sulphur are being imported into India for use as sulphur dust, wettable powders and for the preparations of other sulphur formulations. Sulphur is being widely used in a finely ground condition or in the colloidal form in dust preparations. Finely ground sulphur is obtained by grinding coarse sulphur in special types of mills called micronizers and then fractionating it with the air blasts. As the mechanically ground sulphur tends to aggregate, small amounts of inert material like kaolin or lead arsenate are also added to the micronised sulphur. The fungicidal efficiency of sulphur dusts depends on the fineness of the particles. Most of the sulphur sold for fungicidal use is ground to pass at least a 300 mesh sieve and the present trend is towards even finer particles.

Sulphur even though finely ground, is not wetted by water. Hence, wettable powders are made by grinding sulphur with protective colloidal materials such as sulphite lye, casein, bentonite clay, skimmed milk, flour or dextrin. The amount of deposition and retention of sulphur varies with different plant surfaces. Sulphur dusting powders containing materials such as gypsum, bentonite or talc in small amounts make the sulphur flow freely and this form is called "conditioned sulphur".

Though sulphur is used in its natural state, certain other forms of sulphur produced by chemical reactions are also used. Two forms of colloidal sulphur were prepared by De Ong, one a hydrophylic sol made by passing H_2S through a saturated solution of sulphur dioxide in water or by reaction of sodium thiosulphate and sulphuric or mineral acids.

$$2H_2S + SO_2 \rightarrow 3S + 2H_2O$$

$$Na_2S_2O_3 + H_2SO_4 \rightarrow Na_2S_2O_4 + SO_2 + S + H_2O$$

and a hydrophobic colloidal sulphur prepared by acidifying lime- sulphur solution with mineral acids.

$$CaS_x + 2HCl \rightarrow CaCl_2 + H_2S + S_{x-1}$$

As the hydrophobic suspensions tend to aggregate and thus precipitate after a few days, they are stabilised by the addition of hydrophylic colloids such as glue, gelatin and other similar materials. The hydrophylic sulphur preparations were found to have greater toxicity than hydrophobic suspensions, probably due to differences in the particle sizes.

Another form of sulphur called floatation sulphur is a byproduct of the manufacture of fuel gas from coal. When bituminous coal is carbonised to produce coke and gas, the sulphur present goes out as H_2S and this is recovered by the floatation process. The floatation sulphur is in a fine state of division, almost approaching colloidal dimensions. It is marketed both as a paste and as wettable powders. However, floatation sulphur contains certain impurities like hydrocyanic acid and thiosulphates which are removed by washing.

In addition to the free sulphur, number of sulphur compounds have also been tried as fungicides. Among them lime sulphur or calcium polysulphide (CaS_n) is an important one. Lime sulphur was recommended as a fungicide even as early as 1833. It was later known as "Eau Grison". When a suspension of calcium hydroxide is boiled with sulphur, a number of sulphides of calcium are formed. In the subsequent lime-sulphur preparation made in early 1900's, salt was also added, probably to elevate the boiling point. Lime sulphur is usually prepared by boiling 1.8 parts of sulphur, 1 part of lime and 10 parts of water for one hour. Then the solution is filtered and the filtered solution is diluted with water to specific gravity of 1.003 and 1.035 for summer and dormant sprays respectively.

When sulphur and calcium hydroxide are suspended in water and boiled, the reaction products go into true solution and at the same time the colour of the solution passes from a light yellow through deeper shades of orange until it finely becomes deep orange. The composition of the final product depends upon the initial concentration of the hydroxide, duration and temperature of the reaction and the relative amounts of the reactants. The reaction probably takes place in number of steps. Possibly the first reaction is

$$3S + 3H_2O \rightarrow 2H_2S + H_2SO_3$$

which proceeds at extremely slow rate. The hydrolysis of sulphur in the presence of an alkali proceeds as follows:

$$2S + 4NaOH \rightarrow Na_2S + Na_2SO_2 + 2H_2O$$

From the insecticidal and fungicidal activity point, polysulphide sulphur is the most important constituent of lime sulphur solutions. The older method of measuring fungicidal and insecticidal value of lime-sulphur preparations was by the specific gravity, which was expected to be slightly more than 1.283. Obviously the presence of soluble materials added unknowingly or as adulterants, will increase the specific gravity of the solution. Recently also a high degree of correlation between specific gravity and polysulphide sulphur was also observed.

When lime-sulphur solutions are boiled for considerable time, the calcium sulphides are completely broken down with the formation of calcium thiosulphate and hydrogen sulphide. When exposed to air, lime-sulphur solutions undergo decomposition at temperatures below boiling

$$CaS{:}\ S_x + 3O \rightarrow CaS_2O_3 + S_{x-1}$$

$$CaS_2O_3 \rightarrow CaSO_3 + S$$

$$CaSO_3 + (O) \rightarrow CaSO_4$$

The effect of lime sulphur is based upon the reaction with CO_2 of the air or upon oxidation of finely dispersed sulphur that is deposited on the plant.

$$CaS_n + H_2O + CO_2 \rightarrow CaCO_3 + H_2S + S_{n-1}$$

$$2\,CaS_n + 3CO_2 \rightarrow 2\,CaS_2O_3 + 2\,S_{n-2}$$

Possibly other reactions also take place in the decomposition of lime-sulphur solution. Even simple dilution with water was found to decrease the amount of polysulphide present after one hour of stirring although in solutions more concentrated than 1:50, the polysulphide sulphur was present. Possibly hydrolysis of polysulphides takes place in dilute solution.

When the temperature of the solution is raised from 23 to 32°C, the decomposition of the polysulphide sulphur results in a corresponding increase in the thiosulphate sulphur. Continued stirring of the dilute lime sulphur solution was also found to hasten the decomposition, with increase in thiosulphate sulphur. When lime sulphur and lead arsenate are mixed, the white colour of the lead arsenate is changed to brown in a few seconds and it becomes progressively darker with time. After mixing, there is an increase in the concentration of thiosulphate and sulphite salts in solution and also of soluble arsenic. When acid lead arsenate and lime-sulphur solution containing mainly calcium polysulphides are mixed, the polysulphides are first broken down to yield H_2S and this reacts with lead arsenate to form black lead sulphide. Among other products of reaction, arsenic acid that is formed from the arsenate, combines with the calcium salts to form calcium arsenate and thioarsenates which are later decomposed. Calcium thiosulphate and calcium sulphide are also probably formed during the course of reaction.

As the increase in the soluble arsenic increases the phytotoxicity, while an increase in polysulphides results in increase in the fungicidal efficiency, one has to attempt to reduce the soluble arsenic content and simultaneously either increase or at least maintain the amount of polysulphide already present. To achieve this a number of agents called correctives or adjuvants are used. Lime has been used as a corrective to reduce the amount of soluble arsenic that is set free and preserve the polysulphide content of the fungicide. However, this cannot retain the sulphide-sulphur in solution at as high a level as in a corresponding solution of lime sulphur alone.

Proteinaceous materials like gelatin and casein have been suggested as correctives. Additions of casein or dried milk powder caused an increase in the amount of soluble arsenic formed in lime sulphur-lead arsenate spray mixture. The increase is apparently proportionate to the amount of milk powder added. Casein materials have been found to decrease the extent of decomposition of polysulphide content. A combination of lead arsenate, lime sulphur, lime and skimmed milk (the later two in equal amounts) gave lower amounts of soluble arsenic and at the same time, higher amounts of polysulphide sulphur than those without skimmed milk. Use of gelatin alone increased the soluble arsenic but had no effect on the reactions of calcium sulphide. Addition of nicotine sulphate has little effect on the decomposition of mixed lead arsenate-lime sulphur spray. The order of mixing the ingredients also has considerable effect on the composition of the resulting mixture. Generally lime sulphur is first diluted nearly to its final volume and then lead arsenate is mixed with a small amount of water just before application.

Many metallic salts like manganese sulphate, ferrous sulphate, ferric oxide and zinc sulphate have been used as corrective but their mode of action has not yet been understood. They are supposed to produce elemental sulphur from the polysulphide present in the lime sulphur and hence prevent interaction between the polysulphides and lead arsenates. Sometimes spreaders are used with lime-sulphur solution to increase the efficiency of the spray. Possibly the spreader causes an amorphous precipitate of lime sulphur to be deposited on the leaves which probably results in greater adherence.

Many other forms of lime sulphur have also been prepared. When lime sulphur concentrate is evaporated with sucrose as a stabiliser, probably the polysulphides present as the tetra- and pentasulphides are replaced by a mixture of tri and tetrasulphides in approximately equal proportions. This form of lime sulphur is called "dry lime sulphur" and is slightly less effective than lime sulphur. In another form of lime sulphur, called "self boiled lime sulphur", the temperature is allowed to rise due to the slaking of calcium hydroxide and the sulphur present in the suspension reacts with calcium hydroxide in an accelerated rate due to the increase in temperature. The self boiled sulphur is mainly used for sulphur-sensitive plants. However, now it is rarely used.

Other compounds of sulphur tried are dry mix sulphur-lime, calcium monosulphide (CaS), ammonium polysulphide (APS), barium sulphide (Solbar), potassium and sodium sulphides. In the dry mix sulphur-lime the ingredients are mixed in dry conditions. However, this is rarely used now, as formulations with greater effectiveness are available. Calcium monosulphide was first used as a spray material by Hurt and Schnerderhan in 1929. The fungicidal efficiency of the mixture which contains between 60 and 65% calcium monosulphide, and varying amounts of calcium sulphate, starch, charcoal and coal was found to depend on the amount of calcium sulphide present. Crude calcium monosulphide is a powder varying in colour from pale grey to yellow and is readily soluble in water. It has better adhering properties than sulphur-lime dry mix and does not undergo decomposition on storage. This is also not used now.

Ammonium polysulphide is prepared by saturating concentrated ammonia (28%) with hydrogen sulphide and then dissolving the excess of sulphur in the resulting solution of ammonium sulphide. this is less stable than calcium salts and due to the volatility of the ammonium radical, more free sulphur is deposited on decomposition. This is also not widely used.

Barium polysulphide (BaS_x) is prepared by saturating barium hydroxide solution with hydrogen sulphide or mixing ground sulphur with barium sulphite powder. This is highly soluble in water and is rather stable in storage in the absence of air and water. This preparation is widely used in Russia and is now being marketed in India. However, solbar is not better than lime sulphur in its fungicidal properties and at the same time more toxic to man. Its LD_{50} is 375 to 500 mg/kg. It is compatible with parathion. metsystox, systox, gusathion, diptrex, and chlorthion but these mixtures must be applied at once. However, it should not be combined with organomercurials, DNOC and TMTD.

The sulphides of potassium and sodium have long been used as fungicides. These are prepared by boiling sulphur and alkalihydroxide together similar to the lime sulphur solutions. There is no need to apply any heat treatment. The active principle of this mixture appears to be alkali, mono- and polysulphides and is probably similar to lime sulphur. As this is not widely used, not much work has been done on it. Sodium sulphide unlike the corresponding calcium salt is soluble and therefore remains intimately mixed with the active constituents. As the sodium and potassium hydroxides are more caustic than calcium hydroxide, the sulphides of sodium and potassium have never been used widely.

Another new. form of fungicides is calcium tetrathionate, a white crystalline substance, poorly soluble in water and organic solvents. It is prepared by passing sulphur dioxide into sulphuryl chloride (S_2Cl_2).

$$2SO_2 + 2H_2O + S_2Cl_2 \rightarrow H_2S_4O_6 + 2HCl$$
$$H_2S_4O_6 + CaCl_2 \rightarrow CaS_4O_6 + 2HCl$$

This has been reported to be at least as effective as colloidal sulphur. It is used in the form of aqueous suspensions prepared from 50% wettable powders.

Fungicidal Action of Sulphur

To explain the fungicidal action of elemental sulphur, various theories like sulphur vapour theory, sulphur dioxide theory, pentathionic acid theory and hydrogen sulphide theory have been proposed. Sulphur is volatilised even at relatively low temperature and later condensed under suitable conditions. The rate of volatilisation increases with the decrease in particle size of the sulphur particles. However, as the volatilised sulphur has been shown to be non-toxic to fungus spores and hyphae, the sulphur vapour theory is not accepted now. Similarly the theory that sulphur dioxide plays an important part in the toxic action of sulphur has also been refuted. Young in 1922 suggested that pentathionic acid formed by the oxidation of sulphur, was the toxic factor in sulphur fungicides. Pentathionic acid is probably formed by the following reactions:

$$S + O_2 \rightarrow SO_2 \xrightarrow{+H_2O} H_2SO_3 \xrightarrow{+S} H_2S_2O_3$$
$$5H_2S_2O_3 \rightarrow 2H_4S_5O_6 + 3H_2O$$

Formerly the acidity of sulphur dusts was also attributed to pentathionic acid and sulphuric acid. However, later it was found that the amount present was very negligible. As none of the compounds suggested were more toxic than the hydrogen ion concentration of their solutions in water, the above theories have not been accepted.

Later the "hydrogen sulphide theory'' was proposed. The H_2S produced was found to be 6 to 200 times as toxic to the fungi as pentathionic acid and sulphuric acid. However, later even this theory was refuted as the colloidal sulphur was 5 to 50 times more fungi- toxic on molar basis than the hydrogen sulphide and also the formation of hydrogen sulphide from elemental sulphur was not detrimental to the germination of certain spores.

The latest theory is called the "direct action theory" according to which sulphur acts as a hydrogen acceptor in the metabolic system and by doing this disturbs the normal hydrogenation and dehydrogenation reactions in the cells. In this theory, it is assumed that sulphur being next to oxygen in periodic table, acts much like oxygen and forms H_2S instead of H_2O in the fermentation process, thus causing a disturbance in the Kreb's cycle. The process by which fungus tissues reduce sulphur to hydrogen sulphide is not fully understood. Practically all plant tissues reduce small amounts of sulphur to hydrogen sulphide possibly due to a direct chemical reaction between sulphur and glutathione, to form the disulphide of the latter. The work of Tweedy indicates that the cytochrome system is also involved in the reduction of sulphur. Owens has postulated that the polysulphides free radicals produced in metabolic breakdown of 8-membered rings of elemental sulphur to H_2S might be responsible for the toxicity. He suggested that the reactive polysulphide—free radicals, might form stable cross linkages with protein and other cellular compounds. Thus the basis of toxic action of sulphur in the cell is still not clear. Sulphur fungicides emit sufficient vapour to prevent the growth of fungus spores at a distance of several millimeters from deposits on leaves. This vapour phase activity ensures protection of gaps on leaf surfaces resulting from weathering or expansion of leaf surface.

Acute injury of sulphur is rare in temperate climates but in hotter climates (>80°C), severe burning is sometimes caused on cucurbits and also on the sun exposed side of apple fruit trees. Some fruit trees are sulphur sensitive. Lime sulphur when applied on green leaves, seems to reduce the photosynthesis considerably.

ORGANIC SULPHUR COMPOUNDS—DITHIOCARBAMATES

Investigations of the Du Pont Company (U.S.A.) showed that some of the derivatives of dithiocarbamic acid $H_2N\underset{\underset{\displaystyle S}{\|}}{C}SH$ had insecticidal and fungicidal properties. However, dithiocarbamic acid itself is not known to exist in the free state. When the substituted primary and secondary aliphatic and aromatic amines are treated with carbon disulphide in alcoholic solutions, substituted dithiocarbamates are produced

$$\begin{matrix} R_1 \\ R_2 \end{matrix}\!\!>NH + CS_2 \rightarrow \begin{matrix} R_1 \\ R_2 \end{matrix}\!\!>N\underset{\underset{\displaystyle S}{\|}}{C}SH$$

R_1 and R_2 represent aryl or alkyl organic radicals or in the case of primary amine, one of them may happen to be hydrogen. The hydrogen attached to sulphur dissociates and may be replaced by a metal or any salt-forming organic radicals producing a variety of derivatives with a considerable range of physical properties. The group $>N\underset{\underset{\displaystyle S}{\|}}{C}S$ is considered to be essential for insecticidal and fungicidal action.

Dithiocarbamate fungicides were introduced by the patent of Tisdale and Williams which disclosed Thiram (tetramethylthiuram disulphide) and their sodium, iron and cadmium salts, all of them prepared by reacting dimethylamine and carbon disulphide under alkaline conditions. Later Diamond, Heuberger and Horsfall discovered Nabam (disodium ethylene bisdithiocarbamate) produced by the reaction of ethylene diamine and carbon disulphide. Subsequently the heavy metal complexes of dithiocarbamates like ziram and ferbam and complexes of ethylene bisdithiocarbamates like maneb and zineb were developed.

The esters of thiocarbamic acids like eptam, vernam, tiilam, avadex and triallata are mainly used in weed control. A large number of derivatives of dithiocarbamic acid possess fungicidal properties.

They can be classified as follows:

(i) Metallic dithiocarbamates— Ziram and Ferbam

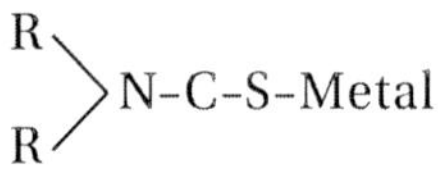

—Ziram

—Ferbam

(ii) Thiuram disulphide (Thiram)—Sulphur to sulphur linkage.

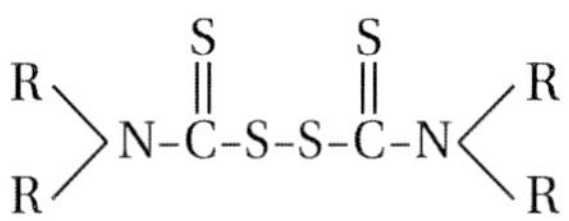

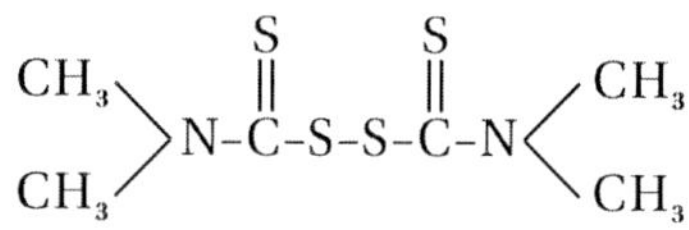

Thiram

(iii) Bisdithiocarbamates

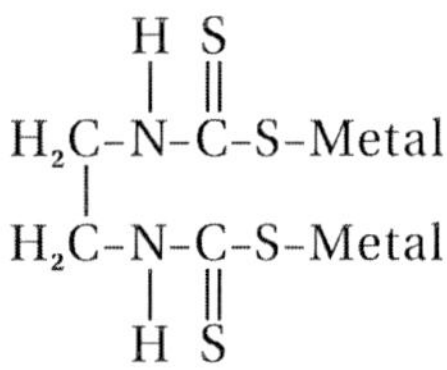

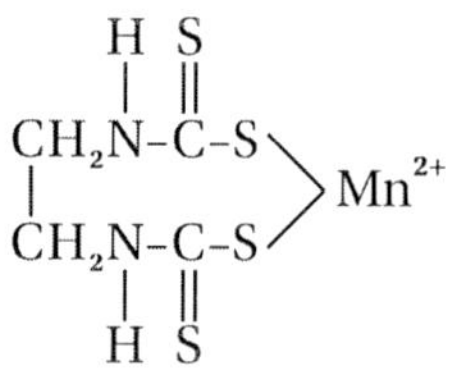

Maneb

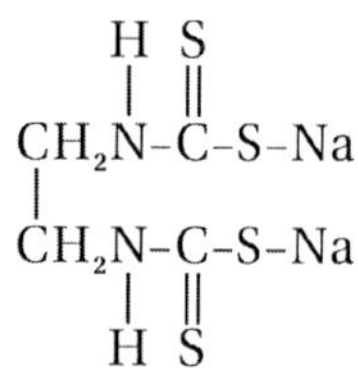

Nabam

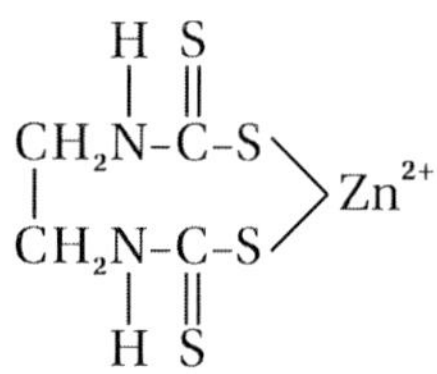

Zineb

Based on the mechanism of action dithiocarbamates can also be divided into two groups:

(i)  Nabam, Maneb, Vapam, Zineb

(ii) 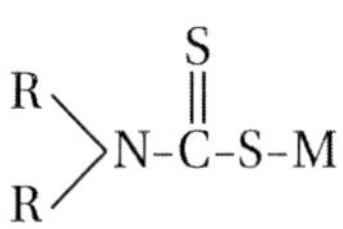 Thiram, Ziram, Ferbam

The first group of monoalkyldithiocarbamates includes compounds like sodium-N-methyldithiocarbamate (vapam) and sodium methylene bis dithiocarbamate (nabam) which are derived from monoalkylamines. The second group, the dialkyldithiocarbamates, includes compounds like sodium dimethyldithio-carbamate and TMTD which are derived from dialkylamines.

The salts of methyl-, dimethyl-, and ethylene bis dithiocarbamic acids like maneb, zineb, ziram, carbothion, polycarbazine, and TMTD are widely used because of their relatively low cost, simplicity of production and availability of raw materials. The alkali salts of alkyldithiocarbamic acids are produced easily and in practically quantitative yields by reacting the corresponding amine, alkali hydroxide with carbon disulphide.

$$R_2NH + NaOH + CS_2 \rightarrow R_2N\overset{\overset{\displaystyle S}{\|}}{C}SNa + H_2O$$

When this water soluble alkali salt reacts with zinc, iron or manganese, the corresponding salts of dithiocarbamic acids are formed. These are practically insoluble in water.

$$2\{R_2NCSNa\} + ZnSO_4 \rightarrow (R_2N\overset{\overset{\displaystyle S}{\|}}{C}S)_2\,Zn + Na_2SO_4$$

However, it is better to use ammonium salts of substituted dithiocarbamic acids, as the waste water containing ammonium salt can be used as fertilizer after evaporation.

Salts of dithiocarbamic acids are also produced by reacting oxides of the corresponding metals with amines and carbon disulphide.

$$2R_2NH + CS_2 + ZnO \rightarrow (R_2N\overset{\overset{\displaystyle S}{\|}}{C}S)_2\,Zn + H_2O$$

By this process the waste water is avoided as pure water is obtained. This water could be used for subsequent operation. However, the salt obtained may have zinc as impurity but this is not of much importance since it does not lower its effectiveness.

Ziram (Zerlate, Methasan, Coruzate, Fuklasin, Karbam white, Milbam, Niagara-C, Zirberk, Cuman, Zinc dimethyldithiocarbamate)

Ziram is prepared by any one of the above methods. It is crystalline white powder with a m.p. 240-246°C, tasteless and odourless. Its molecular weight is 305.8. It is sparingly soluble in water, slightly soluble in most of the organic solvents. On prolonged heating at 170-180°C violent decomposition takes place with the formation of a carbonaceous product. This fact has to be considered during the drying of the compound in a dryer. Ziram is rather stable in storage. Dilute acids do not decompose it at room temperature but concentrated sulphuric acid, phosphoric acid and other acids break the compound completely with the evolution of carbondisulphide and a small amount of hydrogen sulphide. Ziram is also decomposed when it is heated with caustic alkalies.

It is used mostly in the form of wettable powder containing 70-90% of active ingredient. Among the dithiocarbamates fungicides, ziram is the most stable one. It is compatible with most of the other pesticides except those containing either lime or heavy metals like copper, iron, or mercury. The oral toxicity of ziram for rats is > 5200 mg/kg. The chronic toxicity is very slight and it is only slightly irritating to nose, throat, and skin. This compound is not highly toxic to animals though it may cause irritation to mucous membrane. It is very slightly toxic to plants except to those sensitive to zinc.

Ferbam (Fermate, Coromet, Fermate-D, Fermocide, Ferradow, Karbam black, Ferric dimethyldithiocarbamate)

Ferbam was one of the first dithiocarbamates made for commercial use. It is prepared by reacting a ferric salt in solution with sodium dimethyl dithiocarbamate.

$$3(CH_3)_2N\overset{\overset{\displaystyle S}{\|}}{C}SNa + FeCl_3 \rightarrow \{(CH_3)_2\,N\overset{\overset{\displaystyle S}{\|}}{C}S\}_3\,Fe + 3NaCl$$

Ferbam is a dark brown crystalline substance with a molecular weight of 416.5. It is soluble in chloroform, pyridine and acetonitrile, slightly soluble in water. It is somewhat unstable to heat, light and moisture. In cold and in absence of moisture, it is rather stable but when it is heated in moist conditions, it may decompose. It is also decomposed by the action of alkaline agents. It is compatible with most insecticides but has to be used with caution when mixing with calcium arsenate, TEPP, quarternary ammonium compounds, organomercurial compounds, Bordeaux mixture, fixed copper, lime sulphur, zinc sulphate, and lime. However, its fungicidal efficiency is reduced when used with copper, lime or mercury compounds. Skim milk powder has been found to prevent double decomposition of ferbam when mixed with basic copper salts. Ferbam is considered to be less phytotoxic than Bordeaux mixture but leaves a visible deposit on foliage and fruits. Ferbam is relatively non-toxic to both plants and animals. However, it has been found to cause unusually high chromosomal aberrations in roots of *Allium cepa,* which indicates the possibility of injurious effects on certain host plants. The LD_{50} of ferbam is 17,000 mg/kg.

Thiram {Thiuram-D, TMTD, Arasan, Nomersan, Panormats, Pomasol, Spotrete, Tersants, Thylate, Tetramethylthiuram disulphide, bis (dimethyl thiocarbamyl) disulphide, α, α-dithio bis (dimethylthio) formamide}

Thiram differs chemically from the dithiocarbamate, in that it contains two dithiocarbamates nuclei connected together through the sulphur to sulphur linkage. When two molecules of substituted dithiocarbamates are oxidised, two atoms of hydrogen are eliminated and a condensation takes place.

$$R_2'N\underset{\underset{\displaystyle S}{\|}}{C}SH + HS\underset{\underset{\displaystyle S}{\|}}{C}NR_2'' \rightarrow \begin{matrix} R' \\ R' \end{matrix}\!\!>\!N\underset{\underset{\displaystyle S}{\|}}{C}\text{-}S\text{-}S\text{-}\underset{\underset{\displaystyle S}{\|}}{C}\text{-}N\!<\!\!\begin{matrix} R'' \\ R'' \end{matrix} + H_2$$

The tetraalkylthiuram disulphide containing all R groups as methyl radicals is widely used. Thiram is produced in good yields by the oxidation of alkali salts of dimethyldithiocarbamic acids with hydrogen peroxide or other oxidising agents like sodium nitrites.

Pure thiram is a yellow crystalline compound with a m.p. of 155- 156°C. It is almost completely insoluble in water, slightly soluble in ether and alcohol and freely soluble in chloroform. It is quite stable chemically and is not phytotoxic at ordinary doses at which it is used. Thiram appears to be more versatile in its fungicidal action than other thiocarbamates. It is an effective seed protectant also. The toxicity to higher animals is low but it may be irritating to the skin and mucous membrane when used in dust form. Thiram is compatible with most commonly used pesticides except those containing copper and lime.

Tetramethylthiuram monosulphide also shows fungicidal properties but is weaker than the disulphides. Thiuram disulphides when treated with a sulphur acceptor such as potassium cyanide (KCN) is converted into thiuram mono-sulphide.

$$(C_2H_5)H\underset{\underset{S}{\|}}{C}S\underset{\underset{S}{\|}}{S}CN(C_2H_5) + KCN \rightarrow (C_2H_5)_2N\underset{\underset{S}{\|}}{C}S\underset{\underset{S}{\|}}{C}N(C_2H_5) + KCNS$$

The tetraethylthiuram monosulphide appears to be highly toxic to fungus infection on human skin and hence soaps containing this compound have been used successfully in treatment of scabies. Like other carbamate fungicides, thiram is of low mammalian toxicity. The LD50 value for rats is 780 mg/kg.

SALTS OF ETHYLENE BIS DITHIOCARBAMIC ACID

Nabam (Dithane-14, Chembam, Dithane A-40, Dithane D-14, Parsate liquid, Disodium ethylene bisdithiocarbamate.

Nabam is prepared by reacting sodium hydroxide, carbon disulphide, and ethylene diamine in aqueous medium.

$$\begin{array}{l} CH_2NH_2 \\ | \\ CH_2NH_2 \end{array} + 2CS_2 + 2NaOH \rightarrow \begin{array}{l} CH_2NH\overset{\overset{S}{\|}}{C}SNa \\ | \\ CH_2NH\underset{\underset{S}{\|}}{C}SNa \end{array} + 2H_2O$$

The dry salt is white while the water solution is yellow. Its molecular weight is 256.3. It is soluble in water to about 20% and insoluble in most hydrophobic organic solvents. The anhydrous salt is also unstable in storage. Dry salt is unstable to heat, light, moisture. Nabam could be stabilised with the addition of zinc sulphate—lime mixture due to the formation of zineb. Used alone it is phytotoxic but combined with zinc sulphate, it is similar to zineb.

Zineb (Dithane Z-78, Parazate, Du Pont fungicide A, Lonocol, Zinc ethylene bisdithiocarbamate)

Zineb is one of the most important fungicides used in agriculture. It is prepared by precipitation of the zinc ethylene bisdithiocarbamate from aqueous solution of diammonium or disodium ethylene bisdithiocarbamates by zinc sulphate.

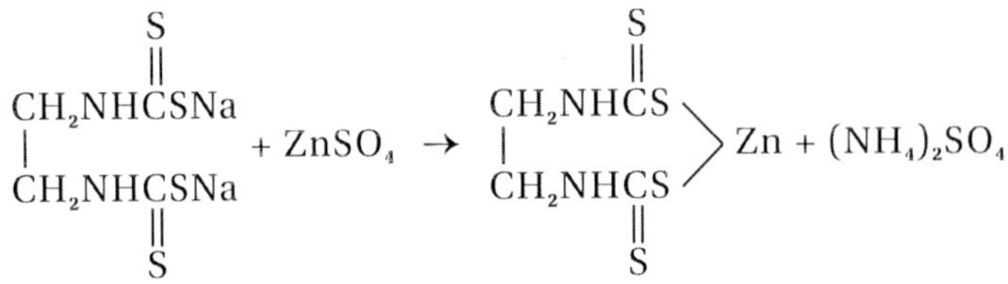

This reaction is carried out in dilute solution (<7%) as polymeric zineb of lower toxicity precipitates from concentrated solutions or by reacting ethylene diamine with zinc oxide and carbon disulphide

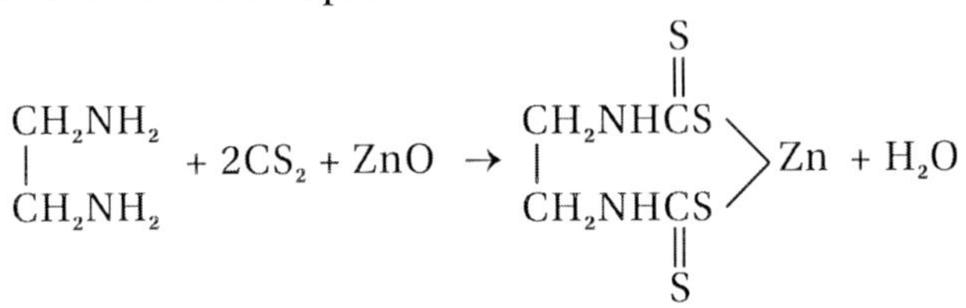

Zineb is a white crystalline substance with molecular weight 275.5, almost insoluble in water and in most organic solvents but moderately soluble in pyridine. It is unstable in the presence of moisture, heat and light. The moist compound (>4%) under unfavourable storage conditions may decompose even up to 50% in a year. To reduce the chances of explosive evolution of CS_2, zineb has to be stored in well ventilated places at the lowest possible temperatures.

Zineb also breaks down relatively rapidly (10 to 15 days) on plants with the final formation of easily volatile products. The breakdown takes place considerably faster in warmer and clearer weather. The final product of this decomposition is zinc sulphate which may play the role of zinc micronutrient. In compatibility, zineb is similar to ferbam. It can be combined with parathion, diptrex, gusathion, metasystox, systox, chlorthion and wettable sulphur. In the presence of strong alkaline substances it may be decomposed and also under certain conditions, combining copper salts may also result in decomposition. It may cause irritation of nose and throat and skin. It is formulated as wettable powders and is safe to use on plants, except those sensitive to zinc. Sometimes it may cause defoliation. The residual effect of zineb depends to a great extent on the weather and the season. It is reported that zineb in comparison to Bordeaux mixture, gives better control of late blight of potato in plains of India than in hills, where the reverse is true. Possibly in hills during summer there are frequent rains and therefore the residual action of zineb is reduced, whereas Bordeaux mixture has a tenacity of its own. Zineb can be of great use on zinc deficient soils as plants like tomato and paddy can absorb zinc from zineb. Its LD_{50} value for rats is >5200 mg/kg.

Proprietary mixed formulations containing zineb and fungicides like copper oxychloride (copper-lonocol, miltox) are being marketed now. In addition to these, certain other substitution compounds like zinc-2-aminoethylene dithiocarbamate which is formed as a product in the preparation of zineb is also used as a fungicide. It surpasses sometimes even zineb in its fungicidal activity. But it is less stable in storage. Other compounds like ethylene bis (thiuram polysulphide) also called Polycarbacine or Polyram also possess fungicidal value. Polyram is insoluble in water and most of the organic solvents. Its composition is

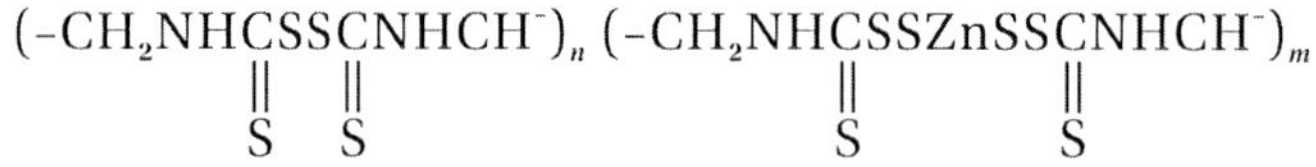

where, n : m is usually 1 : 3.

Maneb (Manzate, MEB, Mn EBd, Dithane M-45, Manganese ethylene bisdithiocarbamate)

Maneb, the manganese salt of ethylene bisdithiocarbamic acid, is similar to zineb in most of its physical and chemical properties. Similar to zineb, it is produced by reacting manganese sulphate with diammonium or disodium ethylene bisdithiocarbamate in aqueous solution.

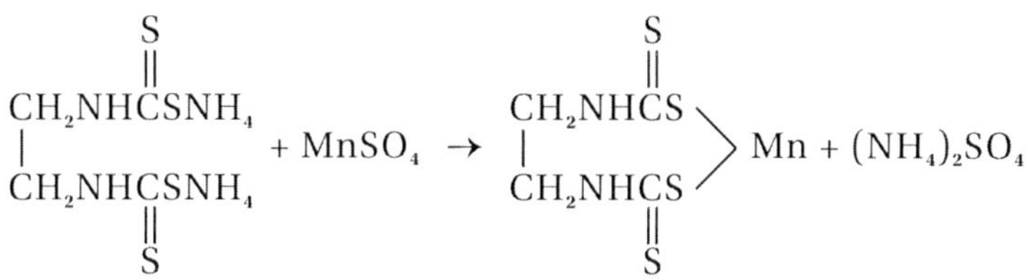

It is a yellow crystalline substance with a molecular weight of 265.3. It is insoluble in water and organic solvents. The pure compound breaks down at about 120°C while the technical grade product breaks down below 100°C. When stored in bulk, maneb is capable of spontaneous decomposition with charring. However, this hazard is reduced when the compound is diluted with an inert diluent. At the same time preparations with a low content of active material decompose very rapidly at an elevated temperature and in the presence of moisture. As such, drying of the compound should not be done at temperatures higher than 45°C. Regarding compatibility, maneb is similar to ferbam. In addition to maneb, various analogues and combinations like Dithane M-45, (10% zineb plus 70% maneb), Dithane C-31 (maneb plus nickel sulphate), Propineb (zinc 1,2-propylene bis dithiocarbamate), Carbathene (or Thioneb) (mixture of 80% polyethylene bis thiuram disulphide and 20% polyethylene bis thiuram monosulphide) have been used as fungicides.

Vapam (SDMC, VPM, Chemrape, Sodium N-methyldithiocarbamate)

Vapam was introduced by the Stauffer's Chemical Co. of U.S.A. in 1954. Besides fungicidal properties, it has nematicidal as well as herbicidal properties. It is soluble in water and only moderately soluble in alcohol. In the soil, vapam releases methylisothiocyanate CH_3 N=C=S. The mode of action of vapam is similar to ethylene bisdithiocarbamates though structurally it is similar to dialkyl-dithiocarbamates.

Mode of Action

Dialkyldithiocarbamates: As dialkyldithiocarbamates are strong chelating agents, they were thought to act by depriving the cell of the needed metal. Recent information however indicates that a heavy metal ion is required for the high toxicity of these fungicides. For example two complexes are formed when copper ions interact with dimethyldithiocarbamates. The complexes are the unsaturated, positively charged 1:1 complex and the saturated uncharged 1:2 complex. According to Goksyn, the 1:1 complex is responsible for the toxicity of dialkyldithiocarbamates at low concentrations as this presumably attacks the vital components of the cell and thus prevents growth. Owens suggested that free radicals rather than the dithiocarbamate ions may be the active forms of ferbam and thiram.

It has been suggested that these are toxic by virtue of their reaction with cellular *thrills* which in turn results in ion interference of the activities of sulphydryl requiring enzymes. The diethyldithiocarba-mate and other toxic chelating agents catalyse incorporation of copper into the diethylester of mesoporphyrin. Lowe and Phillips suggested that the mode of action of these compounds might be associated with the incorporation of copper into a prophyrin precursor of an essential haemintype pigment. Bates and Tweedy reported that thiram has no effect upon glycolysis but did inhibit the activity of the pentose pathway and acetate metabolism.

According to Sisler, the toxicity of dialkyldithiocarbamates in the cell might involve one or more of the following:

(i) Chelation of required heavy metals;

(ii) Attachment of a 1:1 complex of metal and dithiocarbamate ions to enzyme;

(iii) Attachment of dithiocarbamate ion to metals bound to protein:

(iv) Reaction of free radical intermediate with cellular compounds;

(v) Lethel catalysis as suggested by Lowe and Phillips.

In spite of all the above possible mechanisms the actual mode of action of dialkyldithiocarbamate in fungus cells remain obscure.

Monoalkyldithiocarbamates: Introduction of hydrogen at the N-atom in a dithiocarbamate structure considerably reduces its chemical stability. The dithiocarbamates having free H and the nitrogen can split off H_2S or the HS ion with the formation of the isothiocyanate group.

$$R\text{—}\underset{H}{N}\text{—}\underset{S}{\overset{\|}{C}}\text{—}SH \rightleftharpoons H_2S + R\text{—}N{=}C{=}S \rightleftharpoons HS^- + S + R\text{—}N{=}C{=}S$$

In the presence of ions of heavy metals, the equilibrium is forced to the right due to the formation of insoluble metal sulphides. Among the two types of decomposition products H_2S (or the HS^- or S^{--}) and the isothiocyanates, only the latter was found to show antifungal activities and spectra similar to those of the parent dithiocarbamates.

The toxicity of these dithiocarbamates is possibly carried out via an isothiocyante reacting with free SH group in the following manner

$$\mathrm{R{-}N{=}C{=}S + HS{-}Enzyme \rightleftharpoons R{-}\underset{\displaystyle H}{\underset{|}{N}}{=}C{=}S{-}Enzyme}$$

Ludwig and Thorne found ethylene thiuram monosulphide (ETM) among the decomposition products of nabam and thought that ETM might account for the toxicity of nabam solution. Further it was postulated that the compound was formed via an isothiocyanate intermediate and that the two compounds were in equilibrium. It has been suggested that oxidative pathways of carbohydrates metabolism may be involved since ETM at 10^{-5} M produced 77% inhibition of glucose-6-phosphate dehydrogenase and 80% inhibition of 6-phosphogluconate dehydrogenate. Such a plausible explanation is lacking for thiram and other dimethyl dithiocarbamates that do nor obviously form isothiocyanates. However, thiram at 10^{-5} M solution, inhibited glucose-6-phosphate dehydrogenase up to 77% and 6-phosphogluconate dehydrogenase up to 72%, being approximately as active as this *in vitro* test as the ETM. Therefore it is difficult to separate the action of the two types of dithiocarbamates. Thiram can be reduced to dimethyldithiocarbamic acid and it is not possible that the latter may react with the -SH group of the essential enzymes to form disulphides. Formation of volatile mesoisothiocyanate from Vapam also supported the isocyanate theory. However, there is evidence that the isothiocyanates may not entirely be responsible for the toxicity of monoalkyldithiocarbamates. Hence the isothiocyanate theory has been disputed. Much more work needs to be done to understand the mechanism of action of monoalkyldithiocarbamates.

Structure-activity relationships: Since the dimethyldithiocarbamates and the ethylene bisdithiocarbamates have different modes of action, it is difficult to make any generalisations regarding the effects of structural changes. Methyldithiocarbamic acid is a weaker fungicide than dimethyldithiocarbamic acid, while in the corresponding ethylene bis dithiocarbamates replacement of the hydrogen atom by the methyl group greatly decreases the fungicidal activity. Increasing the size of the alkyl groups on the nitrogen atom of dialkyldithiocarbamates greatly decreases the toxicity. Similarly increasing the number of methylene groups between two thiocarbamyl nitrogens of bisdithiocarbamates also decrease the activity. Replacement of C=S by C=O in dimethyldithiocarbamates decreases their activity.

CHAPTER XII

COPPER COMPOUNDS

Among the fungicides used for the control of diseases, copper and sulphur fungicides occupy the prime position. As early as 1761, Schultess used copper sulphate ($CuSO_4.5H_2O$) for treating wheat seeds. Though Prevost recognised the fungicidal properties of copper sulphate as early as 1807, it was used for the protection of foliage only in 1885. Widespread use of copper resulted from an accidental observation by Millardet of France in 1882 about the efficiency of copper sulphate for the control of downy mildew of grapes caused by *Plasmophora viticola.* Most of the roadside vines were being sprayed with copper sulphate to prevent stealing of grapes. At the same time, the roadside vines was found to be free of the disease. This gave the clue to Millardet and his colleague Prof. David that these mixtures could be responsible for the control of mildew disease.

Bordeaux Mixture

A mixture of copper sulphate and lime was initially applied as a paste and this gained wide recognition under the name "Bovillie Bordelaise" (Bordeaux mixture). As the initial mixture containing 8 kg of blue stone dissolved in 100 litres of water and 30 litres of lime suspension prepared from 15 kg lime, was found to cause foliar injury, they tried various combinations. Now a 4-4-50 mixture (copper sulphate in lb : lime in lb : water in gallons) is used but the concentration of the ingredients is varied, depending on the purpose. However, the proportion of copper sulphate to lime usually remains constant, to give a slight excess of lime over the chemical equivalence. In India Bordeaux mixture is being made by preparing a solution of copper sulphate and quicklime (or hydrated lime) in finely ground form in separate containers and then mixing them simultaneously into a third container with constant agitation. The containers should be such that they do not react with copper-sulphate like copper, wooden and earthern vessels. Air slaked quick lime is unsuitable for the preparation of this mixture. Wide variation in the composition of the resulting mixture will be observed due to the chemical reactions between calcium hydroxide and copper sulphate in solution when the ratio between the components is changed. Earlier Bordeaux mixture was divided into three types: (i) acid Bor-deaux-containing small amounts of soluble copper and having a faintly acid reaction, (ii) neutral Bordeaux—the ratio of copper sulphate to calcium oxide being 1 : 0.3 and (iii) alkaline Bordeaux in which the ratio of copper sulphate to calcium is 1 : 0.5 or greater. However, this system is not being used now.

The active constituents consist largely of a voluminous precipitate of tetracupric sulphate ($4CuO.SO_3$) and pentacupric sulphate ($5CuO.SO_3$). However, others have suggested the formation of series of compounds with the general formula $Cu_4(OH)_6 SO_4 x Ca(OH)_2$ since the composition of the precipitate was dependent on the amount of calcium hydroxide added per unit of copper sulphate. It was found that when the ratio of copper sulphate to calcium oxide was 1:0.166—a double salt having composition $[Cu(OH)_2]_3 CuSO_4$ was formed, with a ratio of 1 : 0.181—$[Cu(OH)_2] CuSO_4$, 1 : 0.20— $[Cu(OH)_2]_9 CuSO_4$, 1 : 0.269—$[Cu(OH)_2]_9 Ca(OH)_2 CuSO_4$ and 1 : 1.00—$[Cu(OH)_2]_9 (CuSO_4) [Ca(OH)_2]_3 CaSO_4$.

Initially Millardet and Gayon thought the reaction to proceed according to the following equation:

$$CuSO_4 + Ca(OH)_2 \rightarrow Cu(OH)_2 + CaSO_4$$

However, the mechanism does not appear so simple. Possibly number of reactions take place simultaneously:

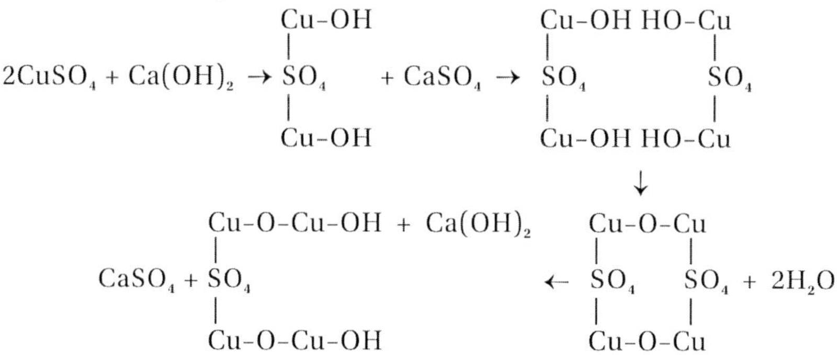

The basic copper sulphate is formed and this remains in suspension. Lime neutralises the acid of the copper sulphate solution, complete neutrality occurring when 0.75 equivalent of calcium hydroxide combines with one equivalent of copper sulphate. This results in the formation of a basic sulphate and the addition of further $Ca(OH)_2$ may decompose the basic sulphate and form a series of copper compounds. The cupric hydroxide formed from the basic sulphate by the reaction of calcium hydroxide is stabilised by the absorbed sulphate ion and if washed repeatedly with CO_2-free water, it undergoes dehydration to form a brown cupric oxide.

In mixtures containing 0.125% copper sulphate and above, the gelatinous precipitate becomes crystalline after standing for a short time. The crystallisation occurs even when an excess of calcium hydroxide is present and obviously affects the adhesiveness and the rate of settling of the precipitate. Crystallisation is retarded by the presence of such impurities as ferrous sulphate, calcium carbonate and magnesium oxide or carbonate. The physical state of the precipitate of Bordeaux mixture depends to a large extent on the dilution of the salts used and the manner in which they are brought together. The most readily suspended precipitate is formed when a dilute solution of copper sulphate is added to a concentrated suspension of lime.

Freshly made Bordeaux mixture is of a gelatinous nature. The colour is a deep blue to slightly greenish. The presence of calcium carbonate in the lime hastens the rate of settling and reduces the volume of the precipitate. Therefore, the proportion of calcium carbonate should not be more than 25%. After preparing the Bordeaux mixture, it should be tested for free copper as it is phytotoxic. This is done by adding a few drops of 10% solution of potassium ferrocyanide to a small quantity of the prepared Bordeaux mixture. If a reddish brown deposit is formed immediately, more lime should be added to the mixture. If potassium ferrocyanide is not available, a clear iron blade could be dipped inside the mixture for sometime and then the presence of copper deposit observed.

Bordeaux mixture should be sprayed while fresh. On standing, it looses its fungicidal property. However, the mixture may be stabilised and kept for several days by adding sugar or jaggery at the rate of 1/2 lb in 50 gallons of mixture. When Bordeaux mixture is applied to the plant surface and allowed to dry, the deposit does not undergo visible decomposition. A reaction between carbon dioxide of the air and dried deposit of the basic sulphate and hydroxide results in the formation of the less highly hydrated brown form of copper oxide. A brown copper oxide is generally formed when the Bordeaux mixture precipitate is first dried in an atmosphere free from CO_2 and then washed with CO_2-free water. The exact nature of the products formed by the reaction has however not been determined. Possibly the reaction between the precipitate and atmosphere CO_2 causes the formation of a highly resistant deposit.

Sprayed films of Bordeaux mixture undergo continuous change in composition under the influence of rain and dew. When deposits of Bordeaux mixture were exposed to rainfall, the toxicity of the deposits was gradually lessened while those stored in the laboratory for the same period did not loose their toxicity. In freshly deposited Bordeaux mixture residue, a component having available copper is present and this is exhausted by washing with rains but not with ageing. After it is lost, conidia that come in contact with the residue are inhibited in germination but are not killed and can germinate freely when removed from contact. While it is still present, conidia take up copper before any visible evidence of vital activity is observed. Possibly the conidial cells are not capable of secreting a substance which will dissolve the inert copper of these washed residues.

The inherent solubility of the Bordeaux mixture is quite low as it yields a solution containing only 4 ppm of copper when equilibrated with water. After the materials are thoroughly dried, as in a spray film, the solubility decreases to 0.2 to 0.3 ppm. As the concentration of copper is less than 1 ppm, it is too low for any substantial toxicity. A number of mechanisms have been suggested to explain the possible ways in which copper is brought into soluble form in quantities sufficient enough to be toxic to spores. One of the theories suggested the action of atmosphere CO_2 and ammonium salts dissolved in the rain water and dew on the deposits of Bordeaux mixture. Millardet and Gayon observed the carbonation of excess of lime and soluble copper by the action of the CO_2 and also by the slight traces of ammonium salt present in rain, water and dew. However, this view is not accepted now. The second mechanism suggested is the action of the secretions of fungus in solubilising the copper. McCallan and Wilcoxon demonstrated the presence of maleic acid and of the amino acids in the washings from the spores of *Neurospora sitophila.* As the excretions of these spores are almost neutral, possibly the salts of hydroxy acids such as malate and of amino acids present in the spore excretions act on the Bordeaux deposit to form soluble toxic copper hydroxy and copper amino salts. Another theory suggested the bringing of copper into soluble form by the host plant secretions. Barker and Gimingham showed that substances exuded from the mechanical injuries to the leaf are capable of dissolving copper from Bordeaux precipitates.

However, the solubilisation hypothesis as a whole has recently been questioned. Somers showed that the formation of copper complexes from the reaction between leaf or spore exudates with a Bordeaux deposit reduces the uptake of copper by the spores of *Neurospora crassis,* suggesting that these water soluble complexes are less effective in penetrating spores than is the free cupric ion. So, he suggested that simple accumulation of ionic copper is the mechanism by which copper from Bordeaux mixture becomes toxic to the spores.

Spraying with Bordeaux mixture deepens the green colour of the foliage possibly due to an increase in the total amount of chlorophyll produced or to the slow removal of products of photosynthesis. Enough copper is also absorbed by leaves sprayed with Bordeaux to kill the nymphs and adults of potato leaf hopper. Bordeaux mixture sprays affect transpiration through both the cuticle and the stomato of the leaves, accelerating the former and retarding the latter. The normal transpiration of a plant may thus be either stimulated or depressed according to the location where it is grown and also the various physical factors.

Due to the presence of an excess of calcium hydroxide, Bordeaux mixture cannot be combined with soaps or organic insecticides such as pyrethrum which are decomposed by free alkalies. However, Bordeaux mixture is compatible with nicotine, lead and calcium arsenate, DDT and oil emulsion for dormant trees unless soap has been used as the emulsifiers. If an arsenical is desired in the spray mixture, either lead arsenate or calcium arsenate may be added. On long standing sulphur will form calcium polysulphide and calcium sulphide when combined with Bordeaux mixture containing excess hydrated lime. Bordeaux mixture is a fairly efficient emulsifier and has been used in combination with various oils at different times.

The "Pickering type of Bordeaux mixture" substitutes saturated lime water for either slaked calcium oxide or hydrated lime that is used in the standard type of bordeaux mixture. Bordeaux mixture is preferred because of its natural adhesiveness or tenacity, its relative cheapness, its ability to control a wide variety of diseases and comparitive safety in handling. However, it exhibits phytotoxicity on certain crops, particularly those like apples and peaches. In certain cases it has a tendency to delay ripening. Further, the preparation of the mixture is laborious and its corroding action on metallic spraying containers has made workers switch over to other copper compounds.

The fungicidal action of Bordeaux mixture is more protective than eradicative. This is due to the insolubility of the basic compound formed by the interaction of copper and lime and the slow release of active fungicidal material from the spray deposit. Certain difficulties experienced in the preparation and use of Bordeaux has resulted in the modification of Bordeaux.

Dried Bordeaux mixture has been found to be inferior to that made immediately before use. The addition of casein in small amounts to lime has helped in the preparation of a stable mixture. Later, Bordeaux mixture was made in the form of a paste. Another attempt is the "Oil-Bordeaux mixture spray" prepared by the addition of edible cotton seed oil of low acid content and crude oil of mustard. The addition of oil (which contains glycerides—a combination of glycerol and fatty acid forming vegetable oils) to the Bordeaux mixture improved the retention of copper on the foliage and this coupled with the fungicidal value of the oil itself, has considerably helped in reducing the copper in the formula to 1/2 the usual amount. The modified Bordeaux mixture could be applied with greater ease and rapidity than the standard formula and is also a better carrier for insecticides. In certain cases, zinc sulphate was also combined with Bordeaux mixture for use on citrus trees for the control of *Septora citri.* To stablise the oil-water emulsion and also avoid injury, a weak Bordeaux mixture can be added to petroleum oil emulsion. The Bordeaux mixture being of a colloidal nature when combined with soap solution, forms an emulsion. The copper and lime of the Bordeaux mixture may react with the soap emulsifier but when aided by agitation in the spray tank, no appreciable separation of oil occurs. However, a certain amount of oil is absorbed by the lime in the Bordeaux mixture, which in weak emulsions may appreciably retard the kill. To compensate for such loss of oil, a small additional amount of oil emulsion should be used.

Plants sensitive to copper however may show foliage injury and fruit russetting due to Bordeaux. This led to a study of substitutes that are less corrosive. The fixed (insoluble) coppers like basic copper sulphate, copper oxychloride, copper silicates, cuprous oxide and mixtures of these have come into general use because of the higher copper content, compatibility with many insecticides and decreased phytotoxicity. However, the fixed coppers lack the tenacity of Bordeaux mixture and hence an additive is usually added to increase the adhesion.

Fixed copper compounds are usually marketed at 50 to 53% copper as metallic. They are usually formulated at the same copper content as Bordeaux mixture. They may be applied as dust also, although it will increase adhesion. Fineness of division of the pesticide both from the standpoint of adhesion and that of fungicidal value is important. The fixed coppers are largely used for treating vegetables. They may also be alternated with organic fungicides such as ziram. The concentration of the diluted mixture ranges from 1 to 5% copper as metallic. A series of copper fungicides termed as insoluble, low soluble or fixed coppers were tried. In all these, the copper was stabilised and rendered less soluble to reduce the phytotoxicity of copper. These low-soluble copper fungicides like copper oxychloride are much less injurious to plants.

Burgandy Mixture

Burgandy mixture or soda mixture was prepared by Masson by allowing one part of copper sulphate pentahydrate to react with one or two parts of crystalline sodium carbonate in solution ($Na_2CO_3.1O\ H_2O$). It gives a water soluble precipitate, the composition of which varies depending upon the amounts and proportions of the reacting compounds. The reaction between copper sulphate and sodium carbonate results in the formation of a basic carbonate $\{Cu(OH)_2\}_3(CuCO_3)_2$, sodium sulphate and CO_2 which combines with sodium carbonate to form sodium bicarbonate ($NaHCO_3$). The dried residue from this mixture varies from blue to greenish colour but since it is not opaque, does not shade the leaf as does the Bordeaux mixture. The ratio, one part of copper sulphate to one part of sodium carbonate, is satisfactory for all purposes.

Acid burgandy mixtures are produced whenever less than 0.86 parts of crystallised sodium carbonate is combined with one part of copper sulphate and such mixtures contain both basic carbonate and basic sulphate. The basic carbonate is stable as it is converted into malachite $Cu(OH)_2CuCO_3$. An excess of sodium carbonate retards the formation of malachite. In actual practice a large excess of sodium carbonate is as disadvantageous as an excess of soluble copper. Plants vary in their sensitivity to burgandy mixture. The presence of dew and rain drops particularly before drying greatly increases the possibility of injury. Dried burgandy mixture whether they are acid, alkaline or neutral differ very little in their effect on plants. Calcium arsenate can be safely combined with burgandy mixture and this tends to neutralise any soluble copper which may be present. There is some doubt about the safety of combining lead arsenate with the burgandy mixture. Burgandy mixture was prepared when good quality lime was not available. Since it is available now, burgandy mixture has become a thing of the past.

Chestnut Compound

Chestnut compound, suggested by Bawley in 1921, contains two parts of copper sulphate and 11 parts of ammonium carbonate. These two compounds are well powdered, thoroughly mixed and the dry mixture is stored in airtight containers for 24 hours before use. If left exposed, it gradually loses ammonia and becomes less effective.

Copper Oxychloride (Basic chlorides of copper, Blimix 4%, Blitox 50, Cupramar, Micop D-06, Mico P.W-50, Fytolan, Vaigfru Tamraghol, Blue copper 50)

Copper oxychloride is one of the low soluble copper fungicides produced by the action of air on cupric chloride solution or scrap copper.

$$4Cu + O_2 \rightarrow 2Cu_2O$$

$$2Cu_2O + 2HCl \rightarrow 2CuCl + H_2O$$

$$2CuCl + O + 2HCl \rightarrow 2CuCl_2 + H_2O$$

$$CuCl_2 + Cu \rightarrow 2CuCl$$

$$4CuCl_2 + 3CaCO_3 + 3H_2O \rightarrow 3Cu(OH)_2\,CuCl_2 + 3CaCl_2 + 3CO_2$$

It is marketed in the form of wettable powder containing 50 and 90% copper oxychloride and dusts containing 4 to 12% metallic copper. The 50% formulation contains a diluent, usually kaolin, and a surface active agent.

This basic copper carbonates vary in their composition depending upon the ratio and nature of the ingredients. Carbonates are preferred to hydroxides for such combinations. The basic sulphates are marketed under various names like Tribasic, Basicop, Copper hydroxysulphate, Spraycop and Coppatone.

Copper Sulphate ($CuSO_4.5H_2O$, Blue stone, blue vitriol)

Copper sulphate is blue in colour and is readily soluble in water. The solution in slightly acid and too dangerous to the foliage and fruits if used alone. Monohydrated copper sulphates are also used. Copper lime dusts are well adapted to power application and compare favourably with standard Bordeaux mixture in effectiveness.

Copper Oxides

Horsfall in 1932 suggested the use of red oxide of copper as a seed treatment fungicide. The fungicidal properties of both cuprous and cupric oxides is influenced by the particle size. On equal particle size basis cuprous oxide inhibits spore germination to a greater degree than cupric oxide of the same copper content.

Cuprous oxide (Fungimar, Perenox, Cu_2O) is produced by the precipitation from the water soluble salt of copper in alkaline-reducing solution. The reducing agent is usually formaldehyde or glucose. Cuprous oxide is a bright red coloured powder, practically insoluble in water and organic solvents. It is rather stable in storage. Wettable powders of cuprous oxide are prepared by drying in a spray dryer an aqueous suspension of cuprous oxide containing sulphite waste liquid as a surface-active agent. Usually a light dusty form, free from oil is preferred for use as a fungicide. The finer the particle, the more toxic is the compound. The fungicidal and protective values were found to increase as the colour of the cuprous oxide changes from red through orange to yellow, possibly due to an increase in the area of chemically active surface per unit weight and a corresponding increase in the rate of release of soluble copper.

Copper carbonate preparations are produced by combining soda ash with hot copper sulphate solution.

$$2CuSO_4 + 2Na_2CO_3 \xrightarrow{H_2O} 2Na_2SO_4 + CO_2 + CuCO_3.Cu(OH)_2$$

When copper sulphate reacts with chalk, a product consisting of the basic sulphate and basic carbonate of copper with gypsum as a contaminant, is obtained. It was used as a seed disinfectant. Wheat treated with copper carbonate dust was not injured by the chemical even when held in storage for an indefinite period. Another preparation of copper carbonate is sold under the name of Chaubattia paste in U.P. and it is made by mixing 800 g each of copper carbonate and red lead with one litre of lanolin or raw linseed oil. The paste is reported to be much more resistant to washing by rain.

Miscellaneous Copper Compounds

Number of other copper compounds like copper ammonium zeolite, copper oxychloride sulphate, copper phosphate, copper silicate (Coposil), copper ammonium fluoride, neutral acetate of copper, copper naphthanate, copper resinate, double salt of copper sulphate and dihydrazine sulphate have been tested as fungicides. Of these, neutral acetate of copper $\{Cu(C_2H_3O_2)_2H_2O\}$ and basic acetate of copper $\{Cu(C_2H_3O)_2CuO.6H_2O\}$ containing 31.8 and 34.4% copper as metallic copper have been found to be less injurious to the plant and have been rated as very near to Bordeaux mixture in safety to host plants, and in the fungicidal properties.

Organic forms of copper salts when used in oil emulsion are decidedly superior both in penetration and wetting ability to inorganic forms such as copper sulphate whether in Bordeaux or burgandy mixture. The organic copper salts usually form a true solution particularly in pine tar oils and hence are carried directly into the leaf and stem tissues wherever the oil penetrates. However, these have limited commercial uses.

Phytotoxicity. Copper in Bordeaux mixture sometimes has an injurious effect on plants. This is characterised by chlorosis followed by brown or purple flocks on the leaves and fruits. In the case of leaves, the leaf veins develop purpling and the leaves fall off. The effect of frost damage was found to be more severe on potatoes and tomatoes sprayed with copper fungicide like Bordeaux mixture than those sprayed with non-copper fungicides like zineb. Compared to the Bordeaux mixture, copper phosphate and basic copper sulphate are more inferior in toxicity while copper oxychloride, copper oxalate and copper zeolate are intermediate. Among the various copper fungicides tested, no copper compound has high fungicidal value with low phytocidal properties. Bordeaux mixture is the best possible compromise available now.

CHAPTER XIII

MERCURY FUNGICIDES

Mercuric Chloride

Mercuric chloride was used as bactericide during the early nineteenth century. Kellerman and Swingle in 1878 tested mercuric chloride ($HgCl_2$) for the first time as a fungicide, for seed treatment on cereals. Hiltner, a German worker, successfully used mercuric chloride for the control of *Fusarium* disease on rye in 1910. However,, its high toxicity to man, animals and plants made it necessary to seek more active and safer compounds for use in agriculture. Mercuric compounds can be divided into inorganic mercurials and organic mercury fungicides. Among the inorganic mercury compounds, mercuric chloride and mercurous chloride have been widely used in India for protecting potato tubers and propagative materials of other root crops from disease during germination. In India, Burns in 1914 used mercuric chloride for controlling *Rhizoctania* in potato tubers.

Mercuric chloride ($HgCl_2$) is a corrosive sublimate. It is a white crystalline substance with a specific gravity of 5.42, m.p. of 265°C and b.p. 303°C. It is soluble up to 5.7 and 54 parts in cold and hot waters respectively. It decomposes in the presence of hydroxides including sodium, calcium and potassium. Generally the seed materials are dipped in 1 : 2000 or stronger solution of mercuric chloride for a few minutes and then planted. Sterilisation of seed beds with mercuric chloride has been widely practiced. In view of the availability of organomercurials, the use of these compounds have declined.

Mercurous Chloride

Mercurous chloride ($HgCl$) also known as Calomel, is much safer for plants than mercuric chloride. Mercurous chloride is very insoluble in water and hence easier to handle than the mercuric form.

Mercury also known as quick silver, is a metallic liquid, very slightly volatile, insoluble in water but soluble in acids. Mercury vapour when confined, is toxic to germinating seeds and to infesting insects.

Though inorganic mercury compounds have been in use for a long time, because of their extreme toxicity to plants, seeds and human beings, they are being replaced by organomercury compounds. The organomercurials are best for controlling externally seed borne smuts. The first important organomercury compound was suggested by Wesenberg of I.G. Farben Industries in Germany during 1913 as a seed disinfectant. The first commercial product called "Uspulam" containing 18.8% mercury in the form of chlorophenylmercury $\{Cl(OH)C_6H_3.Hg).SO_3.Na\}$ was introduced in 1915 by Bayer and Co. Later, a number of compounds like Germisan containing 16% metallic mercury in the form of Crysyl mercuric cyanide $\{(HO)(CH_3)\ (C_6H_2)HgCN\}$ and Tillantion have appeared in the market. Organic mercury dusts for dry seed treatment were introduced in 1924 with 0-nitro-phenol mercury $\{HOC_6H_3.NO_2.HgOH\}$.

The organomercurial compounds considerably surpass inorganic mercury compounds in bactericidal and fungicidal action. Many of the organomercury compounds at the concentrations used are not only harmless to plants but even stimulate their growth. Sometimes the stimulation is so great that it leads to a substantial increase in yield. The insecticidal effect of organomercury compounds is negligible. Organomercurials can be represented by the general formula RHgX and R_2Hg where R is an aryl, aryloxy, alkyl or alkoxy ethyl groups and X is an organic or inorganic anionic radical like chloride, acetate, lactate, urea carbamate, hydroxy or related structures. Strong microbiological activity is shown by the first group of compounds RHgX. While symmetrical compounds of mercury R_2Hg are practically inactive as fungicides, the nature of R group appears to regulate the transport and stability of the compound and the anionic group determines its solubility. Booer (1944) has described ethyl- and methoxyethyl mercury radical as resembling sodium ion in that they are strongly alkaline and form highly ionised salts that are generally water soluble and appreciably volatile.

Organomercurials can be further divided into compounds of aliphatic series and aromatic series, the former being more powerful fungicides than the later. In the aliphatic series, the fungicidal activity of a compound decreases with an increase in the size of the hydrocarbon radical bound to the mercury. A change in the structure of the *cis* group also affects the toxicity of the compound since its physical properties, like solubility in water and organic solvents, are modified. The fungicidal activity of the compounds depends upon functional groups in an aliphatic radical, a carbonyl group in 2-position decreases the fungicidal activity much more than the methoxy group in that position. An increase in size of the hydrocarbon radical also lowers the activity of the compounds. Polycyclic derivatives of mercury are less active. Addition of various acid groups to the mercury atom also modifies the toxicity of the compound to mammals.

ORGANOMERCURY COMPOUNDS OF THE ALIPHATIC SERIES

When the number of mercury compounds used in agriculture is considered, the compounds of the aliphatic series occupy a dominant place. Generally these compounds RHgX are produced by reacting a dialkyl mercury with the appropriate mercury salt:

$$R_2Hg + Hg(Ac)_2 \rightarrow 2RHg\ Ac.$$

The methyl and ethylmercury derivatives have the highest fungicidal activity. Most of these compounds are white crystalline compounds, soluble in water, moderately soluble in most organic solvents. Most of the compounds produce blisters when they come in contact with the skin.

Among the various compounds used as fungicides, the following are widely used: ethyl mercuric chloride, ethyl mercuric phosphate, methyl mercuric sulphate, methyl mercuric cyanamide, Ceresan-M, Panogen, Agalol, Ceresan—Universal Trackenbeize, Emmi, Metazol and Memmi.

Ethyl Mercuric Chloride (C_2H_5HgCl)

It is sold as Granosan (2% ethyl mercuric chloride), Mercuran (2% ethyl mercuric chloride plus 12-14% of γBHC), and mercurhexan (1% ethyl mercuric chloride, 20% BHC, and 20% γBHC). It is resistant to the action of acids but readily reacts with the reducing agents, forming diethyl mercury and metallic mercury. The LD_{50} value of ethylmercuric chloride is 18-30 mg/kg.

Ethyl mercuric phosphates surpasses mercuric chloride in fungicidal property. Usually the total mercury content of the formulation does not exceed 10%. Its LD_{50} value is about 30 mg/kg. Methyl mercuric sulphate is also a seed disinfectant and decomposes on heating to 260°C. It is marketed as formulations containing 0.8 and 1.2%.

The methyl mercuric cyanide (also called Chipcote) is much more toxic than methyl mercuric dicyanaodiamide. It is marketed as aqueous solution in ethylene glycol and with a mercury content of 1.53 and 4.5%. All the above mercury fungicides are produced by reacting dialkylmercury with the appropriate mercury salts.

Ceresan-M {N-(ethyl mercuric)-p-toluene sulphonanilide}

Ceresan-M is widely used as a grain seed disinfectant in the U.S.A. It is produced by reacting ethylmercuric salts with p-toluene sulphonanilide in the presence of NaOH.

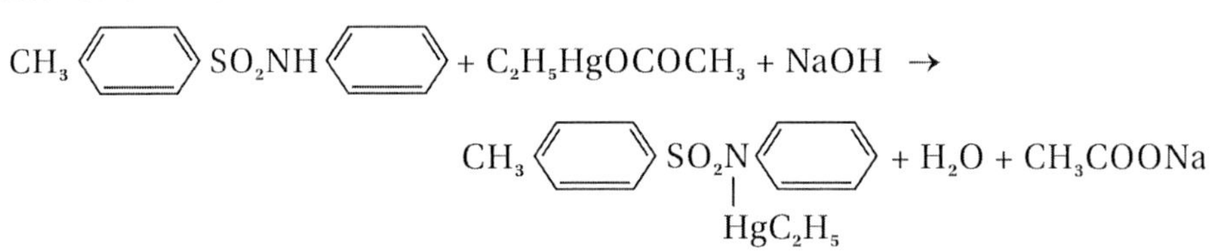

When ethyl mercuric chloride or ethyl mercuric acetate is used, the yield is about 85%. It is also prepared by reacting acetic acid with tetraethyl lead and mercuric oxide. The ethyl lead acetate that is formed is removed by filtration and the aqueous solution of ethylmercuric acetate is reacted with p-toluene sulphonanilide.

$$(C_2H_5)_4Pb + HgO + 2CH_3COOH \rightarrow$$

$$C_2H_5HgOCOCH_3 + (C_2H_5)PbOCOCH_s + H_3O$$

Its LD_{50} for rats is about 50-70 mg/kg.

Panogen (Methylmercuric Dicyanodiamide)

Panogen is being sold as Panogen 15 (2.2% methyl mercuric dicyanodiamide) and Panogen 42 (6.3% methyl mercuric dicyanodiamide) for disinfecting seeds of various crops. It is produced by the action of methyl mercuric hydroxide on dicyanodiamide.

$$CH_3HgOH + NH_2\text{-}\underset{\underset{NH}{|}}{C}\text{-}NHCH \rightarrow CH_3HgNH\text{-}\underset{\underset{NH}{|}}{C}\text{-}NHCN + H_2O$$

In chemical properties it is similar to ethyl mercury salts. Its LD_{50} for rats is 45 mg/kg.

Agalol (Ceresan-Universal Nassbeize, Aretan, Methoxy ethyl mercuric chloride)

Agalol is produced by precipitation of the chloride salts from solutions of methoxy ethyl mercuric acetate and sodium chloride.

$$CH_3O.CH_2CH_2HgO.CO.CH_3\text{+}NaCl \rightarrow CH_3OCH_2CH_2HgCl + CH_3COONa$$

It is decomposed by acids with evolution of ethylene.

$$CH_3CH_2CH_2Hg\ C1\text{+}HCl \rightarrow HgCl_2 + CH_3OH + CH_2 = CH_2$$

Its LD_{50} value is 50 mg/kg.

Similar to Agalol, Ceresan-Universal Trockenbeize (Methyl ethyl mercuric silicate) is produced by precipitation from a solution of methoxy ethyl mercuric acetate with sodium silicate

$$3CH_3O.CH_2CH_2HgOCOCH_3Na_3HSiO_4 \rightarrow$$
$$(CH_3OCH_2CHHg)_3HSiO_4\text{+ } 3CH_3COONa$$

In chemical properties, it is similar to those of methoxy ethyl mercuric chloride. Ceresan-Universal is sold as a formulation containing 1.75% mercury for disinfecting seeds.

Other mercury compounds of the aliphatic series are Memmi {N- (methyl mercuric) 3,4,5,6,7,7-hexachloro-3,6-endomethylene-l,2,3,6-tetrahydro-phthali-mide}, Emmi {N-(ethyl mercuric) 3,4,5,6,7,7-hexachloro-3,6-endo methylen-l,2,3,6-tetrahydrophthalimide} and Metazol (methyl mercury-(3-hydroxy quinolite). Memmi and Emmi are synthesised by the action of respective sodium derivatives of the imide of chlorendic acid (3,4,5,6,7,7-hexachloro-3,6-endomethylene- 1,2,3,6-tetrahydrophthalic acid) with mercuric chloride in acetone or in polar solvents. Both are marketed as 10% wettable powders and emulsive concentrates. Their LD_{50} values are around 160 mg/kg. Metazol is synthesised from sodium 8-hydroxy quinolate and methyl mercuric chloride. It is a golden yellow crystalline compound and is formulated as a water-alcohol solution containing 1.3% Hg with a red dye added. The LD_{50} for mice is 72 mg/kg.

ORGANOMERCURY COMPOUNDS OF THE AROMATIC SERIES

The first organomercury compound "Uspulan" suggested for seed treatment belongs to the aromatic series. These are produced by the mercuration of phenols, cresols, chlorophenol and similar compounds. Only some of them are used now. The mercuric phenols have been replaced by other aryl mercury derivatives.

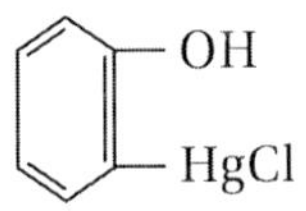

Uspulan

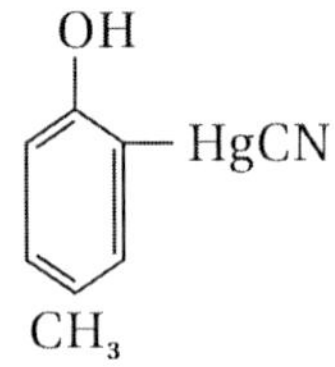

Germisan

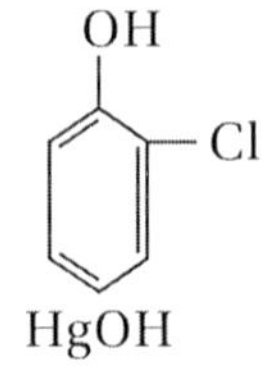

Semesan

Phenyl mercury derivatives have the greatest effectiveness of all the aromatic mercury compounds. These are produced by simple methods like direct mercuration of benzene when mercuric acetate or other mercury salts of carboxylic acids are heated with benzene at temperatures higher than 130°C in the presence of some catalysts.

$$C_6H_6 + (CH_3COO)_2Hg \rightarrow \underset{\text{Ceresan}}{C_6H_2HgOCOCH_3} + CH_3COOH$$

One of the important mercury compounds of the aromatic series, Ceresan (Phenyl mercuric acetate, PMAC, FMA, TAG, Kwissan) is produced by the above method. It is a white crystalline compound soluble in water up to 24.7 g/ litre, more soluble in alcohol, acetone, benzene and ethylalcohol. The compound is resistant to the action of mineral acids and alkalies at room temperatures. On heating to 200-250°C, it breaks down to yield metallic mercury, biphenyl and phenyl acetate. The LD_{50} for experimental animals is 17-20 mg/kg. Like most mixed organomercury compounds PMAC also causes painful blisters.

Phenyl mercury triethanol ammonium lactate (Puratized) has been used practically as curative fungicide. It is produced by reacting phenyl mercuric lactate with triethanolamine.

$$C_6H_5HgOCOCH(OH)CH_3 + N(CH_2CH_2OH)_3 \rightarrow C_6H_5HgN(CH_2CH_2OH)^+ \, CH_3CH(OH)COO^-$$

This compound is also a white crystalline substance, highly soluble in water and hydrophilic organic solvents. It reacts with reducing agents to form diphenyl mercury and metallic mercury and also breaks down quickly in sunlight. Its LD_{50} value for rats is 30 mg/kg.

In India the organomercurial fungicides are based on one or more of the following compounds currently manufactured in the country namely ethyl mercury chloride, phenyl mercury acetate, and methoxy ethyl mercury chloride. The seeds are treated with either dry or wet or in slurry form for controlling seed borne smuts and other seed and seedling diseases. The slurry treatment is preferred as there is less handling hazard. Generally, commercial preparations recommended for dry seed treatment contain 1% metallic mercury and the rate of application is 0.25%. For wet seed treatment, preparations containing 2.5 to 6% Hg and at rates of applications varying from 0.2 to 0.5% are used.

In soils, all organomercury compounds are decomposed to mercury salts or to metallic mercury which are active fungicides. Booer suggested that this decomposition takes place through base exchange to form organomercury clays which subsequently form mercury salts by further base exchange. These mercuric salts arc then reduced to mercurous salts and then to mercury.

$$(C_2H_5Hg)_2\text{—Clay} \rightarrow (C_2H_5)_2Hg + Hg\text{=Clay}$$
$$Hg\text{=Clay} + CaO \rightarrow Ca\text{=Clay} + HgO$$

$$2CH_3OCH_2CH_2HgOC(=O)CH_3 + \begin{matrix}H\\H\end{matrix}\!\!>\text{Clay} \rightarrow (CH_3OCH_2CH_2Hg)_2\text{=Clay} + 2CH_3COOH$$

$$(CH_3OCH_2CH_2Hg)_2\text{=Clay} + 2CH_3COOH \rightarrow Hg\text{=Clay} + 2CH_3OH + 2C_2H_4 + Hg(OC(=O)CH_3)_2$$

Possibly the metallic mercury liberated in the soil is ultimately converted to mercury sulphide by reaction with H_2S liberated by soil microorganisms.

The reducing agents produce symmetrical compounds and metallic mercury:

$$2RHgX + 2H \rightarrow R_2Hg + Hg + 2HX$$

Reaction occurs in the soil under the influence of microorganisms and as a result, the duration of fungicidal effect of organomercury compounds is not very great. Depending on the nature of the soil and concentration of the organomercury compounds, the decomposition takes place over a period of 1 to 4 months.

Oxidising agents are also capable of breaking down organic-mercury compounds with splitting off of hydrocarbon radical and formation of inorganic mercury compounds. In the soil the final product of the reaction is mercuric oxide which is further converted to mercury sulphide.

$$RHg + O + H_2O \rightarrow ROH + Hg + OH$$

The mercury sulphide is the least toxic mercury compound for both animals and microorganisms.

At room temperature, mixed organomercury compounds can be kept without decomposition for an unlimited time. Only on prolonged heating with concentrated mineral acids, the organomercury compounds breakdown with the formation of corresponding hydrocarbon slowly.

Mode of Action

The mechanism of toxicity of mercury is not very specific and the mercurial fungicides all owe their activity to Hg^{2+} moiety. In organo-mercurials, the alkyl or aryl portion serves to conduct the Hg^{2+} to the site of action by reason of lipoid solubility and also determines the stability and rate of release of the mercury ion. The biochemical action of mercury fungicides is related to the affinity of Hg^{2+} for the sulphydryl groups of essential respiratory enzymes. Thus the mercury treatments decrease the oxygen uptake of treated fungus spores and poisoned spores can be revived by subsequent application of sulphydryl compounds such as glutathione or cysteine. The specific mercury chelating agent BAL or 2,3-dimercapto-1-propanol is also effective in protecting fungus spores against mercury.

Organomercurials are generally more toxic than the inorganic mercurials which facilitate diffusion through spore membranes to the site of action within the spores. Organomercurials act as good contact fungicides because of their volatility. Mercurials cover the seeds in two steps: (i) they get distributed on the seed in the treating equipment and (ii) they get redistributed by evaporation and resorption on the seeds. In the second process, the chemical tends to leave locations where it is abundant to locations where it is scarce so that there is uniform coverage. Among the two kinetic processes, viz. evaporation and diffusion, the latter is the rate controlling step in the movement of mercurials.

Seed injury is considered to be one of the hazards in the use of organomercurials. It is more pronounced at higher rates since it increases in direct proportion to increasing rates of application. Seed injury caused by organomercurials to cereals shows an abnormal germination. The effect is a characteristic hypertrophy of the roots and the coleoptile of cereal seedlings when higher dosages of fungicides are used or storage conditions are faulty. The primordia undergo extension, thickening and develop irregular carnation and lobes and in the apical meristem of the plumule, cell division is inhibited and extreme enlargement of existing cells takes place. The cells of the hypertrophied organs become multinucleate containing nuclei ranging from micronuclei to polyploid, giant nuclei, the number of chromosomes in these often exceed 200. Sass considered the malformation to be due to incomplete meiosis.

The organomercury compounds are of high to moderate acute toxicity to animals. On chronic exposure, the mercury compounds are highly toxic and the nature of organic radical determines to some extent the amount of tissue storage and damage in the liver and kidney, (inorganic mercury <aryl mercury> alkyl mercury). In man long exposure to organomercurials causes nervous symptoms like tremors, incoordination, disturbance of hearing, irritability, and severe mental disturbance. Mercury compounds can enter the body through skin absorption, by mouth and the respiratory tract. The appreciable volatility of some alkyl mercury compounds increases the hazard of inhalation and the threshold limit for organic compounds in air is 0.01 mg/m^3. Organic mercury compounds are severe skin irritants and may cause severe dermatitis. These compounds also pose a severe environmental hazard through their concentration and accumulation in food chains, as biologically formed dimethyl mercury.

CHAPTER XIV

MISCELLANEOUS ORGANIC FUNGICIDES

In addition to the various fungicides described earlier, a number of other fungicides have been used in the control of various diseases of plants. Many of them cannot be classified under any one of the previously described groups of fungicides as they do not fall under them. In each of the other chemical groups, number of compounds have been tested for their fungicidal properties. But only a handful of them in each group have been found to possess fungicidal properties. Hence they have been described in this chapter, under different groups as Quinone compounds, heterocyclic N-compounds, organo-tin compounds and miscellaneous fungicides.

QUINONE FUNGICIDAL COMPOUNDS

Among the quinones, two compounds 2,3,5,6-tetrachloro-1,4-ben-zoquinonc (Chloranil, Spergon) and 2,3-dichloro-1,4-naphthoquinone (Phygon, Dichlone) have been used as seed protectants. Benzoquinones show comparatively low fungicidal activity but when halogens and/or hydrocarbons are introduced into the ring, the activity is strongly increased. The p-benzoquinone is less active than chlorobenzoquinone and toluquinone and similarly 2,3-dichloro-1,4 naphthoquinone is less active than 2-chloro-naphthaquinone and 2,3-dichloro-1,4-naphthoquinone. However, further addition of chlorine atoms in the molecule of 1,4-naphthoquinone lowers the fungicidal properties and 2,3,5,6,7,8-hexachloro-1,4-naphthoquinone is completely without fungicidal activity. Activity is improved by halogen substitution in the order Cl>Br>I which decreases the water solubility and phytotoxicity. Alkylation of ortho to quinone reduces activity.

Chloranil or tetrachloro-p-benzoquinone was first synthesised by Erdman in 1843 but it was introduced as a fungicide by Terhorst in 1938 only. It was found to be a successful seed protectant and also a foliar fungicide for downy mildew. Later Dichlone (2,3-dichloro- 1,4-naphthoquinone) and Dithianon (2,3 dicyano-1,4-dithio-anthra-quinone) were developed as effective fungicides for seed treatment.

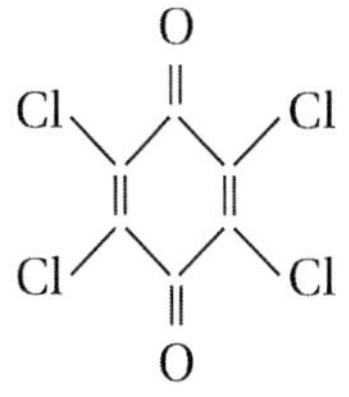

Chloranil

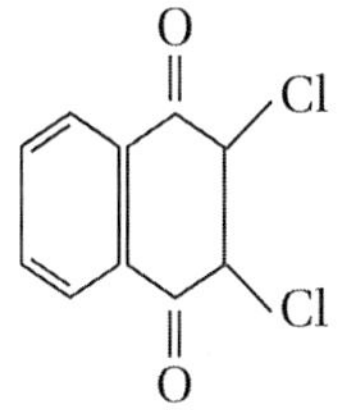

Dichlone

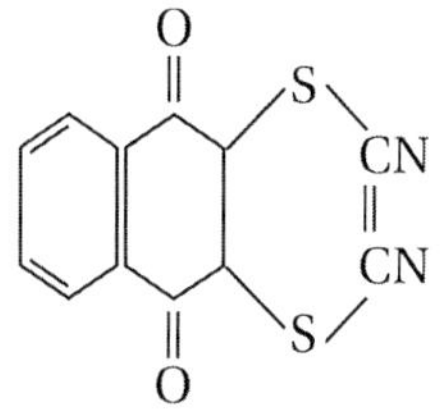

Dithianon

Tetrachlorobenzoquinone (Chloranil) is produced by the oxidative chlorination of phenol or chlorophenols and also by oxidative hydrolysis of pentachlorophenol. Dichlone is used much more than Chloranil. It is produced by various methods like: (i) chlorination of 1-naphthol in sulphuric acid in presence of iron salts, (ii) chlorination of naphthionic acid, and (iii) chlorination of 1,4-naphthoquinone or 1-nitronaphthalene in aqueous, weakly alkaline medium. Dichlone is usually prepared by dissolving naphthionic acid in 50% sulphuric acid and small amount of ferric sulphate and then chlorinating it at a low temperature. After chlorination, the precipitate is filtered off, washed with water and dried.

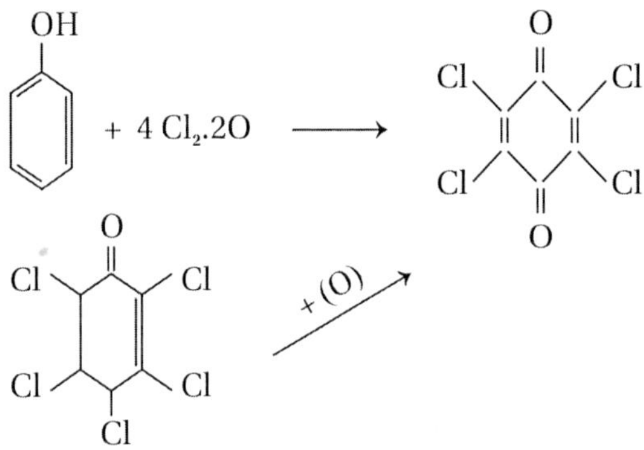

Chloranil

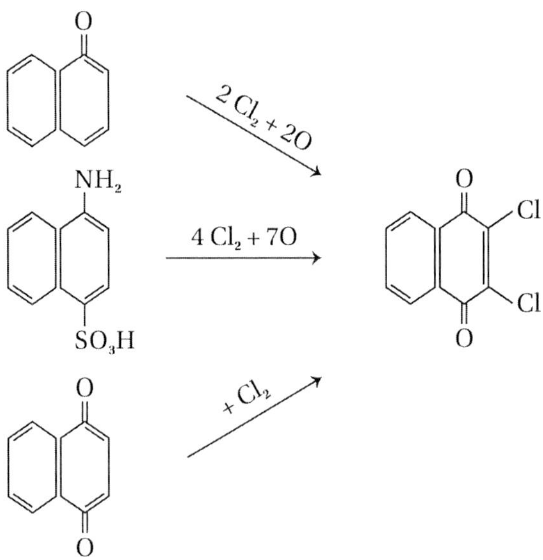

Dichlone

Both these compounds are yellow crystalline substances. They are practically insoluble in water but soluble in organic solvents such as xylene, chloroform and ether. Both are unstable in the presence of alkalies and are compatible with most of the seed dressing pesticides. They are incompatible with alkalies. In the presence of sunlight and moisture chloranil changes to chloranilic acid while dichlone is much more stable. Because of the phytotoxicity, chloranil has not been used successfully. Though these compounds are resistant to further oxidation, they arc highly reactive because of the presence of a ketone group adjacent to the C=C double bond and thus resemble the aliphatic α-β unsaturated ketone. Their reactivity is also related to their fungicidal action. The LD_{50} of chloranil, dichlone and dithianon for rats is 4000, 1300 and 1015 mg/kg respectively.

Recently another compound Dithianon (2,3-dicyano-1,4-dithio- anthraquinonc) has been developed. However, its phytotoxicity has been a major difficulty in developing it into a foliar spray. Another compound Cerenox (benzoyl hydrazone of quinone oxime) has been prepared by reacting p-nitrophenol with aqueous solution of benzoylhydrazinc in an acid medium. A 10% preparation of cerenox is being used for seed treatment. The LD_{50} of ccrenox for rats is about 100 mg/kg.

Two mechanisms which appear to be responsible for the fungicidal activity of chloranil and dichlone are the binding of the quinones nucleus to the sulphadryl (—SH) and amine groups (—NH_2) in the fungus cells by substitution or addition at the double bond and a change in redox potential and secondly the disturbance of the electron transport system. The inhibition of the enzyme by fungicides could be reversed by cysteine or glutathione.

HETEROCYCLIC NITROGENOUS COMPOUNDS

Kittleson in 1952 developed Captan (N-trichloromethylthiotetrahydrophthalimide) as a versatile fungicide for foliar application. Captan is a very efficient fungicide in plant disease control and also does not have any adverse effect on the appearance and quality of fruits. Later analogs of Captan like Folpet (N-trichloromethylthio-phthalimide), Difoltan (N-1,1,2,2-tetrachloro-ethylthiophthalimide) and Mesulfan (N-trichloromethylthio-4-chloroaniline of methane sulphonic acid) were developed. As all of the 18 compounds of fungicidal activity developed contained $=NSCC1_3$ group, Kittleson attributed the fungicidal activity to that group.

Captan (Esso fungicide 406, Orthocide 406, Vancide 89)

Captan is a heterocyclic nitrogenous compound. It is produced by the condensation of perchloro (methylmercaptan) with tetrahydrophthalimide in aqueous alkaline medium with constant stirring, at the lowest possible temperature to avoid hydrolysis of both captan and the perchloromethylmercaptan.

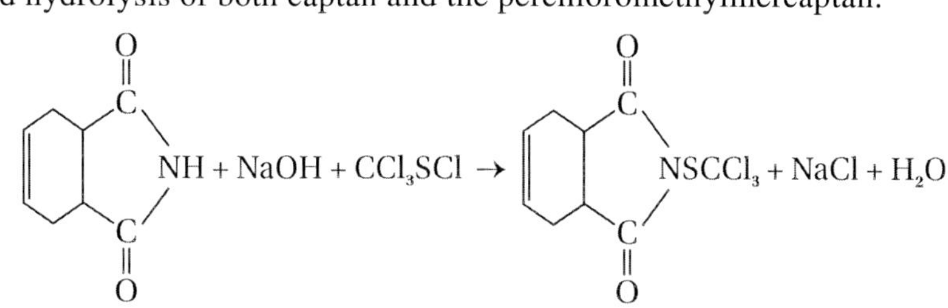

An excess of NaOH is necessary because a part of the perchloro (methylmercaptan) is hydrolysed and ties up alkali.

In pure state, captan is a white crystalline substance, and almost odourless with m.p. at 172°C. The technical grade is a yellow or grey one with characteristic odour of perchloro (methylmercaptan) and thiophosgene. It is practically insoluble in water and slightly soluble in most of the organic solvents. In moist conditions captan is hydrolysed to tetrahydrophthalimide.

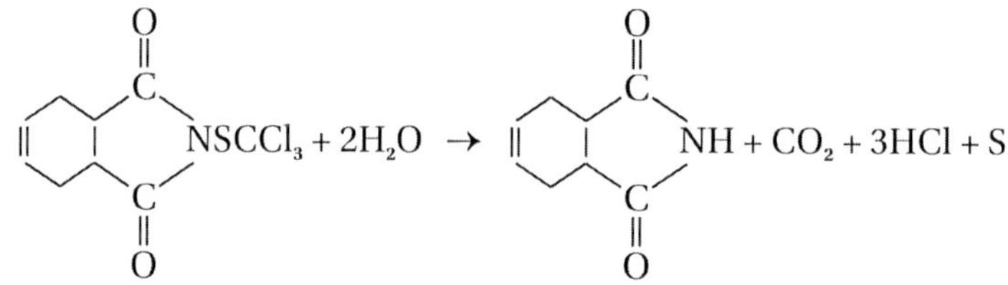

Presence of alkali and higher temperatures increase the rate of reaction. Hence captan is incompatible with all compounds of alkaline nature like lime sulphur and Bordeaux mixture. It is however compatible with other pesticides like lead arsenate, foliage oil, DDT, TDE, parathion, TEPP, chlordane, toxaphene and phenylmercuric acetate.

Kittleson attributed the fungicidal activity to the $=NSCCl_3$ group. Later it has been shown that captan reacts with —SH group within the fungus cells, to produce thiophosgene which reacts with the free sulphydryl, amino and hydroxyl groups of enzymes. Another theory suggested that captan may form the transitory N-chlorotetrahydrophthalimide which is highly reactive and may combine with the reactive groups within the fungus cell. Captan has been found to decrease oxygen uptake by the fungus and the inhibition could be reversed by the addition of thiamine, suggesting a competition between captan and cocarboxylase for sites on the apoenzymes. It has also recently been shown that part of the sulphur of carbondisulphide given off by the cells of *Saccharomyces pastorianes* after treatment with captan was partly derived from the sulphur of the cells as seen with S^{35}. Possibly the sulphur from the cells was derived from thiols and depletion of this sulphur might be responsible for the toxic action of captan. Using labelled captan, Richmonds and Somers in 1968 concluded that fungitoxicity was due to the irreversible changes following oxidation of the protein thiols to disulphide.

It is still not clear whether the activity of captan is due to captan itself or its degradation products of the chemically unstable compound. The total amount of captan on plants decreases with time. The degradation of captan depends upon the environment as it gets degraded when exposed to sunlight. Though captan has been reported to be extremely safe, still caution has to be exercised, as high sensitivity has been reported to oral toxicity of captan by rats kept on low protein diets, and pregnant animals.

Folpet (Phaltan, Phthalan, N-trichloromethylthiophthalimide)

Folpet is an analog of captan introduced by the California Spray Chemicals Corporation. It is slightly more effective than captan against certain diseases. Folpet is produced similar to captan by reacting perchloro (methylmercaptan) with phthalimide in the presence of NaOH in aqueous medium. The main contaminant in the technical grade product is the starting phthalimide that has not reacted and that has been formed as a result of hydrolysis.

Folpet is a white crystalline substance, m.p. 177°C, insoluble in water and poorly soluble in the usual organic solvents. Like captan it is hydrolysed by water but the rate of hydrolysis is higher than that of captan.

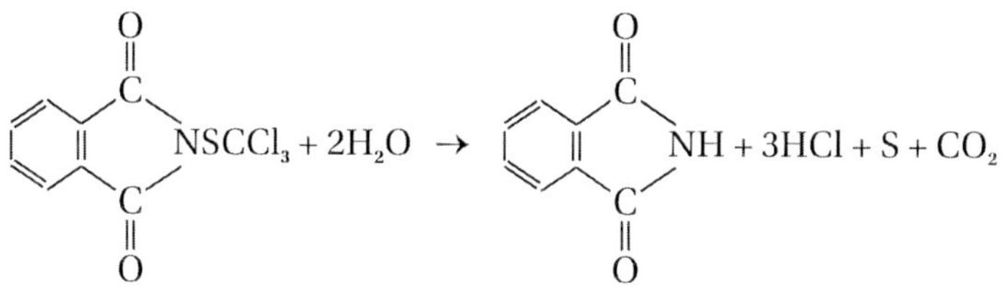

As a large amount of HCl is produced on hydrolysis, folpet like captan cannot be packed in paper containers.

The differences in the imide structure of captan and folpet may account for their differences in the fungitoxic spectra. Captan has mobile hydrogen to carbonyls while folpet does not. Infrared spectrometry show that the mobile hydrogen allows captan to enolyse, hydroxylating the carbonyl groups. The hydroxylation could reduce electronegativity of the carbonyls, thereby strengthening the N—S bond. A strengthened N—S bond in turn may reduce reaction of captan near the potential sites of action in tolerant fungi. Secondly hydroxylation would increase water solubility of captan and reduce its permeation into spores. Lukens and Horsfall (1965) feel that enolisation of captan and not folpet should account for the differences in the fungitoxicity of these two structurally similar compounds. Using labelled carbon C^{14} and sulphur S^{35} Siegel and Sisler suggested that binding of sulphur containing fragments of folpet to ethanol insoluble entities of the cells was primarily responsible for the fungicidal effects of the toxicant.

Difolatan (N-1,1-2,2-tetrachloroethylthiophthalimide)

Difolatan is produced by the reaction of tetrahydrophthahmides with pentachloroethylmercaptan in the presence of NaOH.

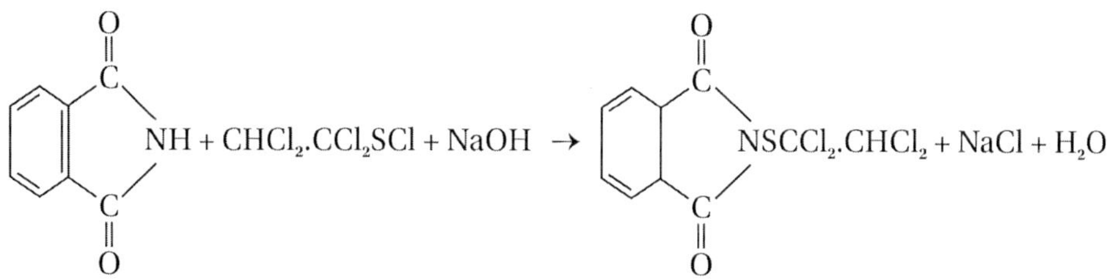

Difolatan is a white crystalline substance, m.p. 160-161°C, practically insoluble in water, poorly soluble in most of the usual organic solvents but somewhat more soluble than captan. The compound is slowly hydrolysed by water and more rapidly by caustic alkalies. It is more resistant to hydrolysis than captan and folpet.

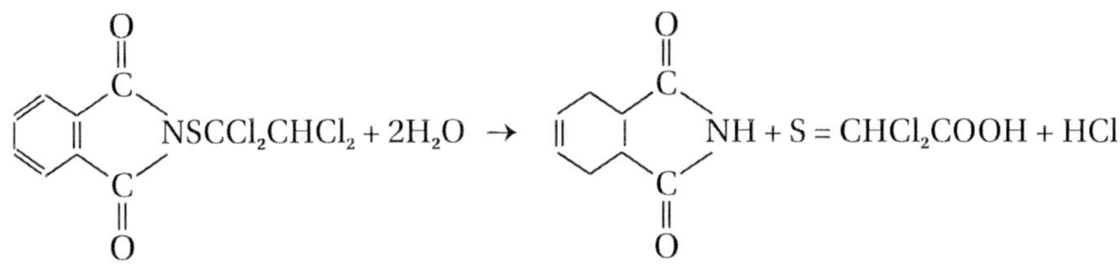

Mesulfan (N-trichloromethylthio-4-chloroanilide of methane sulpho- nic acid)

Mesulfan is an analog of captan but is less active than captan. Mesulfan is produced by reacting perchloro (methylmercaptan) with 4-chlorophenylmethane sulphonimide in the presence of NaOH at the lowest possible temperature.

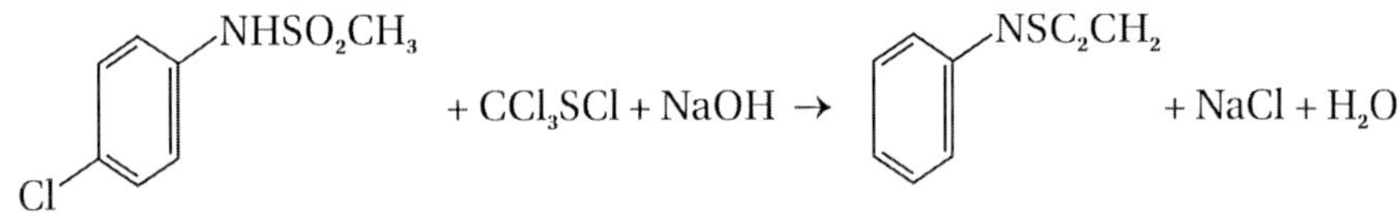

Mesulfan is white crystalline substance, m.p. 113-114°C. In chemical properties it is similar to captan. Mesulfan is marketed in the form of a 50% wettable powder.

ORGANOTIN COMPOUNDS

The biocidal activity of organotin compounds seem to have been studied towards the end of the nineteenth century but the insecticidal properties were discovered only during 1929. However, the studies on fungicidal activity of the tin compounds was begun only in 1950's by van der Kerk and his co-workers.

Although inorganic tin compounds, stannous and stannic chloride were inactive, certain organic tin compounds were found to be most effective fungicides. When the chloride atom in the stannic chloride is replaced by alkyl or aryl groups the fungicidal effect gradually increased up to trialkyl and triaryl tin acyls and decreased with longer chain. Increasing the length of hydrocarbon chain of the radical attached to the tin in trialkyl tin acyl leads to an increase in the fungicidal activity of the compound up to a certain limit. The compound containing a total of 12 carbon atoms has the maximum activity. Introduction into the aromatic radical of triphenyl tin acetate, any substituent like chloride lowers the fungicidal activity of the compound. For example tri(p-chlorophenyl) tin acetate is 10-100 times less active than triphenyl tin acetate.

The effect of anion in the trialkyl tin acyl on fungicidal activity of the compound is very little though it varies considerably with the alkyl group. The toxicity (LD_{50}) of the trialkyl tin salts decreases with an increase in the number of carbon atoms in the molecule, for example LD_{50} of triethyl tin chloride is less than 10 mg/kg, while tributyl tin acetate is 500 mg/kg. The anion has a great influence on the toxicity of trialkyl tin salts. Thus the toxicity of triphenyl tin hydroxide is almost four times less than that of acetate.

An important problem in the development of organotin fungicides has been the high phytotoxicity of many of the compounds. The trialkyl tin compounds have in general proved too phytotoxic for foliar use. However, the triphenyl tin compounds are substantially less phytotoxic and triphenyl tin acetate or Brestan has proved effective as a foliar fungicide.

The phytotoxic effect of triphenyl tin $(C_6H_5)_3Sn$ X is related to the nature of X and is the highest for chloride and sulphate. Among the safest of all the triphenyl tins are triphenyl tin sulphide $(C_6H_5)_3$ SnSSn $(C_6H_5)_3$ and triphenyl tin disulphide $(C_6H_5)_3SnS{-}SSn(C_6H_5)_3$ which are of more limited spectrum of fungicidal activity than the triphenyl tin acetate. Because of the relatively high toxicity of organotin compounds for animals, only tributyl tin and triphenyl tin salts are used in agriculture. At present tributyl tin hydroxide $(C_4H_9)_3$ SnOH is used as an anti mildew agent in wood and paints. Triphenyl tin chloride is used in anti fouling paints and triphenyl tin hydroxide and acetate are used as fungicides.

Brestan (Triphenyl tin acetate)

Brestan is a white crystalline substance, m.p. 124-125°C, slightly soluble in water and in most organic solvents, it is produced by reacting trialkyl tin hydroxide with acetic acid.

$$(C_6H_5)_3SnOH + CH_3COOH \rightarrow (C_6H_5)_3SnOCOCH + H_2O$$

Brestan is marketed in the form of 20 and 60% wettable powder. Though it is phytotoxic, still at concentrations normally used, it is safe for most crops. Though it is similar to copper compounds, it is used in much smaller doses than copper. Recently it has been combined with Maneb. Brestan causes irritation of the mucous membrane. Its LD_{50} value for rats in 125 mg/kg.

Duter (Triphenyl tin hydroxide, TPTH)

Duter is a white crystalline compound prepared by reacting the triphenyl stannous chloride with potassium hydroxide or by

$$(C_6H_6)_3SnCl + KOH \rightarrow (C_6H_5)_3SnOH + KCl$$

reacting tin halide with Grignard's reagent. The compound is practically insoluble in water and most organic solvents. Its m.p. is 120°C. It does not show any phytotoxic effect. The LD_{50} is about 500 mg/kg.

MISCELLANEOUS FUNGICIDES

Though sulphur is one of the best fungicides against powdery mildew, it could not be used on sulphur-sensitive plants like certain varieties of cucurbits and apples. Hence Rohm and Hass of the U.S.A, introduced fungicides like Dinocap, Karathane, Arathane, Iscothan and Mildex in 1934. Dinocap (Dinitro-(1-methyl heptyl) phenyl cortonate) is a brown liquid produced by nitrating 2-sec-octyl phenol.

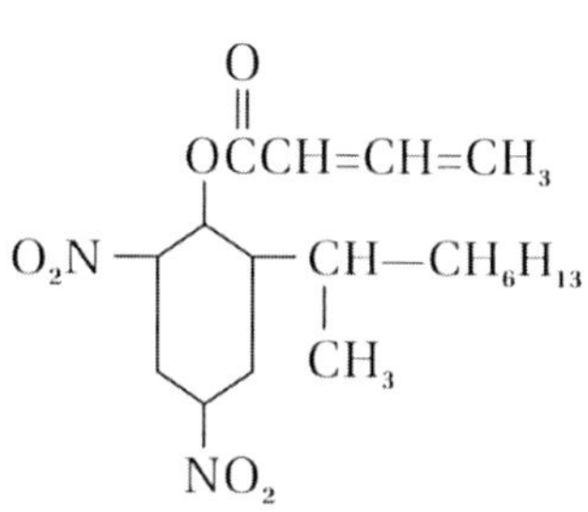

The dinitrophenol formed is further treated with crotonyl chloride in the presence of HCl acceptor like organic tertiary amines in a hydrophobic solvent. It is soluble in organic solvents and its boiling point is 138-140°C. Dinocap is an excellent substitute to sulphur for controlling certain species of mites. Dinocap is marketed in the form of 25% wettable powder. Its LD_{50} for rats is 980-1100 mg/kg.

Dodine (Cyprex, N-dodecyl guanidine acetate)

Dodine was Introduced by the American Cyanamide Co. as a fungicide in 1959 for the control of fruit diseases. This belongs to the guanidine group. It is produced by reacting cyanamide with dodecylamine.

$$C_{12}H_{25}NH_2 + NH_2CN + CH_3COOH \rightarrow C_{12}H_{25}NH\underset{\underset{NH}{|}}{C}NH_2CH_3COOH$$

It is a white crystalline substance with m.p. 136°C. It is poorly soluble in cold water, more soluble in hot water and alcohol but insoluble in most organic solvents. It is stable under moderately acid or alkaline conditions. Because of the presence of a radical with a long chain of carbon atoms in its molecule, dodine lowers the surface tension of water which automatically ensures a better coverage. Dodine in alkaline medium breaks down releasing the free base. In acid medium, the compound is more stable. Upon boiling with acid or alkalies, it decomposes with the formation of ammonia and dodecylamine. Dodine penetrates the host tissues and exhibits eradicative properties. It exhibits adverse effects on certain fruits. Dodine is marketed in the form of 70 per cent wettable powders. It is used at concentrations of $< 0.1\%$. Higher concentration of the compound is phytotoxic. The toxic effect of the fungicide may be due to the physical disruption of the membrane or inhibition of enzymes located in the fungus. Its LD_{50} for rats is 1000 to 2000 mg/kg.

Glyodin (Glyoxalidine, Glumide, Cragglyodin, 2-Heptadecyl-2- imidazoline acetate)

Glyodin is a derivative of glyoxalidine and possesses fungicidal properties. As the compound has the maximum fungistatic activity when the side chain is 13-17 carbon long, and maximum phytotoxicity when the side chain is 11-13 carbon long, especially if unsaturated, the heptadecyl derivative $C_{17}H_{35}$ has been chosen to give it a high fungitoxicity and low phytotoxicity. It is produced by the condensation of ethylene diamine with stearic acid at elevated temperature.

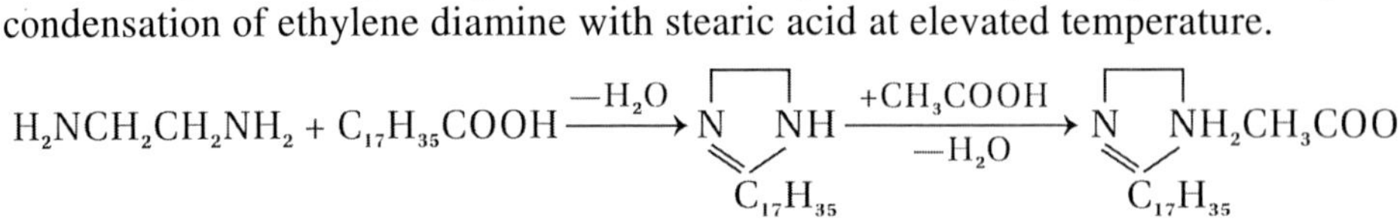

The amide that is formed is dehydrated on heating in vacuum. Glyodin is a light orange coloured powder, practically insoluble in water but soluble in propylene glycol and ethylene dichloride. When it is heated or allowed to stand for a prolonged period in a solution of alkali hydroxide, it is hydrolysed to 2-aminoethylamide of stearic acid. Like dodine, it is also highly surface active.

Glyodin is rather a protective than an eradicative fungicide. In certain trials, in addition to its fungicidal value, it appears to have increased the chlorophyll content of apple leaves and also prevented the build-up of the red mites. However, it was observed that the later build-up was much more rapid following the use of glyodin than with most organophosphorus insecticides and sulphones. It is toxic to mammals when consumed orally. Also, it causes irritation to skin and eyes. The fungitoxicity of glyodin is reversed by the purines, guanine and xanthine suggesting that glyodin is a competitive inhibitor of biosynthesis reaction, involving purines. It is compatible with lead arsenate, nicotine sulphate and most of the phosphate insecticides. Care should be taken in combining with zinc arsenate, cryolite, rotenone, pyrethrum, dormant oils, nitros, toxaphene and chlorodane. The LD_{50} of glyodin for rats is 3720 mg/kg.

In the 1930's the chloronitrobenzenes were introduced as a fungicides by I.G. Farbin Industries based on the findings that 1,3,5-trichloro-2,4,6-trinitrobenzene and 1,2,4-trichloro-3,5-dinitrobenzene were highly toxic to specific fungi. One of the successful compounds of chloronitrobenzene is 1,2,3,4,5-pentachloronitrobenzene (PCNB) and 2,3,5,6-tetrachloronitrobenzene (TCNB). Though not much work was done up to 1950s, interest in use of PCNB was revived in 1956 in the U.S.A. It was recommended for the control of soil borne diseases because of its long persistence in soil. PCNB is synthesised by nitrating pentachlorobenzene or chlorinating various chloronitrobenzene in the presence of a catalyst iron-iodine.

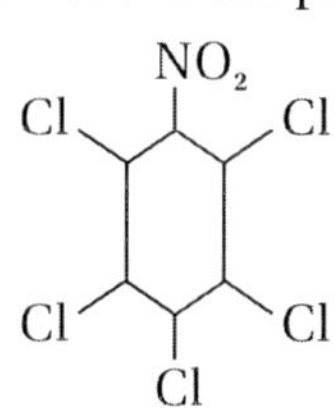

PCNB

As the rate of reaction is sharply decreased in the presence of even a trace of moisture, sometimes chlorosulphinic acids are added to the reaction mixture. PCNB is a fairly yellowish brown crystalline substance, m.p. 140-143°C, readily soluble in benzene and toluene and sparingly soluble in alcohol. It is stable in storage and does not burn. With alkali aicoholates, it gives an ether of pentachlorophenol.

$$C_8Cl_5NO_2 + CH_3OK \rightarrow C_6Cl_5OCH_3 + KNO_2$$

The activity of PCNB as soil fungicide seems to be related largely to the vapour pressure. PCNB is most persistent in soil among nitrocompounds. When applied to the soil, PCNB appears to decrease competition for nutrients in soil by suppressing PCNB- sensitive actinomycetes and fungi, thus permitting PCNB insensitive fungal pathogens like *Fusarium* sp. to increased selectivity. Soil moisture appears to influence persistence of PCNB in soil as PCNB disappeared more rapidly from the submerged soils than from moist soils. Disappearance of PCNB in submerged soil was found to be accompanied by an increase in pentachloroaniline (PCA) which was inhibitory to soil actinomycetes and fungi.

PCNB acts largely by fumigation, by retarding germination and colony growth and also by suppressing sporulation. Certain adverse changes like off-colours and off-flavours were observed in potatoes grown in soils treated with PCNB. Changes in starch and total solid contents were also reported. However, no change in the tyrosine, total phenols, specific gravity, pH and titratable acids was reported in potatoes grown in PCNB treated soils. The LD_{50} for PCNB is 1650 ing/kg.

Antibiotics

Antibiotics are chemicals produced by living, organisms that arc selectively toxic to other organisms. The successful use of antibiotics against animal diseases led to their use for the control of plant diseases. The first material prepared from *Streptomyces* and termed as Antimycin A was used as a protective fungicide for the control of apple scab and tomato early blight *Alternaria solani.* The use of antibiotics has been restricted to chemicals produced by microorganisms that prevent the growth and development of bacteria and fungi. Most antibiotics are of complex structure and are generally highly unstable. Hence their use has been restricted.

A number of antibiotics like penicillin, gliotoxin, streptomycin, blastmycin, blastocidin S, cycloheximide, auerofungin, and griseo- fulvin have been tested for the control of diseases. Penicillin and gliotoxin have not been widely used, either due to their limited activity or their unstable nature. Streptomycin isolated from *Streptomyces griseus* in 1942 by Waksman has been used widely as a systemic fungicide. This is isolated as trichloride having m.p. 190- 200°C. It is readily soluble in water, less soluble in most organic solvents. This is a basic substance that is irreversibly hydrolysed under either acidic or alkaline conditions.

Blastmycin produced from *Streptomyces blastmycetius* is insoluble in water but soluble in a number of organic solvents. Blastocidin S isolated from *S. griseochromegenes* is soluble in water and glacial acetic acid. Both have been found to be effective against paddy blast *Piricularia oryzae.*

The antibiotics Cycloheximide, Auerofungin and Griseofulvin have been found to possess systemic action against a number of fungi. Cycloheximide (Actidione, 3(2-(3,5-dimethyl-2-oxocyclohexyl)- 2-hydroxyethyl)-glutarimide) has been isolated from *Streptomyces griseus* by Whiffen *et al.* in 1946. It is a colourless crystalline compound, m.p. 119.5-121°C, very slightly soluble in water but soluble in organic solvents. This is relatively stable to heat and acid conditions but is decomposed by alkali. It forms stable derivatives such as acetate, oxime and semicarbazone that are active as fungicide. Actidone has a wide antifungal spectrum but is costly to produce and there is a narrow margin between the effective and phytotoxic concentrations. Possibly cycloheximide might interfere with the glutamine and protein synthesis. Commercially it is available under various trade names such as Actipray, Actidion RZ (5% cycloheximide and 75% PCNB).

Another antibiotic Auerofungin having systemic action, is produced from *Streptomyces cinnamomeus* var. *terricola* tir. Auerofungin is a golden yellow powder insoluble in water, soluble in alcohols, dimethyl sulphoxide and alkaline solutions. As it is unstable in the presence of moisture and light, this should be stored in a dry and dark place. This antibiotic is non-toxic when consumed orally. Auerofungin has been found to be useful in controlling a number of plant diseases. Griseofulvin, another antibiotic having systemic action was discovered by Oxford *et al.* in 1939 as a metabolic product of *Penicillin griseofulvin.* Griseofulvin is a crystalline solid of low solubility. It gives only a certain degree of control under field conditions. It has been shown to be effective in producing severe stunting, excessive branching, abnormal swelling and distortion of hyphae of fungi that have chitinous cell walls. It is found to be effective against a number of mildews.

The action of antibiotics is generally quite specific for the particular molecular structure, and hence it is difficult to study its influence on the structural modifications. Although a number of antibiotics having fungicidal properties have been proposed, only Streptomycin and Cycloheximide have been used commercially. Streptomycin has an oral LD_{50} to rats of 300 mg/kg and cycloheximide of 1750 mg/kg.

CHAPTER XV
SYSTEMIC FUNGICIDES

Previously, fungicidal treatment for the control of diseases has been based on external application of the fungicides only. However, after the success obtained in the internal protection with 8-hydroxy quinioline against Dutch elm disease, the emphasis has shifted to the use of systemics. Materials that are applied to one part of the plant, say root or leaf and translocated in the fluid system of the plant to other parts of the plant to render the whole plant toxic to the fungus, is called a systemic poison. A systemic fungicide may either be toxic to the pathogen or get converted in the host plant concerned to such a fungitoxicant, or it may alter the metabolism of the plant so that the biochemical and/or physical resistance to the pathogens is increased or decreased. Systemic fungicide should not adversely affect the host plant either by reducing the quality or quantity of the produce in that season or in later years. Further, the toxicant should be stable within the host plant for a considerable period of time to protect it during its growth period. The systemic poison should also have low mammalian toxicity.

In 1966, Von Schmeling in the U.S.A., and Kulka in Canada reported the systemic fungicidal activity of the two 1,4-oxathin derivatives. The uptake and translocation of these compounds were quite similar to those of systemic insecticides and herbicides. Most systemic fungicides developed are absorbed by the roots and transported acropetally in the xylem. The rate of absorption from the soil to root is rapid but the rate-limiting step in the absorption is transportation from the roots to other parts of the plant. Generally, systemic fungicides are not translocated downward and in a number of cases they persist in the harvested produce. These properties, combined with the cost have to a certain extent been responsible for the lack of wider use of systemic fungicides.

Several systemic fungicides have been tested for their fungicidal activity and among them Benalate, Terrazole, Oxathiins, Chloroneb and Methyrimol appear to be promising. Benalate (Methyl-1-(butyl- carbamoyl)-2-benzinidazole carbamate) is a white crystalline substance, non-volatile in nature. It has a faint acrid odour. It is insoluble in water and oils. In certain cases, it may result in poor finish to certain apple fruits. It possesses residual and systemic properties against a number of fungi. The LD_{50} for rats is >9590 mg/kg.

Terrazole (OM-2424, 5-ethoxy-3-(trichloromethyl)-1,2,4-thia-diazole) and PCNB are mixed at 6.1 and 23% respectively and are used as Terrachlor super X, for the control of a number of fungi. Better emergence of maize seeds was also reported with 0.15% terrazole.

The two oxathiins derivatives developed by Von Schmeling and Kulka are DMOC (5,6-dihydro-2-methyl-l,4-oxathiin-3-carboxanilide) and DCMOD (2,3-dihydro-5-carboxanilide-6-methyl-l,4-oxathi- in-4,4-dioxide). Both are fairly water-soluble and do not show any phytotoxic effect on crops. DMOC is a crystalline compound with two crystal structures having a m.p. 91.5-92.5 and

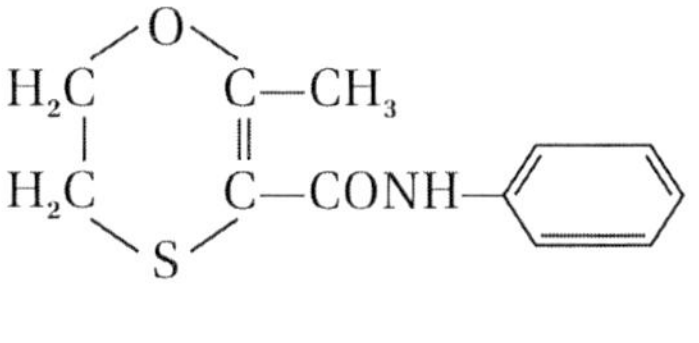

Vitavax (DMOC)

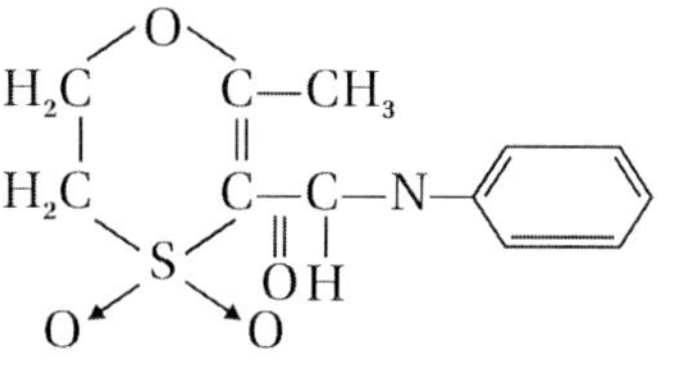

Plantvax (DCMOD)

98-100°C. In solution these two crystal structures revert to one and do not have different biological activity. It is quite stable in acetone and methonal. DCMOD is more water-soluble than DMOC. Both compounds seem to be highly specific in their fungitoxicity. DMOC is specific for Basidiomycetes while DCMOD is highly toxic to Basidiomycetes and also other fungi. DMOC is not very stable when applied to the soil since it is biologically degraded in about 10-30 days.

Chloroneb (Demosan, l,4-dichloro-2,5-dimethoxy-benzene) is relatively less toxic. It is very slightly soluble in water. But it is compatible with most pesticides and can be mixed with organomercurials for seed treatment. Chloroneb apparently moves systemically upward inside the plants when applied as seed treatment. The compounds were either in the original form or as an analogue of chloroneb when it moves in the plant. Metabolic detoxification of chloroneb appears to be associated with an unaltered molecule rather than its conversion product. This has been used for controlling a number of plant diseases.

Methyrimol (PP675, 5-Butyl-2-dimethylamine-4-hydroxy-6-methyl pyrimidine) is a systemic fungicide. It is a white crystalline solid, stable to heat, alkalies, and acid solution. It is slightly soluble in water but highly soluble in chloroform. This has been found to be effective against number of fungi. Methyrimol has been found to move freely in cucurbit plants following the transpiration stream and tends to accumulate at the margin of the leaves. As the persistence of deposits of this compound on leaves is relatively short, for disease control, this fungicide has to be introduced continuously into the transpiration stream and therefore soil application has been found to be satisfactory. The uptake of this fungicide depends upon the soil moisture content and if it is dry, uptake will cease although the plant may be taking water from elsewhere. When the soil is remoistened, the uptake will be again resumed. Methyrimol is loosely adsorbed on to moist soils, particularly more strongly in organic soils with low pH. Hence movement through soil is slow and application has to be done near the absorbing roots as drenches.

In addition to the above, two other fungicides TBZ (2-(4 thiazolyl) benzimidazole) and PP149 (5-butyl-2-ethylamino-4-hydroxy-6- methylpyrimidine) have been found to be promising as systemic fungicides, and have been used against a number of diseases of sugarbeets, potatoes, and oranges.

CHAPTER XVI

COMPATIBILITY OF PESTICIDES AND AGROCHEMICALS

As many of the pesticides cannot control all types of pests and diseases occurring on a plant simultaneously, very often it becomes necessary to use number of them together to control the various pests and diseases occurring on the same crop. As considerable saving also will be effected in labour, time and water in areas of acute water scarcity, if these could be mixed and applied in one single spray, attempts have been made to study their compatibility. As pesticides are also chemicals, they tend to react among themselves as well as with other agrochemicals. This will result in either the pesticide loosing its pesticidal activity or becoming phytotoxic, or adversely affecting the spraying equipment like clogging the nozzle and corroding the sprayer. Many pesticidal firms have been carrying out compatibility studies in respect of their newer pesticidal formulations. A compatibility chart has been prepared by the Farm Chemicals Information in the U.S.A., as a rough guide for mixing of pesticides. In addition to the above, a number of studies have also been published recently.

Combinations of captan with dieldrin and lindane, furadan with ziram, γBHC with rogor, aldrin, carbophenothion and chlorfenvinphos as liquid formulations, thiram with endrin, malathion, metasystox, dieldrin and dithane M.45 have been found to be compatible or have been found to have synergistic effects in the control of pests and diseases. Combinations of fertilisers like urea with various pesticides like thiram, phosphomidon, dithane M.45, thiometon, endosulphan, dichlorovos, parathion, fenthion, carbaryl, and methyl parathion plus DDT, neutral or acidic fertilisers with captan, superphosphate with methyl mercaptophos, anthio, lindane or aldrin, ammonium phosphate with rogor, ammonium sulphate plus farmyard manure with phorate have been found to be compatible and these combinations at least did not reduce the effectiveness of the pesticides or the fertilisers.

Certain other combinations like urea with lime sulphur, karathane, glycdin, phosdrin, sevin or morocide, NP fertilisers with rogor, chlorophos with rogor and υBHC technical, difoltan with parathion plus TDE, were reported to have produced phytotoxicity symptoms or reduced the insecticidal, fungicidal or fertiliser efficiency. As could be seen from many of the compatibility studies, there are many contradictory results. In all the above studies, combinations have been prepared either in the tank or earlier and kept for sometime and used later or different modes of mixing adopted for mixing the pesticides. All these have considerable effect on the efficacy of the combinations. Also, all the formulations are not prepared with the same emulsifiers, wetting agents and other additives. In certain instances, the emulsifiers themselves may not be compatible. In most compatibility studies so far reported, the compatibility or otherwise of the combinations have been assessed by parameters like germination percentage, yield of crops, phytotoxicity effects, and percentage of pests or diseases controlled. In many cases, the physical and chemical changes that are likely to occur in the mixing process and also on storage have not been studied. These studies could yield information of great utility to manufacturers and farmers.

CHAPTER XVII

DEVELOPMENT OF PESTICIDES

Various pests and diseases have been responsible for the catastrophic harvest failures and also for wiping out certain crops in various countries during the past decades. The development of. pesticides like DDT by Muellar in 1938 and phosphoric acid esters by Schrader, later have completely changed the picture. Before a particular compound is released as a pesticide, considerable amount of money and effort would have been expended and a number of vital decisions would have been taken at each stage.

Before embarking on the development of a new synthetic pesticide the availability of necessary facilities, finance and personnel have to be assessed. Dr. Wegler, formerly of Bayer Ltd. estimated that more than 40 million DM are needed for developing one or two compounds. Another important requirement in the development of pesticides is the availability of reliable test methods. In a number of cases, new methods have to be developed based on original research. Once the appropriate test methods are formulated, a chemist attempts to synthesise new compounds. Before attempting to synthesise, the chemist should have a thorough knowledge of compounds already tested in his laboratory and in other laboratories if possible, and as far as possible, see that the compounds sent for trials have not been already described.

Similarity of composition and structure is taken as a guide for synthesis. In most cases, a small proportion of compounds are synthesised for the first time. At this stage, the chemist has to decide purely by speculation on the nature of compounds to be synthesised and get a clue on the possible effect. When some clue on the effects of a few compounds is obtained the nature of groups responsible for those effects are traced, and stage by stage relationship between chemical structure and activity is established. Attempts are made to graft new findings on to, or to coordinate them with the existing knowledge or eliminate any disadvantage which has become known. Often biologists and chemists of different companies reason out along similar lines so that it becomes simply a matter of chance who will be the first to realise the idea, and as such it is very risky one.

About 10% of the compounds sent for biological testing originate from basic product development branch which is an independent group not attached to any particular division like pesticides, pharmaceuticals and dyestuffs. About 200-300 new compounds each about 15 to 25 kg are sent by each chemist for biological testing.

As soon as it is received, the Biological Division conducts exhaustive tests over a period of few months. When a particular compound at low concentration is found highly effective, the chemist is quickly supplied with an interim report of the finding. Upon receipt of this information, the chemist immediately starts studies on modifying the molecule by substituting one or more of the external radicals. Simultaneously an exploratory determination of the toxicity of the potential candidate is also made in the green house studies. Vital decisions based on toxic hazards and biological efficacy of the compounds are made and only a few of them are advanced to the next stage of testing. Also a major decision as to whether chemical research work in that particular group of compounds have to be intensified or not is also taken. If encouraging results are obtained, chemists are asked to prepare 1-2 kg of compounds for the first field trials.

At the field trials stage, it is taken up by the agricultural scientist and advance evaluations are made on the compounds. Before taking the potential products to pilot plant stage, some more important decisions regarding the products are made. As each pesticide is saleable only in relation to other well established products, the new compounds have to be either cheaper or should have very clear recognisable advantages over the other products. Only those having clear distinguishable advantages are moved to the pilot stage. Simultaneously with the introduction of compounds for fields trials, short-term toxicity trials are also taken up.

Before introduction of a pesticide for practical use, the chemical has to be tested for its toxicological nature for at least 2 to 3 years on a variety of warm-blooded animals like mice, rats, rabbits and dogs. The decision to subject a pesticide to long-term toxicity testing is a vital one, as it involves heavy expenditure. At the same time if the testing is not started early, the introduction of the product on a commercial scale could be delayed by 1 or 2 years which may mean in certain cases, loss of the product once for all. After evaluation in the small scale field trials, a small number of highly potential candidates are sent for wider field testing in various countries to assess the influence of various factors like climate, moisture, soil, time of application, intensity and nature of pests and their population and crop phytotoxicity. Such trials normally take another 2 or 3 years or even more if unfavourable climatic conditions exist. Based on these results two or three compounds are selected for pilot plant manufacture.

As soon as the required amount of information becomes available, patent applications have to be filed as otherwise the firm may loose its hold on the compound. Side by side, specific qualitative and quantitative methods of analysis have to be developed for collecting information, like transformation of the pesticide in soil or plant, rate of degradation required for patent applications and registrations. As soon as 2 or 3 products are selected for wider use, the whole method of preparation, scientific and technical details are checked by specialists of the process engineering department and modified wherever necessary. Approximate costs are also worked out at this stage.

The approved active ingredients have now to be formulated by mixing with wetters, emulsifiers, stickers and others. Pesticide formulations are then developed by physico-chemists in cooperation with physicists, engineers and agricultural scientists. Often 10 or more different kinds of formulations of one and the same active ingredient is prepared, taking into consideration the potential areas of marketing and also availability of emulsifiers, stickers and other additives in those areas. These are studied and based on the results, a few of the formulations are approved and about 20 to 50 tons of the pesticide formulation is put into "Restricted trial basis" to assess the opinions of the market and advisory organisations.

Meanwhile, at least two years before the last stage, the Management Division has to chalk out plans for erecting manufacturing plants with provisions for further expansion. Also, they will have to take steps to register the product in different countries. Each country has its own laws regulating registrations and even within each country, each State has its own regulatory laws. For example though there was the Federal Insecticides Act of 1917 in the U.S.A., and later the Federal Insecticides, Fungicide and Rodenticide Act of 1947 in the U.S.A., each one of the States there has its own regulatory laws and also mode of registration. Except in the U.K. all other countries in Europe prohibit the sale of a pesticide without registration for specific uses described on the manufacturers' label. Even for registration for specific uses described, the concerned manufacturers have to furnish complete details in the form of answers to questions. For example, answers to 70 detailed questions have to be given in the Federal Republic of Germany, while one has to answer only 15 questions in the Netherlands. In the U.K. only statutory control is embodied in the poisons legislations. Scientific data presented by the manufacturer is assessed by the interdisciplinary committees, the composition of the same varying in each country. In France, representatives of the industry are also permitted to sit on the Committees. Pesticides could be put into the market only after obtaining clearance from the concerned bodies of the country which may take anything from 6 months to 2 years. In certain cases temporary registration is also granted subject to it being withdrawn at any time. In India, recently the Insecticides Act 1968 was introduced, requiring all manufacturers and distributors to register their products before they are marketed and also to have a check on the quality of the pesticides.

As such, manufacturers also have to take steps to get the registration by the time the chemical is ready for wide scale sale. As it will be costly to formulate all the pesticide and despatch to another country, in many instances steps have to be taken well in advance to erect the formulating plants in countries where a potential market exists, making use of the available local resources of emulsifiers, fillers, wetting agents and materials for the plants. All these take up at least a minimum of 5 to 7 years period and have been reported to cost more than 40 million DM for putting one pesticidal compound in the market.

CHAPTER XVIII

REACTIONS OF PESTICIDES WITH SOILS

Pesticides are generally applied to soil for controlling soil-borne diseases, pests and diseases of plants. In addition to the above, a fraction of all pesticides applied to plants either as spray or dust reach the soil through either runoff from leaves or through wind, rain etc. As soil is a complex physical, chemical and biological system, a number of reactions take place and these modify the efficiency of the pesticides considerably. The activity of pesticides in the soil is controlled largely by the adsorption and desorption of the same by soil components.

Clay is considered to be the most reactive part of the soil. The quantity, nature of clay, surface area, soil structure, organic matter, water content, quality of pesticides, temperature and pH of the pesticides and soil largely influence the efficiency of the pesticide applied to the soil. Some pesticides are cationic in nature and are adsorbed on the clay by exchange reactions. Complexation of the pesticides with clay affects the bioactivity of pesticides to a greater or lesser extent depending upon the energy of adsorption and desorption. The total charge and surface area of the solid components decrease in the order: Organic matter > Vermiculite > Montmorillonite > Ulite > Chlorite > Kaolinitc. On a comparable weight basis, montmorillonite adsorbed about 20 times more of the pesticide than kaolinite. Calcium saturated kaolinite and calcium saturated montmorillonite adsorbed 2.70 and 10.3 mg/g respectively of γBHC. Considering the surface area, kaolinite adsorbed (0.18 mg) far more than the montmorillonite (0.0127 mg/g) which shows that γBHC competes much more effectively with water on the kaolinite hydroxy surfaces than on the oxygen surfaces of montmorillonite. The finer the particles the greater will be absorption of the pesticide, possibly due to the increased surface area and reactive sites. The adsorption, desorption and degradation also depend on the nature of the cations present originally. Adsorption of paraquat was more complete on Na^+ saturated vermiculite than on Ca^{2+} or Mg^{2+} saturated clay. Dieldrin was degraded much more rapidly on a H^+ or Al^{3+} kaolinite than on Na^+ or Ca^{2+} clay. Certain of the chemical reactions of pesticides are catalysed by clay in the soils. For this the two silicate surfaces must be in simultaneous contact with the molecule. The catalytic effect of kaolinite, montmorillinite, some talcs, ferric chloride, aluminium chloride and chromium chloride on the conversion of DDT to DDE at 115-120°C has been demonstrated. Iron oxides in the soil appear to catalyse the reaction of DDT. Thermal degradation of DDT was also promoted by the allophenic soils or clays presumably by their amorphous alumina. High iron and aluminium content in the soils favoured adsorption of DDT, dieldrin and γBHC.

In addition to the mineral fraction of soils, soil organic matter also influences considerably the adsorption of the pesticide in soils. A humic acid adsorbed *ca.* 70 times more of linuron and malathion than the K^+ montmorillonite in comparatively similar conditions. Varying quantities of pesticides were adsorbed depending on the stage of decomposition of the organic matter. Organic matter in this way can be said to retard the degradation of the pesticide. Lambert feels that the active fraction of soil organic matter is a better' predictor of the adsorption of neutral pesticides. In certain cases, fats, cellulose, waxes, nucleic acids etc. contributed very little to the pesticide adsorption while humic acid, lignin and quinalizarin were highly adsorptive. The microbial degradation of pesticides is however increased by the presence of organic matter as it increases microbial activity.

The amount of pesticide adsorbed depends upon the nature of the pesticides and also soil pH. Phenyl mercuric acetate adsorption on montmorillonite, allophane and kaolinite was remarkably pH dependent. At normal pH values, dilute salts slightly increased adsorption while in concentrated solution, adsorption was sharply increased. Acid clays such as attapulgite ($pK_a<1$) rapidly degraded heptachlor while incorporation of oxygen containing chemicals such as diethylene glycol increased insecticidal stability. Significant decomposition of chlordane, toxaphene, heptachlor, DDT, aldrin and endrin were noted on acidic diluents like kaolinite and attapulgite. The adsorption of pesticide depends upon the number, type and relative positions of the functional groups. Adsorption characteristically increased with the functional groups such as R_3N^+, —$CONH_2$,—OH, —NHCOR,—NH_2,—OCOR and NHR. Presence of double bonds in the structure of the pesticide increased adsorption affinities.

Water solubility can be considered an approximate indicator of adsorption as, lower the solubility of the pesticides, the greater is the adsorption. Molecular size of the pesticide also modifies the adsorption of the pesticides. Chemically, organophosphates are less stable in water and soils. Some of them like malathion are base hydrolysed while others like diazinon are acid hydrolysed. The non-enzymatic hydrolysis are catalysed by some metabolic ions like Cu^{2+}, the ease of catalystic hydrolysis being dursban > diazinon > ronnel > zytron.

The above reactions of pesticides in soils are modified by the movement of pesticides through leaching, runoff and volatilisation, by photodecomposition and microbial activity. Pesticides could be moved in the water phase either through molecular diffusion or through mass transfer. Diffusion occurs only for a few centimeters depth and is modified by factors such as diffusion, coefficient of the soil structure characteristics, soil water content and adsorption of the pesticide on soil. The apparent diffusion coefficient of nonvolatile components increases rapidly with increased moisture since the water path becomes less tortuous. The greater the compaction, the significance of the diffusion process will be less.

Another aspect of molecular diffusion is the volatilisation of the pesticide. Volatilisation is increased by higher temperature, lower clay content and organic matter content of soils. Volatilisation is greater from moist soil than in dry soils and gets greatly reduced when the moisture in soil clays dries to less than monolayer thickness.

Mass transfer of pesticides can occur through leaching of the pesticides or through runoff with soil and rainwater or upward movement in drying soils, laterally from wetter to drier soils as well as downwards in the soils. In all above instances, though diffusion takes place simultaneously, the mass transfer is the dominant process. Pesticides applied near the surface may move downwards as a band, and may become diffused due to the diffusion and hydrodynamic processes, the latter due to non-uniform pore-size distribution in the soil. Adsorption of pesticides on soil components will reduce the distance of leaching and will modify the zone of maximum pesticide concentration. Increasing adsorption imparts greater skewing or tailoring if the percolation rate is too great. Pesticides are more likely to be leached in light than in heavy textured soils. The effectiveness of initial soil moisture and water flux is not clear. Microbial degradation will restrict the depth and quantity of the pesticide leached. Lateral movement of the pesticide across soil surface is enhanced by steep topography, low soil permeability and prolonged precipitation. Factors conducive to wind or water erosion are important in cases where pesticides are less mobile and are likely to be removed from bare soil. Leaching protects a pesticide from runoff and also from photodecomposition. Photodecomposition of pesticide occurs at soil surfaces and with only those compounds which are not soil incorporated or moved down to subsoil. Soil DDT slowly degrades in sunlight to DDE, TDE(DDD), pp-dichloro- benzophenone and other products. Solid aldrin yields its epoxide, dieldrin, on exposure to sunlight and they are also converted to photoaldrin and photodieldrin. The latter is approximately twice as toxic as endrin to insects and mices. Pesticides are degraded by microbes to a considerable extent. Warm soil temperatures, adequate moisture, presence of organic matter, neutral pH etc. promote microbial activity and thus increase the degradation of pesticides. Moisture levels play a significant role in the microbial breakdown of pesticides as they control the rate and nature of microbial population in the soil. Thus the accumulation and degradation of pesticides in soils depends on a large number of other soil factors and these control the efficiency of pesticidal action in soils.

APPENDIX

APPENDIX I

STANDARD SOLUTIONS AND SPRAY SOLUTIONS

Solutions are generally prepared by dissolving the solids (solutes) in different solvents. Water is one of the most common solvents used for the preparation of solutions. In cases where the solid is not soluble in water, other solvents like, benzene, chloroform are used to dissolve them. As the density of water is 1 at 25°C and as there is not much difference in the density at room temperatures normally prevalent in tropical countries, it is generally assumed that 1 ml of water is equivalent to 1 gram and therefore in many cases, the weight (g) and the volume (ml) are used interchangeably without much error in final values.

Two types of solutions are prepared for quantitative and qualitative analysis of compounds and also other related uses like preparation of solutions for sprays etc. The first one, "the standard solution" is prepared for the quantitative analysis where accuracy is an absolute requirement. The standard solution is prepared by dissolving a known quantity of a solute in an accurate volume of the solvent. In the preparation of standard solutions, four terms are commonly used viz. primary standards, molecular weight, equivalent weight and normality. The primary standards possess certain properties like stability over a long period of time, high molecular weight, least capacity to absorb moisture from atmosphere, high solubilityin common solvents, particularly solvents like water etc. Usually, primary standards are first prepared after calculating the quantity of the solute required to be dissolved in a known quantity of solvent (usually one litre of solvent). For this purpose, the molecular weight of the compound (including the water of crystallisation) is calculated from the knowledge of atomic weights of the elements that go to make up the compound. The atomic weights of all elements are given in the periodic table as well as in books like Clark's Tables, Seshadri's Tables etc. The molecular weight of a compound is defined as the ratio of the weight of one molecule of a substance to the weight of one atom of hydrogen. For example the molecular weight of hydrochloric acid (HC1) is 1 + 35.46=36.46 (1 and 35.46 being the atomic weights of hydrogen and chlorine respectively). In case there is more than one atom of the same element in the compound the atomic weight of the particular atom has to be multiplied by the number of atoms present (for example in the case of H_2SO_4, the molecular weight is $(1 \times 2) + 32.06 + (16 \times 4)=98.06$. After the molecular weight is calculated, the equivalent weight of the compound has to be calculated. The equivalent weight of any element is defined as the number of parts by weight of the element, which will combine with or displace 1 part (1.008) by weight of hydrogen, 8 parts of oxygen or 35.46 parts of chlorine.

For practical purposes the equivalent weight can be considered to be equal to the molecular weight/ valency. For example the equivalent weights of HC1 and H_2SO_4 are 36.46 (=36.46/1) and 49.03 (=98.06/2) respectively.

After calculating the equivalent weight of the compound, one equivalent weight of the solute is dissolved in one litre of the solvent to prepare one normal solution. If solutions of lower normality are required, then the quantity of the solute is proportionately reduced and then dissolved in one litre of the solvent. Next, this solution is again standardised against a primary standard or a solution that has already been standardised. The preparation of these standard solutions consumes considerable time and hence is used only for accurate, quantitative analytical work.

However, in qualitative analysis only dilute solutions or percentage solutions are used since not much accuracy of results is warranted. A percentage solution is prepared by dissolving a known quantity of substance in 100 ml of the solvent. The solute need not be as accurate as in the quantitative analysis. For example a 10% solution is prepared by dissolving approximately 10 g of the solute in 100 ml of the solvent, 20% solution by dissolving 20 g in 100 ml of the solvent etc. The dilute solutions are usually prepared by knowing the approximate normality of the solution and then by diluting the reagent with the required quantity of another solution like water, alcohol etc.

Pesticide sprays are also prepared using the same method of preparation of percentage solutions. The manufacturers of pesticides recommend that the pesticide may be applied at certain percentage of the concentration of active material or as so many grames of active material over a particular area. Hence the correct quantity of pesticides formulation to be used for giving required concentration of active material or the quantity of formulated material to furnish the required quantity of pesticide over a known area, has to be calculated. This can be done as follows:

Example 1

Calculate the quantity of pesticide formulation required for spraying 40 g of active ingredient in 100 litres of water using 20% wettable powder/emulsion concentrate.

100 g of the 20% formulation will contain = 20 g of a.i.

20 g of active material will be supplied by = 100 g of formulation

1 g of the active material will be supplied by $= \frac{100}{20} = 5$ g of formulation.

40 g of active material will be supplied by $= \frac{100}{20} \times 40 = 200$ g.

Hence 200 g of the formulation has to be mixed in 100 litres of water. The formula can also be used for calculating the quantity of the formulation required to be added in 100 litres of water:

$$F = \frac{AR \times 100}{AP}$$

Where F = Quantity of formulation required to be dissolved

AR = Active ingredient required to be sprayed per hectare

AP = Active ingredient present in the formulation.

Example 2

Calculate the quantity of formulation required for spraying a field with 100 litres of 0.04% active ingredient using 20% wettable powder/emulsion concentrate.

The concentration of spray solution is = 0.04% a.i.

Which means:

100 ml of the spray solution will contain = 0.04 g of a.i.

100 litres (100 × 1000 ml) of the spray solution will contain $= \frac{0.04 \times 100 \times 1000}{100}$

= 40 g of a.i.

(Subsequent steps are as in example 1)

100 g/ml of the formulation will contain = 20 g of a.i.

Hence 20 g of active ingredient will be supplied by = 100 g of formulation

Therefore 40 g of active ingredient will be supplied by $\frac{100 \times 40}{20} = 200$ g.

Hence 200 g of the formulation has to be dissolved in 100 litres of water and sprayed.

Note

In the case of emulsifiable concentrates, where certain manufacturers indicate concentration of the formulation in g/ml or g/100 cc. of the formulation, the above formulae could be used as such. In other cases when the formulations are said to be 5%, 10%, 20% etc. the specific gravity of the formulation has to be taken into consideration for finding out the accurate quantity of formulation needed. However, for practical purposes as the specific gravity of the formulations are around one, not much error will be made by assuming that 1 ml of the formulation is equivalent to one gram of the same, as the manufacturers always recommend the dosages only after providing a safety margin. However, for experimental work, the specific gravity has to be taken into consideration when calculations of the active ingredient are made.

APPENDIX II

NOMENCLATURE OF INORGANIC AND ORGANIC COMPOUNDS

The various compounds present can be divided into inorganic and organic compounds. A chemical compound should contain at least two elements. As a convention in the naming of a binary inorganic compound, the name of the metallic element is written first, followed by the non-metallic element. To distinguish a compound from the elements, a slight change is made in the name of one of the elements. In a binary compound, the suffix '-ide' is added to the second element with a slight mutilation of its name. For example a compound containing sodium and chlorine is called 'Sodium chloride' by adding 'ide' after mutilating the word chlorine.

It will be difficult to write the names of all the three elements in naming a ternary (three) compound. For the sake of reducing the length, a suitable suffix is introduced. Generally most of the ternary compounds are found to contain oxygen and hence it is omitted and a suitable suffix like 'ate' or 'ite' is added. For example when the compound contains sodium, nitrogen and oxygen, the compound may be written as sodium nitrate or sodium nitrite depending upon the number of atoms of oxygen, the former having more oxygens than the latter. A compound with a suffix 'ate' will contain more oxygen atoms than those with the suffix 'ite' for the same number of cations.

In cases where hydrogen is one of the three elements in addition to oxygen and at the same time the compound is an acid, the name is further condensed. Thus a compound containing hydrogen, sulphur and oxygen is called either sulphurous acid or sulphuric acid depending upon whether the compound contains three or four oxygen atoms.

In cases where oxygen is not one of the three elements, different naming procedures are adopted. There are cases wherein the same two elements forming a compound may produce different compounds. In such cases some prefixes or suffixes are added like cuprous oxide (CuO) and cupric oxide (Cu_2O) (the former containing less copper than the latter) or phosphorus trioxide (P_2O_3) and phosphorus pentaoxide (P_2O_6) or barium monooxide (BaO) and barium peroxide (Ba_2O).

The above are the general principles adopted in naming inorganic compounds. However, there are few exceptions. A detailed account can be had from any standard inorganic chemistry book.

Though there is no clear cut demarcation between organic and inorganic chemistry, still organic chemistry has become identified with the chemistry of carbon and its compounds. It can also be said that organic chemistry is the chemistry of hydrogen compounds of carbon and its derivatives. These organic derivatives may contain one or more of the following elements in addition to carbon and hydrogen: oxygen, nitrogen, sulphur, halogens and occasionally other elements.

Organic chemistry is divided into aliphatic chemistry and aromatic chemistry. Aliphatic chemistry can be regarded as the chemistry of methane and its derivatives whereas aromatic chemistry can be considered the chemistry of benzene and its derivatives. Aromatic compounds have certain properties of their own, which are distinct from those of aliphatic compounds. However, there is an intermediate class, the alicyclic compounds which have a mixture of properties of both.

The properties of an organic compound depend mainly on some reactive atom or groups of atoms called Functional Groups and to a certain extent upon the carbon-hydrogen framework. A functional group is a collection of chemically bound atoms that displays a characteristic set of properties. For example all alcohols contain the hydroxyl group (—OH) and the reactions considered to be typical of alcohol are essentially the reactions of the (—OH) group. If one knows the properties of —OH group and also if it is known that it is an alcohol, he can make fairly accurate predictions about the properties of the compound. The following are some of the important functional groups:

Functional group	*Structure*	*Class name*	*Example*
Hydroxyl	R—O—H	Alcohols and phenols	Ethyl alcohol C_2H_5OH
Carbonyl	$R_2C{=}O$ (R, R >C=O)	Ketones	$(CH_3)_2C{=}O$ (CH_3, CH_3 >C=O)
Aldehydes	$R{-}C(=O){-}H$	Aldehydes	Acetaldehyde CH_3CHO
Carboxyl	$R{-}C(=O){-}OH$	Carboxylic acid	Acetic acid CH_3COOH
Amino group	$R{-}NH_2$	Amines	Ethylamine $CH_3CH_2NH_2$
Alcoxyl	R—O—R	Ethers	Ethyl ether $C_2H_5{-}O{-}C_2H_5$
Nitro groups	$-N(=O)\rightarrow O$	Nitro compounds	Nitroparaffins $CH_3CH_2NO_2$
Cyano groups	$-C{\equiv}N$	Cyanides (Nitriles)	Acetonitrile CH_3CN

Functional group	*Structure*	*Class name*	*Example*
Sulphonic	R—SO_2—OH	Sulphonic acids	Benzenesulphonic acid $C_6H_5SO_2OH$
Alkane	RH	Saturated hydrocarbons	Ethane CH_3CH_2
Alkene	RCH=CHR	Olefines	Ethylene $CH_2=CH_2$
Alkynes	RC≡CR	Acetylenes	Acetylene HC≡CH
Amido	$RCONH_2$	Amide	Acetamide CH_3CONH_2
Ester linkage	RCOOR	Ester	Ethyl acetate $CH_3COOC_2H_5$
Acid halide	RCOX	Acid halide	Acetyl chloride CH_3COCl

The symbol R is the hydrocarbon residue that is attached to the functional groups. For example methyl alcohol (CH_3OH), ethyl alcohol (C_2H_5OH) arc represented by the structure R—OH where R=CH_3—, C_2H_5— etc. The aliphatic hydrocarbon groups such as methyl (CH_3—), ethyl (C_2H_5—) etc. are known as alkyl groups. Similarly the aromatic hydrocarbon groups like phenyl (C_6H_5) are called aryl groups. The acyl group is $\overset{\displaystyle O}{\overset{\|}{(RC—)}}$

Nomenclature of Organic Compounds

As there are thousands of organic compounds, it is difficult to provide each compound with a separate distinctive name. However, fortunately most of them form distinct homologous series or families and hence all of them can be identified with a family name. For example all of alcohols like methyl alcohol (methanol), ethyl alcohol (ethanol) etc. end with the family name alcohol in the common system and with '-ol' in the IUPAC system. Similarly compounds of each family end with a characteristic family name or a particular ending The system that is commonly followed, called "Common System" has certain defects like not being able to name all isomers etc. Hence the International Union of Pure and Applied Chemists (IUPAC) have published certain guidelines for naming the organic compounds and this system is known as "IUPAC System".

In subsequent pages, a short summary of the various principles followed in naming organic compounds by both the Common System and the IUPAC System are outlined. For detailed information readers can refer to the standard books in organic chemistry or the Journal of American Chemical Society 82 : 5545-5574 (1960).

Aliphatic hydrocarbons can be divided into saturated hydrocarbons, unsaturated hydrocarbons with either double or triple bonds and hydrocarbons with substituent functional groups. In the Common System, all the isomeric paraffinic hydrocarbons have the same parent name. The various isomers are distinguished by certain prefixes like 'n', 'iso' etc., which indicate the type of branching present in the molecule. If all the carbon atoms are in one continuous chain, the prefix 'n' (normal) is placed before the parent name (e.g.) n-butane, ($CH_3CH_2CH_2CH_3$), n-hexane ($CH_3CH_2CH_2$—CH_2CH_3). The prefix 'iso' is used to denote a branched chain system where a CH_3-group is attached next to the last carbon of a continuous chain of carbons.

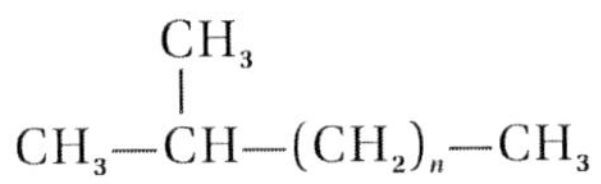

The prefix 'neo' is used for compounds having the same structural feature and at the same time two methyl groups substituting on the same carbon of the main chain:

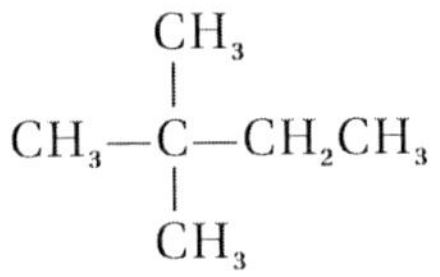

Neo-hexane

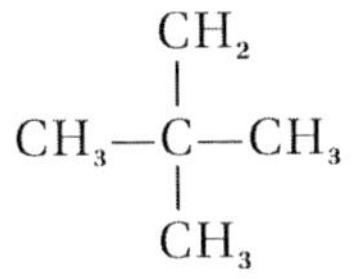

Neo-pentane

However, all the compounds cannot be named by the Common System.

In the IUPAC System, the branched saturated hydrocarbons and the hydrocarbons substituted with simple substituents are named as per the guidelines given below:

1. First the longest continuous chain of carbon atoms is identified in the molecule. The longest continuous chain need not be the portion of the molecule written horizontally.

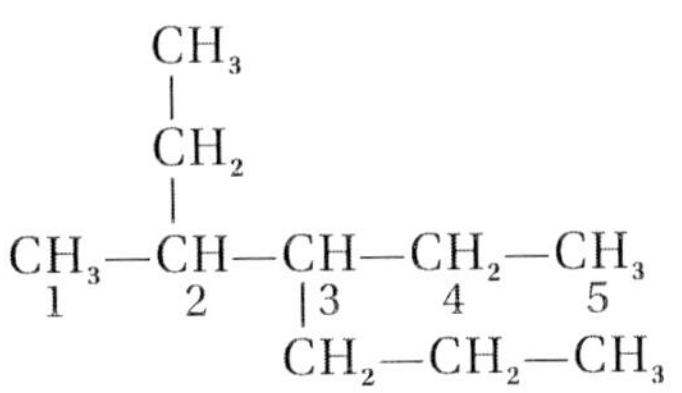

—Heptane

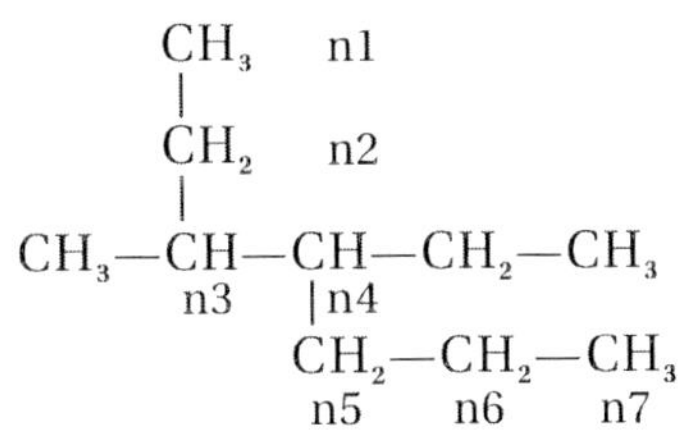

In the above compound heptane is the longest chain and the name of the straight chain alkane is used as the parent name of the molecule.

2. The numbering of the carbon atoms of the continuous chain is done by starting from the end which will give numbers of the smallest value to the carbon carrying a substituent. (A substituent is an atom or group of atoms other than hydrogen that is attached to a carbon atom of the longest chain.)

3. Then position of each substituent is indicated by giving it the number of carbon to which it is attached. For example the methyl group on C_{n3} is written as 3-methyl and the ethyl group on the C_{n5} is written as 4-ethyl. The name of the compound is written as one word. The numbers are separated from the substituent's names by hyphen and numbers are separated from each other by comma. Also the names of the substituents are written as prefixes to the parent name.

4. If the same substituent is present two or more times in a molecule, this is indicated by the prefixes di-, tri-, tetra- etc. attached to the substituent name. The prefixes fluoro-, chloro-, bromo-, and iodo- are added to indicate the presence of different halogens.

Primary, Secondary and Tertiary Carbons and Hydrogens

It is convenient to classify carbons by the number of other carbons attached to them: A carbon attached to another carbon is called a primary carbon and the hydrogen attached to the primary carbon is known as the primary hydrogen. In the 2-methyl butane, the C_{n1} and C_{n4} are primary carbons as they are attached to only one other carbon atom.

$$\begin{array}{cccc} & CH_3 & & \\ & | & & \\ CH_3— & CH— & CH_2— & CH_3 \\ n1 & n2 & n3 & n4 \end{array}$$

The C_{n3} is called a secondary carbon as it is attached to two other carbon atoms (C_{n2} and C_{n4}) while C_{n2} is a tertiary carbon as it is attached to three other carbons (C_{n1}, C_{n3} and C of the CH_3 of the side chain). The hydrogen attached to C_{n2} is called the tertiary hydrogen.

Unsaturated Compounds

In the common system, the name of the compound is obtained by adding the suffix '-ene' to the name of the alkyl group e.g., ethaneethylene, propane—propylene etc.

In the IUPAC System for compounds having double bonds, the longest continuous chain in the molecule's chain is chosen as in the case of paraffins. The only additional restriction imposed is that both the carbons atoms of the double bond should be part of the longest chain.

$$\begin{array}{cccccc} & & n2 & n3 & n4 & n5 \\ CH_3— & CH— & C— & CH_2— & CH_2— & CH_3 \\ & & \| & & & \\ & & CH_3 & & & \\ & & n1 & & & \end{array}$$

2-ethyle-1-pentene

Similar to the numbering of substituents in the case of saturated hydrocarbons, the chain in the unsaturated hydrocarbon is numbered in such a way that the two double bonded C atoms have the lowest possible numbers. In numbering, both the carbons involved in double bonds are also included. The lower bonded carbon is used for indicating the position of the double bond. This number is placed immediately before the parent hydrocarbon name. If there are two, three etc., double bonds, they are indicated by the suffix '-adiene' (for two double bonds), '-atriene' (for three double bonds) etc. The numbers are separated by commas. If there are substituents like halogen or alkyl groups, they are indicated by name and number similar to alkane.

$CH_3—CH=CH—CH_2—CHCl_2$ 5,5-dichloro-2-pentene.

In the case of compounds containing triple bonds, the compound is named by adding '-yne' in the place of '-ane'. The same principle used for naming and numbering of double bonded carbon compounds is adopted here also. If there is more than one triple bond, the suffixes '-adiyne', '-atriyne' etc. are added.

$CH_3C{\equiv}CCH_3$	$HC{\equiv}C{-}C{\equiv}CH$
2-Butyne	1,3 butadiyne

In cases of compounds containing both double and triple bonds, the suffixes '-ene' and '-yne' are added to denote the presence of a double bond and a triple bond respectively. The suffix '-ene' always precedes the suffix '-yne' in the name of a compound even when the double bond is assigned the larger number. When the suffix following 'ene' begins with a vowel or y, the final 'e' of the '-ene' is omitted. Otherwise it is included. Thus a compound containing one double bond and one triple bond is an '-enyne' while one containing one double bond and two triple bonds is an '-enediyne'. The position number for the double bond is placed before the parent hydrocarbon name and the position number of the triple bond is placed between the '-ene' and the '-yne'.

$HC{\equiv}CCH{=}CHC{\equiv}CH$ 3-Henene-1,5-diyne

If the compound contains both a double bond and a triple bond and if the same number results irrespective of the starting point of numbering, the double bond is given the lower number. Otherwise the lowest possible numbers are assigned. The IUPAC System is used for more complex molecules but for simple alkynes, names based on the relationship of the molecule to the acetylene are employed.

Aromatic Compounds

In aromatic compounds, the aryl group is formed by removing a hydrogen from the aromatic hydrogen. Similar to the use of R for representing any alkyl group, Ar is used to represent any aryl group. The group formed by removing one of the hydrogens of benzene is called the phenyl group and represented by ϕ or by Ph also.

In the case of aromatic compounds, many of the common names like benzene, phenol, toluene etc. are often used in the IUPAC System. When a hydrocarbon is substituted, the parent name of the hydrocarbon is used. All ring carbons are numbered in such a way as to give the lowest possible number to the position of substituents. The last named substituent is assumed to be at the 'l' position and the number is not indicated. If the parent hydrocarbons itself is a substituted benzene ring, such as toluene, the carbon carrying. this substituent is assigned number 'l'.

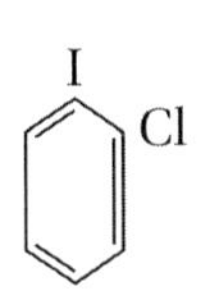

2-chloroiodobenzene

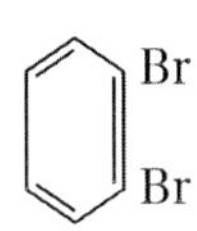

1,2-dibromobenzene

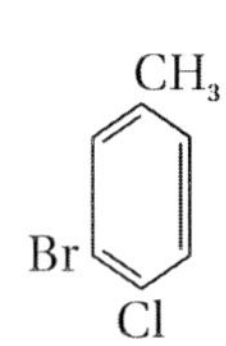

3-bromo-4-iodotoluene

When there are only two substituents on the ring, the relative position of the substituents are often indicated by the terms 'ortho', 'meta' and 'para'. When the two substituents are on adjacent ring carbons, it is called 'ortho', while those on the ring carbons separated by one carbon and two carbons are denoted as para (p) and meta (m) respectively. O, m and p are used to indicate the relative positions of the substituents on a ring where there are only two substituents. If there are more than two substituents, numbering is used.

Naming of Compounds with Substituents

Carboxylic acids

Similar to alkanes, carboxylic acids are named by choosing the longest continuous chain containing the carboxyl group and replacing with the ending '-oic' acid, the final '-e' of the name of the hydrocarbon containing the same number of carbon atoms as are in the chain.

$CH_3CH_2CH_2CH_3$

Butane

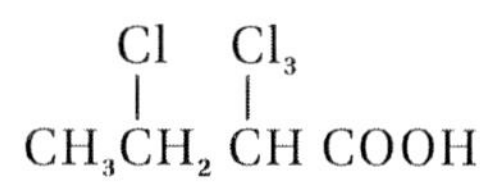

3-chloro-2-methyl butanoic acid

The position of the other substituents or functional groups are indicated by numbering the carbon of the carboxyl group as '1'. The presence of a double or a triple bond in an acid molecule is indicated by the suffiix '-enoil' or '-yenoic' respectively. Numbering starts with the carboxylic carbon and the number of the lower numbered ethylenic or acetylienic carbon is used to indicate the position of unsaturation. This number is placed immediately before the parent name of the compound:

$CH_2{=}CHCH_2COOH$ 3-butenoic acid

If two carboxylic groups are present, the suffix '-dioic' is added instead of '-oic' to the name of the compound.

Carboxylic acid system: The cycloaliphatic acids, aliphatic acids containing three or more carboxylic groups and aromatic acids containing more than one carboxylic group are named by the carboxylic acid system. The ring carbon to which the carboxyl group is attached is given number "1" unless the ring system has a fixed numbering system as in naphthalene. In that case the carboxyl group is given the lowest possible number. In this, the carbon of the carboxyl group is not included when counting the carbons to determine the name of the parent hydrocarbon.

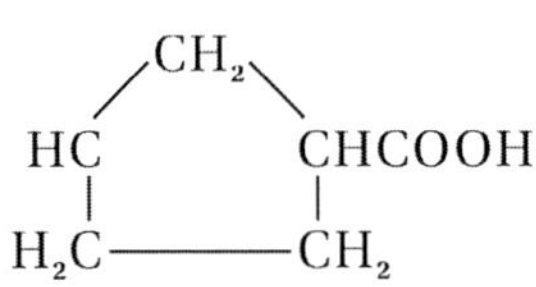

Cyclopentane carboxylic acid

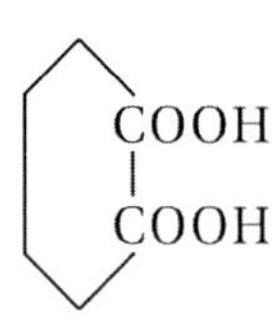

1,2-benzenedicarboxylic acid

Common system: In the common system, the position of the acid molecule, or the substituent or other functional groups are indicated by the Greek alphabets α, β, γ, δ, E, etc. The first Greek letter is used to indicate the carbon to which the functional group is attached and not to the carboxyl carbon itself. Hence α, β, γ etc. will refer respectively to the second, third, fourth etc. carbons of the IUPAC System.

The salts of the carboxylic acids are named in the same way as the salts of inorganic acids by giving the name of the cation followed by the name of the carboxylate anion. The name of the carboxylate anion is obtained by replacing the '-ic' of the acid name with 'ate'.

Alcohols

Alcohols are classified as primary, secondary, or tertiary depending upon whether the hydroxy group is attached to a primary, secondary or a tertiary carbon.

```
   |                 C                 C
C—C—OH               |                 |
   |              C—C—OH            C—C—OH
                     |                 |
                                       C
```

Primary alcohol | Secondary alcohol | Tertiary alcohol

In the IUPAC System, similar to the previous cases, the alcohol is named by choosing the longest carbon chain in the molecule bearing the hydroxy group. Alcohol is named by replacing the final 's' of the parent hydrocarbon with '-ol'. The presence of more than one hydroxy group is indicated by the appropriate Suffixes like '-diol', '-triol' etc., added to the name of the parent hydrocarbon. The position of each hydroxy group is indicated by a number placed before the modified name of the parent hydrocarbon.

$$\underset{\displaystyle OH}{CH_3CH_2\overset{|}{C}HCH_3}$$ 2-Butanol

In the case of unsaturated alcohols, they are numbered in such a way that the hydroxy group rather than the unsaturated point gets the lower number.

$$CH_2{=}CH\ \overset{\displaystyle OH \atop |}{C}H{-}CH_2Cl$$ 1-chloro-3-neten 2-ol

In compounds where the hydroxy group is not the main functional group as in aldehydes, ketones etc. the substituent name '-hydroxy' is used as a suffix to the name of the acid.

$$CH_3\overset{\displaystyle OH \atop |}{C}HCOOH$$ 2-hydroxy propinoic acid

In the Carbinol System the alcohol is named as if it were a substituted methanol. The suffix 'Carbinol' refers to

$$-\overset{|}{\underset{|}{C}}OH$$

In the Common System, alcohols having five and less than five carbons are known by their common names. The name is formed by adding the word "alcohol" to the name of the alkyl group.

$CH_2 = CH\ CH_2OH$	Allyl alcohol

Amines

There is very little distinction between the Common System and the IUPAC System. Amines can be considered to be derivatives of ammonia by replacement of one, two or all the three hydrocarbons by the alkyl or aryl groups. Amines are classified by considering the number of hydrogens replaced by the alkyl or aryl groups.

RNH_2	$R-\overset{R'}{\overset{\mid}{N}}H$	$R-\overset{R'}{\overset{\mid}{N}}-R'$
Primary amines	Secondary amines	Tertiary amines

Classification of amines depends upon the number of R groups attached to the nitrogen atoms and *not* on the nature of the carbon to which the N is attached, as in the case of alcohol or unsaturated hydrocarbons. Monoamines having 2 or 3 different R groups are called primary amines which are substituted on the nitrogen. The largest R group is used to form the name of the primary amine. The other R groups are considered to be substituents on the nitrogen and are indicated by placing on N-before the name of the alkyl or aryl group.

$CH_3NH\ CH_2CH_3$	N-methyl ethylamine
$CH_3\overset{CH}{\overset{\mid}{C}}H-\overset{CH_3}{\overset{\mid}{N}}-\overset{CH_3}{\overset{\mid}{C}}HCH_2CH_2$	N-methyl -N- isopropyl -sec butylamine

In the name of polyamines, the prefixes '-diamine', '-triamine' etc. are added to the hydrocarbons. The position of the amino group is indicated by placing the number before the the hydrocarbon.

Similarly for naming of compounds containing groups like ethers (hydrocarbons substituted with an RO-group, alkoxy group) quinones (cyclic diketones that may be reduced to dihydric phenyls) etc. in the Common and IUPAC systems, standard books on organic chemistry may be consulted. In case of compounds containing different functional groups (polyfunctional compounds), the one higher in the order of precedence of functions is shown as a suffix, while the one lower in that order is indicated by the prefix.

APPENDIX III

ACUTE TOXICITY DATA FOR PESTICIDES

(ARRANGED IN ALPHABETICAL ORDER)

The LD_{50} values given below are for rats except in cases where they arc for rabbits (referred to as Rb indicating rabbits). The values are expressed as mg/kg of body weight.

Name of the pesticides	*Oral*	*Dermal*
Aldrin	40-60	200
Allethrin	680-1000	11,200 (Rb)
Allisan (*see* Dichloran)	1500-4040	—
Amitone oxalate (*see* Tetram)	6	5
Animert (*see* Tetrasul)	6800>14700	—
Arprocarb (*see* Baygon)	95-175	>1000
Aramite	4000-6000	—
Azinophos-ethyl (*see* Ethylguthion)	9	280
Azinophos-methyl (*see* Guthion)	7-13	280
Baygon (*see* Arprocarb)	95-175	>1000
Baytex (*see* Fenthion)	200	1300
BHC (*see* Lindane)	200	500-1000
Bidrin	27-45	—
Binapacryl (*see* Morocide)	150-225	1350 Rb
Birlane (*see* Chlorofenvinphos)	10-39	108
Botran (*see* Dichloran)	1500-4040	—
Bromophos	3750-5050	—
Butonate	1050	—
Captan (*see* Orthocide)	9000-15000	—
Carbaryl (*see* Sevin)	400	> 500
Carbophenothion (*see* Trithion)	7-30	800
Chlordane	283	>1600
Chlorfenvinphos (*see* Birlane)	10-39	108

Name of the pesticides	***Oral***	***Dermal***
Chlorobenzilate	700-3200	>3000 Rb
Chloropropylate	> 5000	>150
Chloroparacide (*see* Chlorobenside)	2000->10,000	—
Chlorothion	625-1500	1500-4500
Cidial	200-300	700->1400
Ciodrin	125	—
Copper salts	700-1000	>1000
Co-Ral (*see* Coumaphos)	15-180	860
Coumaphos (*see* Co-Ral)	15-180	860
Coumithoate (*see* Dithion)	67	>200
Dazomet (*see* DMTT)	6600-9300	—
DDD (*see* TDE)	400-3400	>5000 Rb
DDT	300-500	2500
DDVP (*see* Dichlorovos)	25-30	75-900
Dihydroacetic acid (*see* DHA)	500-1000	—
Delnav (*see* Dioxathion)	20-40	350
Demeton (*see* Systox)	3-5	200
Demcton methyl (*see* Metasystox)	50-75	300-450
Demeton-S-methyl (*see* Meta- isosystox)	40	85
Derris	1500	—
DHA (*see* Dihydroaceticacid)	500-1000	—
Diazinon	300-600	500-1200
Dibrom (*see* Naled)	430	1100
Dicapthon	330-475	800-1250
Dichlone (*see* Phygon)	1300-2250	—
Dichloran (*see* Allisan, Botran)	1500-4040	—
0-Dichlorobenzene	500	—
Dichlorovos (*see* DDVP)	25-30	75-900
Dicofol (*see* Kelthane)	575-1100	1000-1230
Dieldrin	40	>100
Dilan	1100-4000	6000
Dimefox	2	2-10
Dimetan	140-150	—
Dimethoate (*see* Rogor)	200-300	700-1150

Name of the pesticides	*Oral*	*Dermal*
Dimetilan	25-50	600-700
Dimile	500	—
Dinitrocresol (*see* DNOC)	25-40	200-600
Dinitrotrichlorobenzene	500	
Dinocap (*see* Karathane)	890-1190	>9400 Rb
Dioxathion (*see* Delnav)	20-40	350
Diptrex (*see* Trichlorofon)	650	>2800
Disulphoton (*see* Disyston)	4	50
Dithiaon	1015	—
Dithiocarbamates (*see* Maneb, Zineb etc.)	1000-8000	>1000
Dithioquinox (*see* Morestan, Forstan)	2500-3000	>1000
Dithion (*see* Coumithoate)	67	200
DMTT (*see* Dazomet)	500-650	—
DNBP (*see* Dinoseb)	50	80-200
Dodine (*see* Melprex)	1000-2000	—
Dowco 109 (*see* Ruelene)	660-1000	4000 Rb
Dow ET 14 (*see* Fenchlorphos)	1000-3000	1000-2000 Rb
Dow Et-15	710	>1000
Ekatin (*see* Thiometon)	1000	> 200
Endosulphan (*see* Thiodan)	35	74-130
Endothion	23	130
Endrin	3-6	60-120
EPN	8-17	25-230
Eradex (*see* Thioquinox)	1800-3400	>3000
Ethion	13-34	1600
Ethylguthion (*see* Azinphosethyl)	9	280
Ethyl mercuric salts	30	200
Etrolene (*see* Fenchlorphos)	1000-3000	1000-2000 Rb
FAC 20 (*see* Prothoate)	14-25	100-200
Fenchlorphos (*see* Ronnel, Etrolene, Dow Et-14)	1000-3000	>3160 Rb
Fenitrothion (*see* Sumithion)	250-673	>3000
Fenthion (*see* Baytex, Lebyacid)	200	1300
Fluoroacetamide	15	80

Name of the pesticides	***Oral***	***Dermal***
Folpet (*see* Phaltan)	>10,000	—
Formothion	400	400-600
Forstan (*see* Dithioquinox)	2500-3000	>1000
Guthion (*see* Azinphosmethyl)	7-13	280
Haloxon	900-2000	—
Heptachlor	40	200-250
Imidan	147	>3160 Rb
Isobenzene (*see* Telodrin)	5-10	5-30
Isodrin	7-17	23-35
Isolan	12	35-60
Karathane (*see* Dinocap)	980-1190	>9400 Rb
Kelthane (*see* Dicofol)	575-1100	1000-1230
Kilval (*see* Vamidothion)	64-100	1160 Rb
Lead arsenate	10-100	>2400
Lebaycide (*see* Fenthion)	200	1300
Lethane	90-300	125-250 Rb
Lindane (*see* BHC)	200	125-250 Rb
Malathion	1400-1900	>4000
Maneb (*see* Dithiocarbamates)	100-8000	> 1000
Matacil	30	—
Mecarbam	15	380
Menazon	1200-1600	> 500
Mesurol	130-135	—
Meta isosystox (*see* Demeton-s-methyl)	40	85
Metasystox (*see* Demeton methyl)	50-75	300-450
Metasystox-R (*see* Oxydemeton-methyl)	57	100
Metham (*see* Vapam)	820	800 Rb
Methoxychlor	5000-7000	6000
Methyltrithion	98-200	190-215
Mevinphos (*see* Phosdrin)	3-5	90
Mebam	>234	>6000 Rb
Morestan (*see* Dithioquinox)	2500-3000	>1000 Rb
Morocide (*see* Binapacryl)	150-225	135 Rb
Morphothion	200	283

Name of the pesticides	***Oral***	***Dermal***
Naled (*see* Dibrom)	430	1100
Neotran	5800	>1000 Rb
Nicotine	70	140
Orthocide (*see* Captan)	9000-15,000	—
Oxydemeton-methyl (*see* Metasystox-R)	57	100
Parathion	3-6	4-200
Parathion-methyl	12-16	67
PCNB (*see* Quintozene)	1650-12,000	—
Pemphene (*see* Tetrachlorothiophene)	70	—
Pentachlorphenol	280	105-350
Phaltan (*see* Folpet)	>10,000	—
Phenkapton	50	> 1000
Phenyl mercuric salts	60	—
Phorate (*see* Thimet)	2-3	70-300
Phosdrin (*see* Mevinphos)	3-5	90
Phosphomidon	15	125
Phygon (*see* Dichlone)	1300-2250	—
Potasan	20-40	> 300 Rb
Prothoate (*see* FAC 20)	14-25	100-200
Pyrethrins	570	>1880
Pyrolan	50-60	—
Quintozene (*see* Terrachlor, PCNB)	1650-12,000	—
Rogor (*see* Dimethoate)	200-300	700-1150
Ronnel (*see* Fenchlorphos)	1000-3000	1000-2000 Rb
Rotenone	60-132	>940
Ruelene (*see* Dowco 109)	660-1000	4000 Rb
Ryanaia	750	—
Schradan	5	50-100
Sevin (*see* Carbaryl)	400	>500
Strobane	200-250	>20
Sulphotepp	1-5	20 Rb
Sulphenone	>2000	> 1000 Rb
Sulphur	Non-toxic	—
Sumithion (*see* Fenitrothion)	250-673	>3000

Name of the pesticides	***Oral***	***Dermal***
Supracide (*see* Methidation)	20-48	25-400
Systox (*see* Demeton)	3-5	200
TD-183 (*see* Tetrachloro- thiophenene)	70	—
TDE (DDD)	400-3400	> 5000 Rb
Tedion (*see* Tetradifon)	5000-14,700	—
Telodrin (*see* Isobenzene)	5-10	5-30
TEPP (*see* Terrasystam)	0.5	20
Terrachlor (*see* Quintozene)	1650-12,000	—
Terrasystam (*see* TEPP)	0.5	20
Tetrachlorothiophene (*see* pemphene, TD-183)	70	—
Tetradifon (*see* Tedion)	5000-14,700	—
Tetram (*see* Amitone oxalate)	6	5
Tetrasul (*see* Animert)	6800->14,700	—
Thiocron	600-660	—
Thiodan (sec Endosulphan)	35	74-130
Thiometon (sec Ekatin)	100	>200
Thionazin (*see* Zinophos)	12	—
Thioquinox (*see* Eradex)	1800-3400	> 3000
Thimet (*see* Phorate)	2-3	70-300
Thiram (*see* TMTD)	375-865	—
TMTD (*see* Thiram)	375-865	—
Toxaphene	283	>1000
Trichlorfon (*see* Diptrex)	650	2800
Triphenyl tin salts	238	450
Trithion (*see* Carbophenothion)	7 30	800
Vamidothion (*see* Kilval)	64-100	1160 Rb
Vapam (sec Metham)	820	800 Rb
Wepsyn (*see* Triaininophos)	10 20	1500-3000 Rb
Zectran	15-63	—
Zineb (*see* Dithiocarbamates)	1000-8000	>1000
Zinophos (*see* Thionazin)	12	—

GLOSSARY OF TERMS USED IN PESTICIDES CHEMISTRY *

ABSORPTION: As applied to chemical hazards, absorption refers to the process of penetration of a poison into the body. A material is said to be absorbed only when it has gained entry into the blood stream and consequently reaches all parts of the body. This requires that the substance pass through the skin, mucous membrane or air sac of the lungs.

ACETHION: is an analog of malathion having low mammalian toxicity. Its chemical formula is 0, 0-diethyl-S-carbothioxy- methyldithiophosphate (132).

ACETOFOS: is a contact organophosphorus insecticide and is similar to trichlorofon. It is-0,0-diethyl-S-carboxymethyl thiophosphate (112).

ACETOMETAARSENITE: is the first arsenical insecticide to be used in large quantity. It is a compound theoretically containing 55-58% of arsenic trioxide (34).

ACID BURGANDY MIXTURE: is produced by reacting less than 0. 86 parts of crystalline sodium carbonate with one part of copper sulphate (185).

ACID LEAD ARSENATE: *see* dilead orthoarsenate (36).

ACTIDIONE: *see* cycloheximide; 3,2-(3, 5-dimethyl-2-oxocyclohexyl)- 2-hydroxyethyl glutarimide (204).

ACTIDIONE, R.Z.: contains 5% cycloheximide and 75% PCNB (204).

ACTIMYCIN, A. —is the first compound prepared from *Streptomyces* for use as a protective fungicide.

ACTINIC LIGHT: Usually it is the violet and UV light portions of the electromagnetic spectrum (88).

ACTIPRAY: is the market product of an antibiotic cycloheximide (204).

ACTIVE INGREDIENT: is the element, group or groups of the chemical compound which are responsible for insecticidal or fungicidal action of the formulation.

ACUTE: In the medical sense, it means short duration. As applied to materials inhaled or absorbed through the skin, it refers to a single exposure. When applied to materials ingested, it refers to a single dose (208).

AEROSOLS: A colloidal system in which the dispersion medium is a gas (27).

AGALOL: (Ceresan-universal Nassberge, Aretan, Methoxy ethyl mercuric chloride)—is a mercurial *see*d disinfectant. It is sold as a formulation containing 1.75% mercury (190).

ALDICARB: *see* Temik (154).

---

* The Numbers at the end of each para indicates the page number in the text.

ALDRIN: (HHDN; 1, 2, 3, 4, 10, 10-hexachloro-l, 4, 4a, 5, 8, 8-hexahydro- 1, 4-endo, exo-5, 8-dimethano-naphthalene: $C_{12}H_8Cl_6$)—is a chlorinated hydrocarbon produced by the Diels-Alder reaction. It has a minimum of 95% HHDN (active material) (92).

ALKALOIDS: is a naturally occurring heterocyclic, optically active, nitrogenous base of relatively high molecular weight and having marked physiological activity e. g. nicotine (47).

ALLERGY: is a stage in which the cells of the body are hypersensitive to certain substances introduced into it. The reaction of the body is exhibited as oedema, inflammation and destruction of tissue.

ALLETHRIN: (Divinyl pyrethrin I, DVPY-I, Allylrethrum allyl analog of cinerin-S) is a natural organic insecticide and is an ester made from alcohol having the allyl side chain (allethriplone). Allethrum is a mixture of four racemic forms (55).

ALUMINUM OXIDE: is used as a corrective in the formulation of arsenical insecticides to prevent arsenical injury.

ALUMINUM PHOSPHIDE: is one of the insecticides now manufactured in India (3).

AMIDITHION: *see* Thiocron (134).

AMINOPARATHION: (0,0-diethyl-0,4-amino-phenylthiophosphate)—is the compound formed by reacting parathion with reducing agents like metals in acid medium (113).

AMMONIUM POLYSULPHIDE: (APS) is a sulphur fungicide prepared by saturating concentrated ammonia with hydrogen sulphide and then dissolving excess sulphur in the resulting solution and ammonium sulphide (169).

AMMONIUM SULPHAMATE: is a weedicide now manufactured in India (4).

AMMOPHOS: (0,0-dimethyl-S-2-acethyl-amide ethyldithiophosphate)—is a new systemic organophosphorus insecticide.

ANABASINE: *see* Neonicotine (47).

ANTHIO: (Formothion,0,0-Dimethyl-S-(N-methyl-N-formylcarba- moyl methyl)-dithiophosphate)—is an organic phosphorus compound similar to dimethoate but is more stable than dimethoate (135).

ANTIBIOTIC: is a substance produced by microorganisms which inhibit the growth of a poisonous or toxic chemical or radioactive materials (204).

ANTIDOTE: A substance given to a patient to counteract the effect produced by the ingestion of a poisonous or toxic chemical (42).

APS: *see* Ammonium polysulphide (169).

ARAMITE: (2-chloroethyl-l-methyl-2-(p-tert-butyl phenoxy ethyl sulphite)—is a chlorinated hydrocarbon. It is commonly used as an acaricide (83).

ARASAN: *see* Thiram (174).

ARATHANE: *see* Karathane (158).

ARETAN: *see* Agalol (191).

ARSENIC: *see* Arsenic trioxide (34).

ARSENIC ACID: *see* Arsenic pentoxide (33).

ARSENIC PENTOXIDE: is an arsenical insecticide. It is also called arsenic acid (As_2O_5) (34).

ARSENIC SULPHIDE: *see* Realgar (33).

ARSENIC TRIOXIDE: (white arsenic, arsenic, As_2O_3)—is an arsenical insecticide (34).

ARSENIC WHITE: *see* Arsenic trioxide (34).

AUEROFUNGIN: is an antibiotic having systemic action, isolated from *Streptomyces cinnamomeus* var. *terricola tir* (205).

AVENIN: (0,0-dimethyl-N-(isopropoxy carbamoyl) phosphate)—is a systemic organophosphorus insecticide with low activity and narrow selective action (101),

AZINOPHOS METHYL: *see* Guthion (140).

BAL: is one of the specific antidotes for arsenical poisoning since it readily complexes and solubilises trivalent arsenic (42).

BARIUM FLUORIDE: (BaF_2)—is a minor fluorine Insecticide (44).

BARIUM FLUOSILICATE: ($BaSiF_6$)—is a minor fluorine insecticide (45).

BARIUM POLYSULPHIDE: (Solbar)—is a fungicide prepared by saturating barium hydroxide with hydrogen sulphide or by mixing ground sulphur with barium sulphite powder (169).

BARIUM SULPHIDE: *see* Solbar, barium polysulphide (169).

BASIC CALCIUM ARSENATE: is a safe form of calcium arsenate, has the composition of $(Ca_2(AsO_4)2)_3$ $Ca(OH)_2$ with 25% metallic arsenic and 0. 4 to 0. 5% water soluble arsenic (38).

BASIC CHLORIDE OF COPPER: *see* Copper oxychloride (185).

BASIC COPPER ARSENATE: ($Cu(CuOH)AsO_4$)—is an arsenical compound containing 56. 27% copper oxide (44. 8% metallic copper), 40. 6% arsenic pentoxide and 3. 2% of water of constitution (40).

BASIC LEAD ARSENATES: ($Pb(PbOH)$ (AsO_4). $3H_2O$)—are arsenical insecticides. Several compounds of basic lead arsenate exist and the basic lead arsenate of commerce has the approximate composition Pb $(PbOH)(AsO_4)3H_2O$ (37).

BASIC MAGNESIUM ARSENATE: ($Mg_3(AsO_4)_2$. $2H_aO$)—is an arsenical insecticide of minor importance (40).

BASIC MAGNESIUM ORTHOARSENATE: (Mg, $(AsO_4)_2$. MgO)—is an arsenical insecticide of minor importance (40).

BAYER 30687: *see* Eradex (161).

BAYER 39007: *see* Baygon (152).

BAYGON: (PHC, (Japan), Propoxur, Bayer 39007, 2-iso-propoxy-phenyl-N-methylcarbamate)—is a broad spectrum carbamate insecticide with contact action and long residual action (152).

BAYTEX: *see* Fenthion (121).

BENALATE: (Methyl-l-butylcarbamoyl)-2-benzino-diazole carbamate)—is a systemic fungicide and possess residual and systemic properties (206).

BENTONITE: is a colloidal native hydrated aluminum silicate (clay) varying in colour from yellowish white to almost black. It is used as an emulsifier of oil (52).

BENZENE HEXACHLORIDE: *see* BHC (87).

BHC: (Benzene hexachloride, hexachlorocyclohexane, Gammexane, HCH, 1, 2, 3, 4, 5, 6-hexachlorocyclohexane, $C_6H_6Cl_4$)—is a chlorinated hydrocarbon, first synthesised by Micheal Faraday in 1825. At present 8 isomers are known, of which γ-isomer is highly toxic to insects (87).

BI-58: *see* Rogor (133).

BIDRIN: (3-dimethoxy phosphinyloxy)-N,N-dimethyl crotonamide)— is a systemic vinyl phosphate insecticide. Bidrin is the name applied to the pure cis isomer of the compound (101).

BIRLANE: (Chlorfenvinphos: 2-chloro-(2,4-dichlorophenyl) vinyl diethyl phosphate) —is a moderately persistent contact and stomach insecticide. Of the the two isomers the trans isomer is insecticidally most active (101).

2, 2-Bis-(p-chlorophenyl)-l, l-dichloroethylene: *see* DDE (79).

2, l-Bis-(p-chlorophenyl) ethanol: *see* Dimite (82).

2,2, -Bis-(p-chlorophenyl)-1,1,1-trichloroethane: *see* DDT (78).

BIS-(P-CHLOROPHENOXY) METHANE: *see* Neotran (82).

5-Butyl-2-dimethylamine-4-hydroxy-6-methylpyrimidine: *see* Methyrimol (207).

5-Butyl-2-ethylamino-4-hydroxy-6-methylpyrimidine: *see* PPI 49 (207).

BIS-DIMETHYLFLUOROPHOSPHATE: *see* Dimefox (101).

BIS (DIMETHYLTHIOCARBAMYL) DISULPHIDE: *see* Thiram (174).

BLADAFUM: *see* Sulphotepp (144).

BLADEN: *see* Tetraethylpyrophosphate (143).

BLASTIMYCIN: is an antibiotic produced by *Streptomiyces blastimycetius.* It has been found to be effective against paddy blast (204).

BLASTOCIDIN: is an antibiotic isolated from *Streptomyces griseschromegenes.* It has been used against paddy blast (204).

BLIMIX 4%: *see* Copper oxychloride (185).

BLITOX 50: *see* Copper oxychloride (185).

BLUE COPPER 50: *see* Copper oxychloride (185).

BLUE STONE: *see* Copper sulphate (180).

BLUE VITRIOL: *see* Copper sulphate (186).

BORDEAUX MIXTURE: consists of copper sulphate (in lb), lime (in lb) and water in gallons in the ratio of 4-4-50. The solutions of copper sulphate and finely ground quicklime are prepared separately and both mixed simultaneously in a third container with constant agitation (180).

BRESTAN: *see* Triphenyl tin acetate (201).

BROMOPHOS: (0,0-dimethyl-0-2,5-dichloro-4-bromophenylthiophos- phate)—is the 4-bromoanalog of ronnel and is a persistent general purpose organic phosphate insecticide of low acute toxicity to mammals (111).

BULAN: (DNB: 2-nitro-1,1-bis (p-chlorophenyl) butane) (82).

BULK DENSITY: is the weight of a definite volume of dust measured under specific conditions as grams per cubic volume (16).

BURGANDY MIXTURE: (Soda mixture) is prepared by allowing one part of copper sulphate pentachloride to react with one or two parts of crystalline sodium carbonate in solution (185).

BUTYLAMINE STEARATE: (Lovo)—is an additive added to pesticides in aerial spraying for decreasing the evaporation of the drops (26).

CALCIUM ARSENATE: The commercial compound is a mixture of tricalcium arsenate $Ca_3(AsO_4)_2$ with 37.5% arsenic, calcium arsenate ($CaHAsO_4$, excess lime and calcium carbonate (38).

CALCIUM ARSENATE: (Basic)—is $(Ca_3(AsO_4)_2)_3.Ca(OH)_2$ with 25% metallic arsenic and 0.4 to 0.5% water soluble arsenic (generally) (38).

CALCIUM FLUOSILICATE: ($CaSiF_6.2H_3O$)—is a minor fluosilicate insecticide (44).

CALCIUM FLUOSPAR: ($CaSiF_6$)—is a minor fluorine insecticide (44).

CALCIUM MONOSULPHIDE: (CaS)—is a fungicidal powder of varying colours. The fungicidal efficacy of the compound depends on the calcium sulphide present (169).

CALCIUM TETRATHIONATE: is a sulphur fungicide prepared by passing sulphur dioxide into sulphuryl chloride (170).

CALOMEL: *see* Mercurous chloride (188).

CAPTAN: (Esso fungicide 406, Orthocide 406; Vancide 89; N-trichloro- methyl thiotetrahydrophthalimide)—is a versatile fungicide for foliar application. It belongs to the heterocyclic nitrogenous group of fungicide (197).

CARBARYL: (Sevin; Naphthyl carbamate 1-naphthyl-N-methyl- carbamate; methyl isocyanate)—is a good contact carbamate insecticide with occasional systemic action (149).

CARBOFURAN: *see* Furadan (150).

CARBOPHENOTHION: *see* Trithion (130).

CARBOPHOS: *see* Malathion (131).

CASEIN: is a phosphoprotein occurring in milk, in beans and in nuts. The principal protein in cow's milk contains about 3% casein.

CERENOX: (Benzoyl hydrazone of quinoneoxime) is a fungicide belonging to quinone group. A 10% preparation is used for *see*d disinfection (197).

CERESAN: (Phenyl mercuric acetate, PMAC, FMA, TAG, Kwissan)— is an organomercurial fungicide belonging to the aromatic series (190).

CERESAN-M: (N-(ethylmercuric)-p-toluenesulphonilide)—is a *see*d disinfectant (190).

CERESAN-UNIVERSAL NASSBERZE: *see* Agalol (191).

CERESAN-UNIVERSAL TRACHENBERGE: *see* Methyl ethyl mercurous silicate (191).

CHAUBATTIA PASTE: is a copper fungicide used in U.P. (India). It is made by mixing 800 g each of copper carbonate and red lead with one litre of lanolin or lin*see*d oil (186).

CHEMBAM: *see* Nabam (172).

CHEMRAPE: *see* Vapam (177).

CHESTNUT COMPOUND: Well-powdered copper sulphate (2 parts) and ammonium carbonate (11 parts) are thoroughly mixed and the dry mixture stored for 24 hours in air tight containers (185).

CHINOMETHIONATE: *see* Morestan (160).

CHIOLITE: ($3AlF_3.5NaF$)—is fluosilicate of minor importance (44).

CHIPCOTE: *see* Methyl mercuric cyanide (190).

CHLORAL: Trichloroacetaldehyde ($C1_3C.CHO$) (83).

CHLORANIL: (Spergon, 2, 3, 5, 6-tetrachloro-l, 4-benzoquinone-tetra- chloro-p-benzoquinone) is a fungicide belonging to the quinone compounds. It has been used as *see*d disinfectant. It shows comparatively low fungicidal activity (197).

CHLORDANE: (Chlorondin: l, 2, 4, 5, 6, 7, 8, 8-Octachloro-4, 7-methane- 3, 4, 7, 7a-tetrahydroindene Octachloro-4, 7-methane-tetrahydrolin-dene: $C_{10}H_6CL_8$) is a chlorinated hydrocarbon, first prepared by Haymen by the Diels-Alder reaction (90).

CHLORFENVINPHOS: *see* Birlane (106).

CHLORINATED TERPENES: *see* Toxaphene (96).

CHLOROBENZILATE: (Geigy 328, 2-hydroxy-2, 2-bis (4-chlorophenyl) ethyl acetate) —is a chlorinated hydrocarbon and is primarily an acaricide. The technical material contains 90% of active ingredient (82).

2-Chloro-(2, 4-dichlorophenyl) vinyldiethylphosphate: *see* Birlane (106).

2-Chloroethyl-l-methyl-2 (p-tertbutylphenoxy) ethylsulphite: *see* Aramite (83).

CHLORENDIC ACID: *see* Heptachlor epoxide (92).

CHLORONDIN: *see* Chlordane (90).

CHLORONEB: (Demesan: 1, 4-dichloro-2, 5-dimethoxy-benzene)—is a systemic fungicide but is relatively less toxic when utilised as seed treatment (207).

CHLORONIL: (Spergon: 2,3,5,6-tetrachloro-l, 4-benzoquinone) is a fungicide belonging to the quinone group. It was first introduced as a fungicide by Terhost in 1938 (195).

4-Chlorophenyl-4-chlorobenzene sulphinate: *see* Ovotran (83).

4-Chlorophenyl sulphone: *see* Sulphenone (83).

CHLOROPHOS: *see* Trichlorofon (127).

CHLOROTHION: (0,0-dimethyl-0,4-nitro-3-chlorophenylthiophosphate) — is one of the two chlorinated analogs of methyl parathion with more reduced mammalian toxicity (118).

2-Chloro-l-(2,4,5-trichlorophenyl) vinyldimethyl phosphate: *see* Gardona (107).

CHRONIC: in the medical sense, this term is used in contrast to 'acute' and means 'of long duration'. As applied to material inhaled or absorbed through the skin, it means exposure times of long duration. As regards materials ingested, it refers to repeated doses.

CIDIAL: (0,0-dimethyl-S-(l-carboethoxy benzoyl) dithiophosphate)— is a broad spectrum organophosphorus compound (141).

CINERINS I AND II: are esters found in the pyrethrum extracts. However they are of less insecticidal value than pyrethrum I. (53).

COLLOIDAL SULPHUR: There are two forms of colloidal sulphur— one hydrophobic and other hydrophilic (166).

COMPOUND M 74: *see* Disyston (138).

CONTACT POISON: The poisons kill the pests after they have made physical contact with the pesticide (9).

COPPER AMMONIUM FLUORIDE: is a copper fungicide of minor importance (187).

COPPER AMMONIUM ZEOLITE: is a copper fungicide of minor impon tance (187).

COPPER CARBAMATE: is a copper fungicide produced by combining soda ash with hot copper sulphate solution (184).

COPPER NAPHTHANATE: is a copper fungicide of minor importance (187).

COPPER OXIDE: *see* Cupric oxide (186).

COPPER OXYCHLORIDE: (Basic chloride of copper, Blimix 4%; Blitox 50: Cupramar: Mico D-06: Mico P.W. 50: Fytolan: Vaigfru tamraghol: Blue copper 50)—is one of the low soluble copper fungicides produced by the action of air on cupric chloride solution or scrap copper (185).

COPPER PHOSPHATE: Copper fungicide of minor importance (187).

COPPER RESINATE: Copper fungicide of minor importance (187).

COPPER SILICATE: (Coposil) Copper fungicide of minor importance (187).

COPPER SULPHATE: (Blue stone; Blue vitriol: $CuSO_4.5H_2O$)—is used in the preparation of Bordeaux mixture and also as a fungicide (186).

COPOSIL: *see* Copper silicate (187).

COROMET: *see* Ferbam (174).

COUMAFURYL: is a rodenticide, now manufactured in India (4).

CRAGGLYODIN: *see* Glyodin (202).

CRYOLITE: (sodium aluminum fluoride; sodium fluoaluminate; $AlF_3.3NaF$)—is a fluoaluminate containing 54.2% fluorine and 0.06% soluble in water (44).

CRYOLITHIONITE: (2 $AlF_3.3NaF.3LiF$)—is a fluoaluminate insecticide of minor importance (44).

CUMULATIVE EFFECT (of a poison)**:** The effects of certain poisons depend not only on the dose received by the person at a particular time but also on the doses of the poison previously received. Thus, quantity of such a poison, innocuous in the first instalment may in nature be poisonous and is termed as "Cumulative effect". Sometimes such cumulative effects will occur with radiations also.

CUPRAMAR: *see* Copper oxychloride (185).

CUPROUS OXIDE: (Fungimar: Perenox. Cu_2O) It is a copper fungicide produced by precipitation from water soluble salts of copper in alkaline reducing solutions (186).

CYCLOHEX AMIDE: is an antibiotic isolated from *Streptomyces griseus* by Whiffe *et al.* in 1946 (204).

CYOLANE: (Dithiolane, Iminophosphate, Diethyl-N-1,3-dithiolauryl- 2-iminophosphate)—is a stomach and contact organophosphate insecticide with systemic activity in plants through roots and foliage absorption (109).

CYPREX: *see* Dodine (202).

2,4-D: is a weedicide now manufactured in India (3).

DASANIT: (Fensulphothion: Terracur-P: 0,0-diethyl-0, p (methylsul- phinyl) phenyl thiophosphate)—is an organophosphorus compound used as nematocide (121).

DCMOD: (2,3-dihydro-5-carboanilide-6-methyl-l,4-oxathin-4,4,dioxide: Plantvax)—is an oxathiin derivative and is systemic fungicide highly toxic to Basidiomycetes and other fungi. However it is not very stable (207).

DCPC: *see* Dimite (82).

DCPM: *see* Neotran (82).

DDD: (TDE, dichlorodiphenyldichloroethane, tetrachlorodiphenyl- ethane: 1,1-dichloro-2,2-bis (p-chlorophenyl) ethane)—is similar to DDT but differs only in the absence of one chlorine atom in the ethane moiety (81).

DDE: (2,2-bis (p-chlorophcnyl)-1,1-dichloroethylene)—is one of the detoxification products produced by insects from DDT (79).

DDT: (dichlorodiphenyl trichloroethane: 2,2-bis (p-chlorophenyl)- 1,1,1-trichloro-ethane: 1,1,1 -trichloro-2,2-bis (p-chlorophenyl ethane: pp DDT)—is synthetic chlorinated hydrocarbon first synthesised by Zeidler in 1874 (78).

pp DDT: *see* DDT (78).

DDVP: *see* Dichlorovos (102).

DEGUELIN: is derivative of rotenone and has a benzene ring substituting with a α-pyron ring. There are two symmetrical carbon atoms and hence four optical isomers are possible (58).

DELNAV: (Dioxathion; 1,4-dioxano-2,3,5,5-bis-(0,0-diethyl dithiophosphate)—is a general purpose persistent organophosphorus insecticide (141).

DEMETON: (Systox; Mercaptophos: 0,0-diethyl-2-ethylmercapto- ethylthio-phosphate)—is a systemic insecticide and acaricide. The technical material contains 70% thiono and 30% thiolo isomers (124).

DEMOSAN: *see* Chloroneb (207).

DERRIS: The extract from *Derris elliptica* and *D. mallencensis* contain rotenone with certain rotenoid compounds of lesser insecticidal activity than rotenone (56).

DEUTRO DDT: is similar to DDT except that the one hydrogen of DDT is replaced by deuterium.

DFDT: [Gix; 1,1,1-trichloro-2,2-bis (p-fluorophenylethane)]—is a chlorinated hydrocarbon. It is the fluorine analog of DDT (83).

DIANISYL DDT: *see* Methoxychlor (81).

2,2-Di-p-Anisyl-1,1,1-trichloroethane: *see* Methoxychlor (81).

DIAZINON: (0,0-diethyl-0-2-isopropyl-4-methyl-pyrimidyl-6) thiophosphate)—is a persistent general purpose insecticide used for controlling pests on plants and animals (122).

DIBASIC LEAD ARSENATE: *see* Dilead orthoarsenate (36).

DIBROM: *see* Naled (102).

DICAPTHON: (0,0-dimethyl-2-chloro-4-nitrophenylthiophosphate)—is a chlorinated analog of methyl parathion with Cl in 2-position (119).

DICHLONE: *see* Phygon (195).

2,2-Dichloro-1-chloroethoxy vinyl phosphate: *see* Phosphinon (125). 1,1-Dichloro-2,2-bis (p-chlorophenyl) ethane: *see* DDD (81).

2,3-Dichloro-1,1-bis (p-chlorophenyl) ethane: *see* Perthane (82).

1,4-Dichloro-2,5-dimethoxybenzene: *see* Chloroneb (207).

DICHLORODIPHENYL DICHLOROETHANE: *see* DDD (81).

2.3-Dichloro-1,4-naphthaquinone: *see* Phygon (195).

Di-(p-chlorophenyl) ethanol: *see* Dimite (82).

DICHLOROVOS: (DDVP: 0,0-diethyl-0,2,2-dichlorovinylphosphate)- is a derivative of phosphoric acid and is the first representative of enol phosphate series (102).

2,3-Dicyano-1,4-dithioanthra quinone: *see* Dithianon (195).

DIELDRIN: (HEOD: 1,2,3,4,10,10-hexachloro-6,7-epoxy-1,4,4a,5$_1$6,7, 8,8a-octahydro-1,4,5,8 endo, exo-dimethanonaphthene)—is a chlorinated hydrocarbon produced by the Diels-Alder reaction. It contains not less than 85% of the active ingredient HEOD (93).

0,0-Diethyl-0-4-aminophenylthiophosphate: *see* Aminoparathion (113).

0,0-Diethyl-S-carboxymethyl thiophosphate: *see* Acetophos. 0,0-Diethyl-S-carbothioxymethyldithiophosphate: *see* Acethion (132).

0,0-Diethyl-S-(6-chlorobenzolinyl-3-methyl) dithiophosphate: *see* Phosalone (140).

0,0-Diethyl-S-(4-chlorophenyl thiomethyl) dithiophosphate: *see* Trithion (136).

0,0-Diethyl-S-(2,5-dichlorophenylthiomethyl) dithiophosphate: *see* Phenkapton (136).

0,0-Diethyl-0,2,2-dichlorovinylphosphate: *see* Dichlorovos (102).

0,0-Diethyl-S-(2-ethyl mercaptoethyl) dithiophosphate: *see* Disyston (138).

0,0-Diethyl-2-ethyl mercaptoethyl thiophosphate: *see* Demeton (124). 0,0-Diethyl-S-(ethylthiomethyl) dithiophosphate: *see* Thimet (130).

0,0-Diethyl-0-(2-isopropyl-4-methylpyrimidyl-6) thiophosphate: *see* Diazinon (122).

0,0-Diethyl-S-(N-methyl-N-carboethoxy carbamoyl methyl) dithiophosphate: *see* Mecarbam (135).

0,0-Diethyl-0-(4-methyl caumarinyl) thiophosphate: *see* Potsan.

0,0-Diethyl-0-P(methyl sulphinyl) phenyl thiophosphate: *see* Dasanit (121).

0,0-Diethyl-0-nitrophenyl phosphate: *see* Paraoxon (115).

0,0-Diethyl-0,4-nitrophenylthiophosphate: *see* Parathion (112). Di-(p-ethyl phenyl) dichloroethane: *see* Perthane (82). 0,0-Diethyl-0-pyrozinyl thiophosphate: *see* Zinophos (123). 0,0-Diethyl-0-3,5,6-trichloropyridyl thiophosphate: *see* Dursban (122).

DIFOLATAN: (N-1,1-2,2-tetrachloro-ethyl thiophthalimide)—it belongs to the heterocyclic nitrogenous group of fungicides and it is more resistant to degradation than captan and folpet (199).

2.3-Dihydro-5-carboxanilide-6-methyl-1,4-oxathiin-4,4-dioxide: *see* DCMOD (207) .

Dihydrodeguelin: is the optically active form of rotenone and is 10 times as active as the levo deguelin (58).

5,6-Dihydro-2-methyl-l,4-oxathiin-3-carboxanilide: *see* DMOC (207).

Dihydrorotenone: is a natural organic insecticide produced by the catalytic hydrogenation of the side chain of rotenone with elimination of the double bond (57).

N, N'-diisopropyl phosphorodiamide fluoride: *see* Mipafox (108).

Dilan: is a mixture of the two insecticides (53.3% of bulan and 26.7% of prolan). Commercial products contain other related compounds like 0,p-isomers (82).

Dilead Orthoarsenate: (Acid orthoarsenate, dibasic lead arsenate, dibasic lead arsenate, diplumbic hydrogen arsenate, diplumbic arsenate, lead hydrogen arsenate, $PbHAsO_4$)—is an arsenical insecticide prepared by reacting soluble lead with disodium arsenate (36).

Dimecron: *see* Phosphamidon (104).

Dimefox: (Pestox, Hnane, Terrasystam, Tetramethylphosphorodi- amidic fluoride: bisdimethyl fluorophosphate)—is a powerful systemic organophosphorus insecticide with very short residual activity (108).

2,3-Dimercaptoproponal: *see* BAL (42).

Dimetan: (5,5-dimethyl (dihydroresorcinol) N,N-dimethyl-carba- mate)— is a carbamate insecticide (153).

Dimethoate: (Rogor, phosphamide: Bi-55; 0,0-dimethyl-S-(N- methylcarbamoyl methyl) dithiophosphate)—is a systemic organophosphorus insecticide (133).

0,0 Dimethyl-S-(2-acetylamide ethyl dithiophosphate): *see* Amophos (142).

4-(N,N-Dimethyl amino)-3,5-methylphenyl-N-methylcarbamate: *see* Zectran (153).

0,0-Dimethyl-S-(1-carbothoxy benzyl) dithiophosphate: *see* Cidial (141).

0,0-Dimethyl-S-(4,6-diamino-1,3,5-triariisyl-2-methyl) dithiophosphate: *see* Menazon (140).

0,0-Dimethyl-S-l,2-dicarboethoxy ethyldithiophosphate: *see* Malathion (131).

0,0-Dimethyl-0,2,5-dichloro-4-bromophenyl thiophosphate: *see* Bromophos (120).

0,0-Dimethyl-0,2,2-dichloro-1,2-dibromoethyl phosphate: *see* Naled (102).

0,0-Dimethyl-S-(3,4-dichloro-4-keto-l,2,3-benzotriazinyl-3-methyl dithiophosphate): *see* Guthion (140).

5,5-Dimethyl (dihydroresorcinol) N,N-dimethylcarbamate: *see* Dimetan (153).

Dimethyl-S-(2-ethylmercaptoethyl) dithiophosphate: *see* Ekatin (137).

0,0-Dimethyl-S-(2-ethylmercaptoethyl)) thiophosphate: *see* Metasystox-I (127).

0,0-Dimethyl-2-ethylmercaptoethylthiophosphate: *see* Metasystox (127).

0,0-Dimethyl-S,5-(ethylsulphinyl) ethylthiophosphate: *see* Metasystox R (127).

0,0-Dimethyl-(1-hydroxy-2,2,2-trichloroethyl) phosphonate: *see* trichlorofon (127).

0,0-Dimethyl-N (isopropoxy carbamoyl) phosphate: *see* Avenin (108).

0,0-Dimethyl-S(N-2-methoxyethyl carbamoyl methyl) dithiophosphate: *see* Thiocron (134).

0,0-Dimethyl-S-5-(methoxy pyronyl-2-methyl) thiophosphate: *see* Endothion (124).

0,0-Dimethyl S-2-(1-N-methyl carbamoyl ethyl mercapto) ethyl thiophosphate: *see* Vamidothion (128).

0,0-Dimethyl-S (N-methyl) carbamoyl methyl dithiophosphate: *see* Rogor (133).

0,0-Dimethyl-0-(l-methyl-2-carbomethoxy vinyl phosphate: *see* Phosdrin (103).

0,0-Dimethyl-S-(N-methyl-N-formyl carbomoyl methyl) dithiophosphate: *see* Anthio (135).

0,0-Dimethyl-S-(2-methylmercapto) ethyl thiophosphate: *see* Tinox (127).

S,5-Dimethyl-4-methylmercaptophenyl-N methyl carbamate: *see* Mesurol (152).

0,0-Dimethyl-S-morpholine-carbamoyl methyl) dithiophosphate: *see* Ekatin M (134).

0,0-Dimethyl-(S-naphthalimide methyl) dithiophosphate: *see* Imidan (139).

0,0-Dimethyl-0,4-nitro-3-chlorophenylthiophosphate: *see* Chlorothion (118).

0,0-DimethyI-0,4-nitro-3-methyl phenyl thiophosphate: *see* Sumithion (117).

0,0-Dimethyl-0,4-nitrophenyl thiophosphate: *see* Methyl parathion (115).

3(2(3,5-Dimethyl-2-Oxocyclohexyl) 2-hydroxy ethylglutarimide: *see* Actidione (204).

DIMETHYL PARATHION: *see* Methyl parathion (115).

3-Dimethoxy phosphinyl oxy-N,N-dimethyl crotanamide: *see* Bidrin (105).

DIMETHANONAPHTHALENE: *see* Isodrin (95).

1.2.3.4.10.10-Hexachloro-1,4,4a,5,8,8,-hexadydro-1,4-endo exo-5,8-dimethano naphthalene: *see* Aldrin (92).

6.7.8.9.10.10-Hexachloro-1,5,5a,6,9,9a-hexahydro-6,9-methano-2,4,3-Henzodiox -athiopiri-3-oxide: *see* Endosulphan (95).

DIMITE: (DCPC: DMC: l,l-bis(p-chlorophenyl) ethanol: di-(p-chloro- phenyl) ethnol) is a chlorinated hydrocarbon insecticide (82).

DINEX: (DNOCHP: 2,4-Dinitro-6 cyclohexylphenol) belongs to the nitro cresol group of insecticide. It is an effective contact and stomach poison (157).

DINITRO-O-CRESOL: (DNOC: DINOC: 2,4-Dinitro-6-methyl phenol: 4,6-dinitro-O-cresol)—is the earliest synthetic organic compound sold as insecticide in Germany. It is a contact and stomach poison (156).

4,6-Dinitro-O-Cresol: *see* Dinitro-O-cresol (156).

Dinitro-O-cyclohexylphenol: *see* Dinex (157).

Dinitro-(1-methylheptyl) phenyl cortonate: *see* Dinocap (158).

2,4-Dinitro-6-methylphenol: *see* Nitro-O-cresol (156).

2,4-Dinitro-6 sec-butylphenol: *see* Dinoseb (157).

2,4-Dinitro-6-sec-octaphenyl crotonate: *see* Karathane (158).

DINOC: *see* Dinitro-O-cresol (156).

DINOCAP: is a good substitute for sulphur for controlling certain spores and mites (158).

DINOSEB: (DNBP, DNOSBP, DNSBP, 2,4-dinitro-6 sec-butylphenol)—belongs to the nitrocresol group of insecticide. It has insecticidal and herbicidal action (157).

1,4-Dioxano-2,3-5,5-bis(0,0-diethyl dithiophosphate): *see* Delnav (141).

DIOXATHION: *see* Delnav (141).

DIPLUMBIC ARSENATE: *see* Dilead arsenate (36).

DIPLUMBIC HYDROGEN ARSENATE: *see* Dilead arsenate (36).

DIPTREX: *see* Trichlorofon (146).

DISODIUM ETHYLENE BIS DITHIOCARBAMATE: *see* Nabam (175).

DISULPHOTAN: *see* Disyston (138).

DISYSTON: (Dithiosystox, disulphotan. Compound M. 74: 0,0-diethyl-S-(2-ethyl mercaptoethyl) dithiophosphate)—is similar to those of thiometon but considerably more stable. It is also used as *see*d disinfectant (138).

DITHANE, 14: *see* Nabam (175).

DITHANE A-40: *see* Nabam (176).

DITHANE D-14: *see* Nabam (175).

DITHANE Z-78: *see* Zineb (175).

DITHIANON: (2,3-Dicyano-1,4-dithio-anthioquinone)—is a fungicide belonging to the quinone group. However its high phytotoxicity is preventing its use as a foliar spray (195).

α,α- Dithio bis (dimethylthio) formamide: *see* Thiram (174).

DITHIOLANE: *see* Cyolone (109).

DITHIOSYSTOX: *see* Disystox (138).

DIVINYL PYRETHRIN I: *see* Allethrin (55).

DMC: *see* Dimite (82).

DMDT: *see* Methoxychlor (81).

DMOC: (Vitavax, 5,6-Dihydro-2-methyl-1,4-oxathiin-3-carboxanilide):—is an oxathiin derivative developed by Von Schmelung and Kulka. It is highly specific in its fungitoxicity. It is highly toxic to Basidiomycetes (207).

DNB: *see* Bulan (82).

DNBP: *see* Dinoseb (157).

DNOC: *see* Dinitro-O-cresol (156).

DNOCHP: *see* Dinex (157).

DNOPSP: *see* Dinoseb (157).

DNP: *see* Prolan (82).

DNSBP: *see* Dinoseb (157).

N-Dodecyl guanidine acetate: *see* Dodine (202).

DODINE: (Cyprex, N-dodecyl guanidine acetate)—was introduced by the American Cyanamide Company as a fungicide in 1959. It belongs to the guanidine group. It is marketed as 70% W.P. (202).

DOET 57: *see* Ronnel (119).

DOWET 14: *see* Ronnel (119).

DRIED BORDEAUX MIXTURE: This is inferior to the freshly prepared Bordeaux mixture (184).

DRY LIME SULPHUR: When lime sulphur concentrate is evaporated with sucrose as a stabiliser, probably the polysulphides are present as tetra and penta sulphides. This form is called Dry lime sulphur (169).

DRY MIX SULPHUR LIME: In this form of lime sulphur, the temperature is allowed to rise due to slaking of calcium hydroxide. This is used on sulphur-sensitive plants (169).

DUPONT FUNGICIDES A: *see* Zineb (175).

DURSBAN: (0,0-diethyl 0,3,5,6 trichloropyridyl thiophosphate): is an active organophosphorus insecticide used as a stomach poison (122).

DUSTS: These are solid form of pesticides containing a low concentration of the toxicant mixed with a carrier like talc (13).

DUTER: *see* Triphenyl tin hydroxide (201).

DVPY I: *see* Allethrin (55).

DYLOS: *see* Trichlorofon (146).

E 600: *see* Paraoxon (115).

E 605: *see* Parathion (112).

EAU GRISON: *see* Lime sulphur (167).

E.C: *see* Emulsion concentrates (21).

EDCT MIXTURE: Ethylene dichloride-carbon tetrachloride mixture (3).

EKATIN: (Thiometon, Intruthion, 0,0-dimethyl-S-2-ethyi mercapto ethyl dithiophosphate)—is a good organophosphate systemic insecticide with about the same duration of action of metasystox (137).

EKATIN F: *see* Ekatin M (134).

EKATIN M: (Ekatin F: Morphothion): 0,0-diethyl-S-(morpholine- carbamoyl methyl) dithiophosphate (134).

EMERALD GREEN: *see* Paris green (34).

EMMI: (N (ethyl mercuric) 3,4,5,6,7,7-hexachloro-3,5-endomethylene 1,2,3,6-tetrahydrophthalimide)—is a mercurial fungicide and is sold as 10% wettable powders or as emulsion concentrates (191).

EMULSION CONCENTRATES: These are oil based liquid formulations concentrated to extra strength to save transport and storage costs. They contain solvents, emulsifiers, and wetting agents that make them easily soluble in water (21).

ENDOSULPHAN: (Thiodan:6,7,8,9,10,10-hexachloro-l,5,5a,6,9,9a- hexahydro-6,9-methano-2,4,3-benzodioxathiin-3-oxide)—is a chlorinated hydrocarbon containing 4 parts of α-cis isomer and 1 part of β-trans isomer (95).

ENDOTHION: (Phosphoryrone: 0,0-dimethyl-S-(5-methoxy-pyronyl-2-methylthio-phosphate)—is a systemic organophosphate insecticide (124).

ENDRIN: (1,2,3,4,10,10,-hexachloro-6,7-epoxy-1,4,4a,5,6,7,8,8a-Octa hydro-l,4,5,8-endo, endo dimethanonaphthalene)—is a chlorinated hydrocarbon of low persistence. It is a sterio isomer of dieldrin (94).

ENGLISH PURPLE: *see* London purple (35).

EPN: (0-ethyl-0-(4-nitrophenyl). benzene thiophosphanate)—is a powerful acaricide (147).

ERADEX: (Quinoxaline, Bayer 30686, 2,3-trithiocarbonate)—is an active acaricide belonging to the group of derivatives of dithiocarbonic acid (161).

ESSO FUNGICIDE 406: *see* Captan (197).

ETHION: (0,0,0,0,-Tetraethyl-S,S-methylene bis dithiophosphate)—is an organophosphorus insecticide used for the control of mites and aphids (137).

5, Ethoxy-(3-trichloromethyl-1,4,-thiodiazole)—*see* Terrazole (206).

ETHYL MERCURIC CHLORIDE: (C_2H_5HgCl)—is a mercurial fungicide sold under different brand names as Granosen, Mercuron, Mercurhexan (190).

N-Ethyl mercuric-3,4,5,6,7,7-hexachloro-3,6-endo methylene-1,2,3,6- tetrahydrophthalimide—*see* Emmi (191).

ETHYL MERCURIC PHOSPHATE: is a mercurial fungicide better than mercuric chloride in fungicidal property. Its formulation does not exceed 10% of the total mercury (28%) (190).

0-Ethyl-0-(4-nitrophenyl benzene thiophosphate)—*See* EPN (147).

N-Ethyl mercuric-P-toluene sulphonilide: *see* Ceresan M (190).

ETROLENE: *see* Ronnel (119).

EXPOSURE: Substances which under certain conditions of temperature, shock or chemical action can decompose rapidly to evolve either large volumes of gases or so much heat that the surrounding air is forced to expand rapidly; this state is being subjected to the action of a poison or radiation.

FENITROTHION: *see* Sumithion (117).

FENSULPHOTHION: *see* Dasanit (121).

FENTHION: (Baytex, Lebaycid, 0,0-dimethyl-0-(4-methyl mercapto-3- methyl phenyl) thiophosphate)—is a fast killing organophosphate insecticide with long residual action and it approaches chlorophos in its activity (121).

FERBAM: (Coromet, Fermate: Fermates D: Fermocide: Ferradow: Karbam black: Ferric dimethyl dithiocarbamate)—is one of the first dithiocarbamates fungicides produced commercially. It is less phytotoxic than Bordeaux mixture (174).

FERMATE: *see* Ferbam (174).

FERMATE D: *see* Ferbam (174).

FERMOCIDE: *see* Ferbam (174).

FERRADOW: *see* Ferbam (174).

FERRIC DIMETHYL DITHIOCARBAMATE: *see* Ferbam (174).

FERRIC OXIDE: is used as a corrective in the formulation of arsenical insecticides to prevent arsenical injury (25).

FERROUS SULPHATE: is used as a corrective in the manufacture of arsenates (41).

FIXED COPPERS: are formulated at the same content as bordeaux. In all these, copper is stabilised and rendered less soluble to reduce the phytotoxicity of copper (184).

FLASH POINT: is the lowest temperature at which a combustible liquid will give off a flammable vapour which will burn momentarily (63).

FLOTATION SULPHUR: is a form of sulphur fungicide obtained as a by-product of the manufacture of fuel gas from coal. It is recovered by the flotation process.

FLOWABILITY: of the dust is an indicator of the feed rate of dusting equipment. The angle of slope is a measure of flowability (15).

FLUOALUMINATES: is a group of fluorine insecticides of which sodium-fluoaluminate is important (45).

FLUOBORATES: is a group of fluorine compounds (43).

FLUORINE: Elemental fluorine as a gas at 100 to 1000 ppm was effective against number of pests (43).

FLUOSILICATES: is a group of fluorine compounds of which sodium fluosilicate is important (44).

FMA: *see* Ceresan (190).

FOLIDOL 80: *see* Methyl parathion (115).

FOLIDOL E 605: *see* Parathion (112).

FOLITHION: *see* Sumithion (117).

FOLPET: (Phaltan, Phthalan; N-trichloromethylthiophthalimide)—is an analog of captan introduced by the California Spray Chemicals Corporation. It is slightly more effective than captan (198).

FORMOTHION: *see* Anthio (135).

FRENCH GREEN: *see* Paris green (34).

FUKLASIN: *see* Ziram (173).

FUMIGANT: The substance used in the form of vapour and aerosol for destroying microbes and insects (28).

FUNGICIDE: is a chemical used for controlling diseases caused by fungi (20).

FUNGIMAR: *see* Cuprous oxide (186).

FURADAN: (Carbofuran, 2,3-dihydro-2,2,dimethyl-7-benzofuranyl- methyl carbamate) —is a systemic carbamate insecticide of broad activity (150).

FYTOLAN: *see* Copper oxychloride (187).

GAMMA BHC: is one of the eight isomers of BHC. Lindane is a purified form of BHC containing 99% of pure λ isomer (89).

GAMMEXANE: *see* BHC (87).

GARDONA: 2-Chloro-1-(2,4,5-trichlorophenyl) vinyl-dimethyl phosphate) is a low toxicity organophosphorus insecticide with contact action and having moderate persistence (107).

GASOLENE: It is one of the lowest boiling petroleum fractions and is seldom used as insecticide because of its high volatility and fire hazard (64).

GEIGY 328: *see* Chlorobenzilate (82).

GERMISAN: is a mercurial fungicide of the aromatic series. It is a phenyl mercuric derivative (189).

GIX: *see* DFDT (83).

GLUMIDE: *see* Glyodin (202).

GLYODIN: (Glyozalidine: Gluoxide: Craggiyodin: 2-Heptadecyl-2- imidazoline acetate)—is a derivative of glyoxalidine and possesses fungicidal properties. It is rather a protective than eradicative fungicide (202).

GLYOXALIDINE: *see* Glyodin (202).

GRANOSAN: 2% Ethyl mercuric chloride (190).

GRANULES: consist of free flowing granular basic materials impregnated with an insecticide. Basic materials commonly used are clays, bentonite, etc. (17).

GRISBOFULVIN: is an antibiotic having systemic action: It is produced from *Penicillin griseofulvin.* It was discovered by Oxford *et al.*, in 1939 (205).

GUSATHION: *see* Guthion (140).

GUTHION: (Gusathion, Azinophos methyl; 0,0-dimethyl-S-(3,4-dichloro-4-keto-1, 2,3,-benzotrizinyl-3-methyl) dithiophosphate)— is a broad spectrum organophosphorus insecticide (140).

HCCP: (Hexa chlorocyclopentadiene)—is the base produced by the Indian Petrochemical Plants. It is used for the manufacture of all chlorinated hydrocarbons in India (4).

HCH: *see* BHC (87).

HEAVY OILS: Oils having a viscosity greater than 55 seconds saybolt and are usually classified as heavy and used for insecticidal sprays, especially for dormant sprays (66).

HELLABORE: *(Veratrum album or V. viride)*—has been found to be toxic to certain species of insects (59).

HEOD: *see* Dieldrin (93).

HEPTACHLOR: (1,4,5,6,7,8,8-heptachloro-3a,4,7,7a-tetrahydro-l,7- methano-indene) —is a chlorinated hydrocarbon, produced by substantive chlorination of Chlordane at low temperature like 0 to 5° C in carbon tetrachloride solutions (91).

HEPTACHLOR EPOXIDE: (Chlorendic acid: 3,4,5,6,7,7-hexachloro- bicyclo-2,2-heptene-4,5-dicarboxylic 1,2-acid)—is the main metabolite of heptachlor produced mainly by soil microbes. It is more toxic than heptachlor (92).

2-Heptadecyl-2-imidazolone acetate:-*see* Glyodin (202).

HERBICIDES: are chemicals used for the control of weeds (12).

HERCULES 3956: *see* Toxapene (96).

HEXACHLOROCYCLOHEXANE: *see* BHC (87).

1,2,3,4,5,6 Hexachlorocyclo hexnene: *see* BHC (87).

1,2,3,4,10,10, Hexachloro 6,7-epoxy-l,4,4a,5,6,7,8,8a-octahydro-1,4,5,8-endo, exo, dimethane naphthalene: *see* Endrin (94).

l,2,3,4,10,10-Hexachloro-6,7-epoxy-l,4,4a,5,6,7,8,8a-octahydro-1,4,5,8-endo, endo dimethano naphthalene: *see* Dieldrin (93).

HHDN: *see* Aldrin (92).

HNANNE: *see* Dimefox (108).

2-Hydroxy-2,2-bis(4-chlorophenyl) ethylacetate: *see* Chlorobenzilate (82).

HYDROXY PHENOXYTHIAZONE: *see* Thionol (159).

IMIDAN: (Phthelophos: 0,0-dimethyl-S-(naphthalimide-methyl) dithiophosphate)—is an organophosphrus insecticide (139).

IMINOPHOSPHATE: *see* Cyolane (109)

INFUSARIAL EARTH: the siliceous or diatomaceous earth, fossil/,flour, Kiselguhr. It is a white to light grey to pale buff powder insoluble in water, acids or oil, alkalies (91).

INTRUTHION: *see* Ekatin (137).

IODINE VALUE: is a measure of the unsaturated aliphatic compounds in oils (64).

ISODRIN: (1,2,3,4,10,10-Hexachloro-1,4,4a,5,8,8a-Hexahydro-l,4,5,8- endo, endo-dimethano naphthalene)—is a chlorinated hydrocarbon, reminiscent of aldrin in properties. However, it is used mainly for the preparation of endrin (95).

ISOLAN: (1-isopropyl-2-methyl-pyrazolyl-5-dimethyl carbamate)—is an active aphicide with systemic action (153).

ISOPESIOX: *see* Mipafox (108).

2-Isopropoxy-phenyl-N-methyl carbamate: *see* Baygon (152).

1-Isopropyl-2-methyl-pyrozolyl-5-dimethylcarbamate: *see* Isolan (153).

ISOTHAN: *see* Karathane (158).

JASMOLIN II: It is one of the alkaloid compounds found in pyrethrum extracts (53).

KAOLIN: Essentially a hydrated aluminum silicate, white or yellowish white—earthy mass insoluble in water, cold acids, and in alkaline hydroxides.

KARATHANE: (DINOCAP, Arathane, Isothan, 2,4-dinitro-6 sec-octyl- phenyl cretonate)—is a nitrophenol compound, used as an acaricide and also as fungicide (158).

KARBAM WHITE: *see* Ziram (173).

KARBAM BLACK: *see* Ferbam (174).

KORALAN: *see* Ronnel (119).

KWISSAN: *see* Ceresan (192).

LAURYL THIOCYANATE: is a contact insecticide belonging to the thiocyanate group but shows little promise as stomach poison (160).

LD_{50}: The amount of poison which will kill one half of a group of experimental animals (LD = lethal dose).

LEAD ARSENATE: Depending upon the conditions of preparation, different compounds of lead arsenate, like acid lead arsenate Pb (PbOH) $(AsO_4)_3$, monolead arsenate $PbH_4(AsO_4)_2$ etc. are prepared (36).

LEAD HYDROGEN ARSENATE: *see* Dilead arsenate (36).

LEAD HYDROXY ARSENATE: *see* Basic lead arsenate (37).

LEAD METAARSENATE: ($Pb(AsO_3)_2$) is an arsenical insecticide (36).

LEAD PYROARSENATES: are arsenate insecticides with the composition of Pb H_2AsO_7 and Pb_2AsO_7 (36).

LEBAYCID: *see* Fenthion (121).

Levodeguelin: a considerable proportion of the insecticidal activity of derris and cube extractives is due to the nonrotenone portion principally Levodeguelin (59).

Light Qil: Oils with a viscosity between 40 and 65 seconds saybolt are light oils. These are highly refined and contain > 90% unsulfo- natable residues.

Limb Sulphur: (Eau green)—is a fungicide, usually prepared by boiling 1.8 parts of sulphur, 1 part of lime and 10 parts of water for one hour.

Lindane: *see* Gamma BHC (89).

Litharge: PbO (36).

Local: This term refers to the site of action of a poison and means that the action takes place at the point of area of contact. The site may be skin or mucous membrane.

London Purple: (Paris purple, English purple)—a purple residue from the manufacture of dyestuffs. It is a mixture of calcium arsenite, calcium arsenate and other extraneous materials (35).

Lonocol: *see* Zineb (175).

Lovo: *see* Butylamine sterate (26).

Magnesia Mixture: It is used for distinguishing arsenates from arsenites. It forms white precipitate with arsenate. No precipitate is formed with arsenites. Magnesia mixture contains 110 gm of magnesium chloride dissolved in 700 ml of ammonia and diluted to 2 litres (33).

Malachite Green: ($C_{22}H_{25}ClN_2$)—green crystals with metallic lustre (85).

Malathion: (Carbphos, 0,0-dimethyl S-1,2-dicarboethoxy ethyldithiophosphate)—is an organophosphorus insecticide, highly toxic to mites while being generally nonphytotoxic (131).

Maneb: (Manzate; MEB; MnEbD: manganese ethylene bisdithiocarbamate)—is the manganese salt of ethylene bisdithiocarbamic acid and is similar to Zineb in chemical and physical properties (177).

Manganese ethylene bis dithiocarbamate: *see* Maneb (177).

Manganese Sulphate: is used as a corrective in the formulation of arsenical insecticide to prevent arsenical injury (41).

Manzate: *see* Maneb (177).

Maximum Allowance Concentration (MAC): The concentration of a poison or toxic substance in air expressed in suitable units, above which its presence becomes harmful.

Maximum Permissible Dose: for an individual is that dose accumulated over a long period of time or resulting from a single exposure, which in the light of present day knowledge carries a negligible possiblity of severe somatic organic injuries. Furthermore it is such a dose that any effects that ensure more frequently or limited to those of a minor nature that would not be considered unacceptable by the exposed individual and by competent medical authorities.

MEB: *see* Maneb (177).

Mecarbam: (0,0-diethyl-S-(N-methyl-n-carboethoxy carbamoyl methyl)-dithiophosphate)—is an organophosphorus insecticide similar to dimethoate in action (135).

MEMMI: ((N-methyl mercuric) 3,4,5,6,7,7-hexachloro-3,6-endome- thylene-l,2,3,4-tetrahydrophthalimide)—is a mercurial fungicide sold as 10% wettable powder and emulsion concentrate (190).

MENAZON: (Sayfos, 0,0-dimethyl-S-(4,6-diamino-l,3,5-triazinyl-2- methyl) dithiophosphate)—is a good systemic organophosphorus insecticide with low mammalian toxicity for the control of vectors of viruses that cause potato diseases (140).

MERCAPTOPHOS: *see* Demeton (124).

MERCURAN: is a mercurial fungicide containing 2% ethyl mercuric chloride plus 12 to 14% of υ BHC (190).

MERCURHEXAN: is a mercurial fungicide containing 1% ethyl mercuric chloride plus 20% BHC as 20% γBHC (190).

MERCURIC CHLORIDE: is a mercury fungicide. The *see*d materials are dipped in 1 : 2000 or stronger solution for a few minutes (188).

MERCUROUS CHLORIDE: (HgCl, Calomel) is a mercury fungicide and is easier to handle than mercuric form (188).

MESH: is the number of open spaces in a screen per unit area.

MESULFAN: (N-trichloromethyl thio-4-chloroanilide of methane sulfonic acid)—belongs to the group of heterogeneous nitrogenous compounds and is an analog of Captan but less effective than Captan (200).

MESURAOL: (Metamercapturon. 3,5-Dimethyl-4-methyl-mercaptophenyl-N-methyl carbamate)—is an active carbamate insecticide and acaricide with wide spectrum of activity and relatively long residual activity.

METACIDE: *see* Methyl parathion (115).

METALDEHYDE: is a molluscicide, now manufactured in India (4).

METAMERCAPTURON: *see* Mesurol (152).

METAPHOS: *see* Methyl parathion (115).

METASYSTOX: (Methyl demeton, methylmercaptophos, 0,0-dimethyl- 2-ethyl-mercaptoethylthiophosphate)- is a rapidly acting systemic and contact insecticide containing only the thiolo isomer of methyl demeton. It is similar to methyl demeton (126).

METASYSTOX-R: (Oxydemeton methyl: 0,0-dimethyl-S-(2-ethylsul- finyl) ethyl thiophosphate)—is similar to methyl demeton (127).

METASYSTOX-S: Its insecticidal activity is almost the same as oxydemeton methyl (127).

METASYSTOX-I: (0,0-dimethvl-S-ethyl mercaptoethvl thiophosphate) (127).

METATHION: *see* Sumithion (117).

METAZOL: (Methyl mercury-p-hydroxy quinolate)—is a mercury fungicide containing 1.3% Hg with a red dye added (190).

METHAM-N-SODIUM: is a nematicide now manufactured in India (4).

METHANOLYSIS: is an alcoholysis ‘in which the alcohol split is methanol.

METHASAN: *see* Ziram (173).

METHOXYCHLOR: (DMDT; Methoxy analog of DDT, Dianisyl DDT; 2,2-di-p-anisyl-1,1,1-trichloroethane)—is a synthetic organic chlorinated insecticide. It is methoxy analog of DDT (81).

METHOXY ETHYL MERCURIC CHLORIDE: *see* Agalol (191).

METHYL ACETOPHOS: (0,0-dimethyl-S-carboxymethyl thiophosphate) —is the methyl analog of acetofos.

METHYL-1-BUTYL CARBONYL-2-BENZINIDAZOLE CARBAMATE: *see* Benalate (206).

METHYL DEMETON: (Metasystox, methylmercaptophos, 0,0-dimethyl-2-ethyl mercaptoethyl thiophosphate)—is the methyl analog of demeton and the technical grade contains 70% of thiono and 30% of thiolo isomer. It is less toxic than demeton (126).

METHYL ETHYL MERCURIC SILICATE: *see* Ceresan Universal Trackenberge.

METHYL ETHYL PARATHION: *see* Thiophos-ME (139).

0-Methyl-0-ethyl-S-(2-ethyl thioethyl) dithiophosphate: *see* Tetrathion (139).

0-Methyl-0-ethyl 0,4-nitrophenyl thiophosphate: *see* Thiophos-ME (116).

METHYL ETHYL THIOPHOS: *see* Thiophos-ME (116).

METHYL ISOCYANATE: *see* Carbaryl (151).

METHYL MERCAPTOPHOS: *see* Metasystox (126).

METHYL MERCAPTURON: *see* Mesurol (152).

METHYL MERCURIC CYANIDE: (Chipchote)—is a mercurial fungicide and is sold with a mercury content of 1.53 and 4.5% (190).

METHYL MERCURIC DICYANODIAMINE: *see* Panogen (190).

P-Methyl-mercuric-3,4,5,6,7,7-hexachloro-3,6-endomethylene-1,2,3,-6-tetrahydro-phthalimide: *see* Memmi (190).

METHYL MERCURY-β-HYDROXY QUINOLATE: *see* Metazol (190).

2- Methyl -2-methylthiopropional-oxime -O-N-methylcarbamate: *see* Temik (154).

METHYL NITROPHOS: *see* Sumithion (117).

METHYL PARATHION: (Folidol 80,. Metaphos, Dimethylparathion, Metacide, Wofatox, 0,0-dimethyl-0-4-nitrophenylthiophosphate)— is an organophosphorus compound, similar to parathion except having a methyl group in the place of ethyl. It is lower in toxicity than parathion and hence preferred (115).

1-Methyl-2-3-(pyridyl) pyrrolidine: *see* Nicotine (47).

6-Methyl quinoxaline-2,3-dithiocyclocarbamate: *see* Morestan (199) 6-Methyl quinoxaline-2,3-dithiocarbamate: *see* Morestan (160).

METHYL TRITHION: (0,0-dimethyl-S-(4-Chlorophenylthiomethyl) thiophosphate), is a methyl homolog of trithion with less toxicity to mammals and pests than trithion (130).

METHYRIMOL: (PP 675: 5-Butyl-2-dimethylamine-4-hydroxy-6-methyl pyrimidine) —is a systemic fungicide effective against a number of fungi (207).

MEVINPHOS: *see* Phosdrin (103).

MICO D-06: *see* Copper oxychloride (185).

MICO PW-50: *see* Copper oxychloride (185).

MICRON: 10^{-6} cm.

MILBAM: *see* Ziram (173).

MIPAFOX: (Pestox-15: Isopestox; N,N-diisopropyl phosphorodiamide fluoride)—is a strong systemic and contact organophosphorus insecticide. Now it is not used as it causes irreversible paralysis in man and animals (108).

MITIS GREEN: *see* Paris Green (34).

MNEBD: *see* Maneb (177).

MONOLEAD ORTHOARSENATE: ($PbH_4(AsO_4)_2$)—is an arsenate insecticide (36).

MONOMAGNESIUM ORTHOARSENATE: ($MgH_4(AsO_4)_2$)—is an arsenical insecticide (40).

MORESTAN: (Chinomethionate, Oxythioquinox, Quinomethionate, 6- methyl-2,3-quinoxalene dithiocarbamate: 6-methyl quinoxaline-2, 3-dithiocyclocarbonate)—is an acaricide and fungicide (160).

MORPHOTHION: *see* Ekatin-M (134).

NABAM: (Dithane 14 Chembam, Dithane D-14 Disodium, ethylene bisdithiocarbamate)—is a sodium salt of ethylene bisdithio- carbamic acid and is similar to Zineb (175).

NALAD: (0,0-Dimethy 1-0-2,2-dichloro-l,2-dibroinoethylphosphate)— is a powerful organophosphorus insecticide obtained by addition of chlorine or bromine at the double bond of dichloro vinyl group.

NANKAR: *see* Ronnel (119).

NAPHTHA: is a petroleum product falling in the group of 5-10 carbon. These are seldom used as insecticides because of their low boiling point and hence probable fire hazard (62).

NAPHTHYL CARBAMATE: *see* Carbaryl (151).

1-Naphthyl-N-methyl carbamate: *see* Carbaryl (151).

NARBAM BLACK: *see* Ferbam (174).

NEONICOTINE: (Anabasine, 3-(2-piperidyl) pyridine)— occurs in *Anabasine aphylla* Linn. The crude anabasine sulphates of commerce is a mixture of alkaloids (50).

NEOTRON: (DCMP, Bis (p-chlorophenoxy) methane)—is an acaricide, compatible with most common insecticides.

NEUTRALISATION NUMBER: is an expression of the acidity in terms of mgm of KOH required to neutralise the acid in 1 gm of oil, under standard conditions (64).

NIAGARA-C: *see* Ziram (173).

NICOTINE: (1, methyl-2,3-(pyridyl) pyrrolidine)—is a natural organic insecticide occurring in 50 or more species of *Nicotiana.* The economic source of nicotine is *N. tabacum* (47).

2-Nitro-1,1p bis (p-chlorophenyl) butane: *see* Bulane (82).

2-Nitro-1,1-bis (p-chlorophenyl) propane: *see* Prolan (82).

P-NITROPHENOL: is the impurity in parathion and also is the principal metabolic product of parathion (112).

NOMEKSON: *see* Thiram (174).

NORNICOTINE: (2-3 (pyridyl) pyrrolidine, $C_9H_{12}N_2$)—is a natural organic insecticide. It is similar to nicotine except that the methyl group attached to pyrrollidine ring has been replaced with a hydrogen atom (50).

NUVAN: *see* Dichlorovos (102).

OCMA: *see* Schradan (144).

1,2,4,5,6,7,8,8-Octachloro-4,7-methane-3,4,7,7a-tetrahydroindene: *see* Chlordane (90).

OCROCHLORO: 4,7-methanotetrahydrolindane: *see* Chlordane (90).

OCTAMETHYL: *see* Schradan (144).

OCFAMETHYL PYROPHOSPHORAMIDE: *see* Schradan (144).

OCTAMETHYL TETRAMIDE OF PYROPHOSPHORIC ACID: *see* Schradan (144).

OIL BORDEAUX MIXTURE: is prepared by the addition of edible cotton *see*d oil to bordeaux mixture. It improves the retention of copper on foliage (198).

OIL EMULSION: contains lube oil (2 gallons), water (1 gallon), Potash fish oil soap (2 pounds)—was used in the beginning of 20th century for the control of pests (61).

OM 2424: *see* Terrazole (206).

OMPA: *see* Schradan (144).

ORPIMENT: (As_2S_3)—is the arsenical used by the Chinese for control of pests during A.D. 900's (33).

ORTHOCIDE-406: *see* Captan (197).

OVOTRAN: (4-chlorophenyl-4-chlorobenzene sulphonate)—is a chlorinated hydrocarbon. It is primarily an acaricide (83).

OXYDEMETON-METHYL: *see* Metasystox-R (127).

OXYTHIOQUINOX: *see* Morestan (160).

OW TYPE: is a type of emulsion wherein water is the external or continuous phase and oil is the internal or dispersal phase. This is the one type generally adopted for insecticidal purposes (23).

PACHNOLITE: ($AlF_3.CaF_2.NaF.H_2O$)—is a fluoride insecticide of minor importance (44).

PANOGEN: (Methyl mercuric dicyanodiamine)—is sold as Panogen-15 (2.2% methyl mercuric dicyanodimidc) and Panogen 42-(6.3% methyl mercuric dicyanodim ide) for disinfecting *see*ds (190).

PARA-CHLOROPHENYL PHENYL SULPHONE: *see* Sulphenone (83).

PARAOXON: (0,0-diethyl O-nitrophenyl phosphate)—is the oxygen analog of parathion and is more toxic than parathion (115).

PARATHION: (parathion ethyl, thiophos, E-605, 0,0-diethyl-0,4-nitro- phenyl thiophosphate: SNP 3442)—was discovered by Schraden. It is a contact organophosphorus insecticide (112).

PARATHION ETHYL: *see* Parathion (112).

PARATHION METHYL: *see* Methyl parathion (115).

PARAZATE: *see* Zineb (175).

PARAZATE LIQUID: *see* Nabam (175).

PARIS GREEN: (Emerald green, French green, Mitis green, Schweenfurter green)—complex compound of coppermetaarsenite and copper acetate-(3:l to 2:1) is an arsenical (arsenite) insecticide (34).

PARIS PURPLE: *see* London purple (34).

PARTICLE DENSITY: is the actual density of solid materials only (as if there were no air spaces between them) (16).

PCA: (Pentachloroaniline)—this compound is formed when PCNB is applied to the soil in submerged conditions. It inhibits the growth of soil actinomycetes and fungi (203).

PCNB: (1,2,3,4,5-pentachlorpnitrobenzene)—belongs to the chloronitrobenzene group and is recommended for the control of soil borne diseases because of its long persistence in soil (206).

PENICILLIN: is an antibiotic which is not widely used now, due to its limited activity (204).

1,2,3,4,5-Pentachloronitrobenzene: *see* PCNB (206).

PERENOX: *see* Cuprous oxide (186).

PERTHANE: (2,2-dichloro-1,1-bis(p-ethyl phenyl) ethane: Di(p-ethylphenyl) dichloroethane)—is an organochlorinated insecticide with a low toxicity to higher animals (82).

PESTOX-III: *see* Schradan (144).

PESTOX-15: *see* Mipafox (108).

PHALTAN: *see* Folpet (198).

PHC (JAPAN): *see* Baygon (152).

PHENKAPTON: (0,0-diethyl-S-(2,5-dichloro-phenylthiomethyl) dithiophosphate)—is a selective organophosphorus acaricide of prolonged action and is similar to trithion (136).

PHENOTHIAZINE: (thiodiphenyl amine, $C_{12}H_9NS$) (159).

PHENOTHIAZONE: is a highly potent fungicide and is an oxidation product of phenothiazine (159).

PHENYL MERCURIC ACETATE: *see* Ceresan (190).

PHENYL MERCURY TRIETHYL AMMONIUM LACTATE (PURATISED)—is a phenyl mercury fungicide used for curative action (153).

1, Phenyl-3-methyl pyrazolyl-5-dimethyl carbamate: *see* Pyrolan (153).

PHOMASOL: *see* Thiram (174).

PHORATE: *see* Thimet (135).

PHOSALONE: (0,0-diethyl-S-(6-chlorobenzolinyl-3-methyl) dithiophosphate)—is an organophosphorus insecticide used for treating *see*ds to protect the *see*dlings (140).

PHOSDRIN: (Mevinphos: 0,0-dimethyl-0-(l-methyl-2-carbomethoxy vinyl) phosphate)—is a derivative of phosphoric acid. It exists in *cis* and *trans* isomeric forms (103).

PHOSPHINON: (Phosthenon, 0,0-diethyl-0-(2,2-dichloro-l-chloro- ethoxy-vinyl phosphate)—is a derivative of phosphoric acid and is the closest analog of dichlorovos (103).

PHOSPHAMIDE: *see* Rogor (133).

PHOSPHAMIDON: (Dimecron: 0,0-dimethyl-0-(2-chloro-2-N,N,-diethyl carbamoyl methyl vinyl phosphate)—is a derivative of phosphoric acid and is highly soluble in water (104).

PHOSPHORYRONE: *see* Endothion (124).

PHOSTHENON: *see* Phosphinon (103).

PHTHALAN: *see* Folpet (198).

PHTHALOPHOS: *see* Imidan (139).

PHYGON: (Dichlone: 2,3-dichloro-l,4-naphthaquinone)—is a fungicide belonging to the quinone group and is used much more than chloronil (195).

PICKERING TYPE OF BORDEAUX MIXTURE: In this, saturated lime water is substituted for either slaked calcium oxide or hydrated lime that is used in the standard bordeaux mixture (183).

3-(2-Piperidyl) pyridine: *see* Neonicotine (50).

PIROFOS: *see* Sulffotepp (144).

PISCHEMUKA REARRANGEMENT: is the conversion of thiono isomers to thiolo isomers by heat or other reagents.

PLANTVAX: DCMOD (2,3-dihydro-5-carboxanilide-6-methyl-l,4- oxathiin-4,4-dioxide) (207).

PMAC: *see* Ceresan (192).

POISON: Assimilation of these substances in small quantities through inhalation, ingestion or absorption results in systemic injury, acute toxic conditions or death within a reasonable time following exposure.

POLYCHLOROPINENE: ($C_{10}H_{10}CL_8$)—is a chlorinated hydrocarbon obtained by the photochemical reaction or initiated chlorination of bornyl chloride to chlorine content of 64 to 67% (97).

POMASOL: *see* Thiram (174).

POTASSIUM SULPHIDE: is a sulphur fungicide prepared by boiling sulphur and an alkali hydroxide together. However it is not used now.

POTSAN: (Zinophos: 0,0-diethyl-0-(4-methyl coymarinyl-7) thiophosphate)—is a selective organophosphorus insecticide with stomach and contact action.

PP 149: (5-butyl-2-ethylamine-4-hydroxy-6-methylpyrimidine)—is a promising systemic fungicide for use against diseases of sugarbeets, potatoes and oranges (207).

PP 675: *see* Methyrimol (207).

PROLAN: (DNP,2, nitro-1,1 bis(p-chlorophenyl) propane) — is a chlorinated hydrocarbon (82).

PROPOXUR: *see* Baygon (152).

PYRETHRUM: is a natural organic insecticide found in the flower heads of certain species of Chrysanthemum genus. The active principles are Pyrethrins I and II, Cinerins I and II and Jasomolin II (52).

PYROLAN: (1-Phenyl-3-methylpyrazoly-5-dimethylcarbamate)— is a systemic insecticide (153).

PYROPHYLLITE: is the best diluent for most of the dust preparations. It has a lamellar structure and adheres well to plant foliage (14).

QUINONETHIONATE: *see* Morestan (160).

QUINOXALINE: *see* Eradex (161).

Quinoxaline-2,3-trithiocarbamate: *see* Eradex (161).

RALSTONITE: ($2NaF.MgF.6AlF_3(OH)_34H_2O$)—is a fluoride insecticide of minor importance (44).

REALGAR: (As_2S_2)—is an arsenical sulphide used by Chinese for the control of garden insects during A.D. 900's (33).

ROGOR: *see* Dimethoate (133).

RONNEL: (Trolene Etrolene, Dowet 14; Doet 57; Koralan; Nankar; 0,0-dimethyl-0,2,4,5-trichlorophenyl thiophosphate)—is a general purpose organophosphorus insecticide with low mammalian toxicity (111).

ROSIN FISH OIL SOAP: A brown viscous material containing 200-400 oz. of soap/and resin dissolved in 100 gallons of water.

ROTENONE: is a natural organic insecticide, extracted mostly from *Derris elliptica and D. mallenceesis* (56).

SABADILLA: *(Schoenocaulor officinale)*—has toxic insecticidal properties against certain pests (59).

SAYFOS: *see* Menazon (140).

SCHEELE'S GREEN: is an arsenite fungicide containing $CuHAsO_3$-acid copper arsenate (34).

SCHRADAN: (Octamethyl, OMPA, Pestox III, Systam, Octamethyl pyrophosphamide, Octamethyl tetramide of pyrophosphoric acid) —is a systemic organophosphorus insecticide. It is non phytotoxic to plants but toxic to animals (144).

SCHWEENFURTER GREEN: *see* Paris green (34).

SDMC: *see* Vapam (177).

SECONDARY MAGNESIUM ORTHO ARSENATE: ($MgHAsO_4$)—is an arsenical insecticide of minor importance (40).

SEMESAN: is an organomercurial fungicide belonging to the aromatic series. It is a phenyl mercuric derivative (192).

SEVIN: *see* Carbaryl (151).

SILICOFLUORIDE: is a group of fluorine compounds of which sodium fluosilicate is of importance (44).

SNP 3422: *see* Parathion (112).

SOAPS: are alkali salts of fatty acids and are most commonly made from sodium salts (hard soaps) or potassium salts (soft soaps) (74).

SODA MIXTURE: *see* Burgandy mixture (185).

SODIUM ALUMINUM FLUORIDE: *see* Cryolite (45).

SODIUM FLUOALUMINATE: *see* Cryolite (45).

SODIUM FLUORIDE: (NaF)—is a fluorine insecticide with 45.2% of fluorine (43).

SODIUM FLUOSILICATE: (Na_2SiF_6)—belongs to the fluosilicate insecticides (44).

SODIUM METHYLENE BIS DITHIOCARBAMATE: *see* Nabam (175).

SODIUM-N-METHYL DITHIOCARBAMATE: *see* Vapam (177).

SOLBAR: *see* Barium sulphide (169).

SPERGON: *see* Chloranil (196).

SPOTRETE: *see* Thiram (174).

STOMACH POISON: applies to all poisons which act through the stomach (11).

STREPTOMYCIN: is an antibiotic isolated from *Streptomeyces grisons* in 1942, and has been used widely as a systemic fungicide (205).

STROBANE: is a chlorinated hydrocarbon similar to toxaphene and prepared by the chlorination of a mixture of camphene and pinone so as to contain 66% chlorine (97).

SUDAN III: is an indicator which is used for distinguishing the OW and WO type emulsion. Sudan III is soluble in oil and potassium permanganate in water phase (23).

SULFONATION: indicates the degree of refinement of oil in terms of the percentage of oil which does not react with sulphuric acid. This has been used as a measure of safeness for application to plant and animal life (63).

SULFOTEPP: (Pirofos, Bladafum, Tetraethyl dithiopyrophosphate)—is closely related to TEPP and is similar to parathion in insecticidal action (144).

SULPHENONE: (4-Chlorophenylsulphone: para-chlorophenyl-phenyl- sulphone)—is a chlorinated hydrocarbon occurring in two different forms. It is toxic to mites (83).

SUMATROL: is similar to tephrosin but differs in being phenolic. It has the same toxicity as toxicarol (58).

SUMITHION: (Folithion, Metathion, methylnitrophos, fenitrothion, 0,0-dimethyl-0,4-nitro-3-methylphenyl thiophosphate)—is similar to methyl parathion in chemical properties and is an active synergist for many organophosphorus insecticides like parathion, methylparathion, etc. (117).

SYSTAM: *see* Schradan (144).

SYSTEMIC POISONS: are those which are applied to one part of the crop like root or the leaf and are translocated to other parts of the plant in the fluid system, for rendering the whole plant toxic to insects or fungi (10).

SYSTOX: *see* Demeton (124).

TAG: *see* Ceresan (192).

TCNB: (tetrachloronitrobenzene)—belongs to the group of chloronitro benzene. But not much of its insecticidal activity is known (203).

TDE: *see* DDD (81).

TEMIK: (Aldicarb,2-methyl-2-methylthiopropional-oxime-0-N-methyl carbamate)—is a systemic carbamate nematicide (154).

TEPP: *see* Tetraethylpyrophosphate (143).

tepp: *see* Tetraethylpyrophosphate (143).

TEPHROSIN: is a derivative of rotenone and is found along with rotenone, in the Derris and cube roots. It is isolated from the leaves of *Tephrosia vogelii* (58).

TERRACHLOR SUPER: is a systemic fungicide containing Terrazole (6.1%) and PCNB (23%) (206).

TERRACUR-P: *see* Dasanit (121).

TERRASYSTAM: *see* Dimefox (108).

TERRAZOLE: (OMM 2424)(5-ethoxy-3-(trichloromethyl) 1,2,4-thiadia- zole)—is a systemic fungicide; Terrazole and PCNB are mixed and sold as Terrachlor super (206).

TERSANTS: *see* Thiram (174).

2,3,5,6-Tetrachloro-1,4-benzoquinone: *see* Chloranil (195).

TETRACHLORO-P-BENZOQUINONE: *see* Chloranil (195).

TETRACHLORODIPHENYLETHANE: *see* DDD (81).

N-1,1-2,2-tetrachloroethyl-thiophthalimide: *see* Difolatan (199).

2,3,5,6-tetrachloronitrobenzene: *see* TCNB (203).

TETRAETHYLDITHIOPYROPHOSPHATE: *see* Sulfotepp (143).

0,0,0,0-tetraethyl-S,S-methylene bis (dithiophosphate): *see* Ethion (137).

TETRAETHYLPYROPHOSPHATE: (TEPP, tepp, bladen)—Commercial compound contains 40% pure TEPP and is used for the control of aphids. It is highly toxic to warm blooded animals (143).

TETRAMETHYLPHOSPHORODIAMIDOFLUORIDE: *see* Dimefox (108).

TETRAMETHYLTHIURAM DISULPHIDE: *see* Thiram (174).

TETRAMETHYLTHIURAM MONOSULPHIDE: is also a dithiocarbamate fungicide with lower fungicidal properties than the disulphides (175).

TETRATHION: (0-methyl-0-ethyl-S-(2-ethylthioethyl dithiophosphate)—is an organophosphorus insecticide similar to thiometon and disulfotan in chemical properties (139).

THIMET: (Phorate, 0,0-diethyl-S-(ethylthiomethyl dithiophosphate)— is a systemic organophosphorus insecticide of low water solubility. It has long residual activity in plants and at the same time, is highly toxic (135).

THIOCYANATE: Certain thiocyanates like laurylthiocyanate are highly toxic, contact insecticides (160).

THIOCRON: (Amidithion, 0,0-dimethyl-S-(N-2-methoxyethyl carbamoyl methyl) dithiophosphate)—is similar to dimethoate and is a good contact insecticide with prolonged systemic action (130).

THIODIPHENYLAMINE: *see* Phenothiazine (159).

THIODAN: *see* Endosulphan (95).

THIOLO: The radical-SH; the use of the term 'thiolo' commonly indicates the replacement of the oxygen in an—OH group by sulphur.

THIOMETON: *see* Ekatin (137).

THIONAZIN: *see* Zinophos (123).

THIONO: (The radical is=CS) The use of the term thiono commonly indicates the replacement of oxygen in a carbonyl group by sulphur.

THIONOL: (Hydroxy phenoxy thiazone)—is produced by spontaneous oxidation of phenothazine (159).

THIOPHENEN: is found in coal tar and coal gas. It is a liquid having slight aromatic odour, resembling that of benzene.

THIOPHOS: *see* Parathion (112).

THIOPHOS ME: (Methyl ethyl parathion: Methyl ethyl thiophos: Methyl-0-ethyl-0-4-nitrophenylthiophosphate)—is an organophosphorus compound similar to parathion except one ethyl has been replaced by methyl group (110).

THIOPHOS: *see* Parathion (112).

THIRAM: (Tetramethyl thiuram disulphide, Thiram D. TMTD, Arasan, Nomersan, Panoramts, Pomasol, Spotrete, Tersants, Thylate, Bis (dimethyl thiocarbamoyl) disulphide: α,α-dithio-bis (dimethylthio) formamide)— differs chemically from other dithiocarbamates in that, it contains two dithiocarbamates nuclei, connected through sulphur linkage (174).

THIRAM D: *see* Thiram (174).

THRESHOLD LIMIT VALUE: The concentration of a contaminant or toxic substance in air, at which on the basis of present day knowledge, workers may be repeatedly exposed without any preceptable adverse effect on their health during their life time. This value should be used as a guide in the control of health hazards and should not be regarded as a fine line between safe and dangerous concentrations.

THIONAZIN: *see* Zinophos (123).

THYLATE: *see* Thiram (174).

TINOX: (0,0-Dimethyl-2-methyl mercaptoethylthiophosphate)—is an oragno-phosphorus compound. Similar to methyldemeton, the technical product is a mixture of thioloisomer (60-70%) and thionoisomer (30-40%) (127).

TMTD: *see* Thiram (174).

TOBACCO: is a natural insecticide which is used as such, and also extracted for its active insecticidal compounds like nicotine (47).

TOXAPHENE: is a chlorinated hydrocarbon, obtained by the chlorination of the naturally occurring terpenes. A mixture of products formerly called chlorinated camphene, Hercules 3956 and now toxaphene are produced (97).

TOXICAROL: is a derivative of rotenone and occurs along with rotenone and deguelin in derris and cube roots (58).

TBZ: (2-(4-thiozolyl) benzylmidazole)—is a promising systemic fungicide for use against disease of sugarbeets, and potatoes (207).

TPTH: *see* Tributyl tin hydroxide (201).

TRIBUTYLTIN HYDROXIDE: ($(C_4H_9)_3SnOH$)—belongs to the organotin group of fungicides. It is used as an anti mildew agent in wood and paints (201).

TRICALCIUM ARSENATE: is one of the constituents of commercial calcium arsenate. It contains 37.5% arsenic as $Ca_3(AsO_4)_2$ (38).

TRICHLORFON: (Diptrex, Dylos, Chlorophos, 0,0-dimethyl-l-hydroxy- 2,2,2-trichlorophenyl) phosphonate)—is a systemic organophosphorus insecticide (146).

TRICHLOROMETAFOS: *see* Ronnel (119).

1,1,1-Trichloro-2,2-bis(p-chlorophenyl) ethane: *see* DDT (78).

1,1,l-Trichloro-2,2-bis(p-fluorophenyl) ethane: *see* DFDT (83).

N-Trichloromethylthio-4-chloroanilide of methanosulphaonic acid: *see* Mesulphan (197).

N-Trichloromethylthio-phthalimide: *see* Folpet (198)

N-trichloromethylthio-tetrahydrophthalimide: *see* Captan (197).

TRILEAD ARSENATE: ($Pb_3(As_3O_4)_2$)—is an arsenical (arsenate) insecticide (38).

TRIMAGNESIUM ORTHOARSENATE: ($Mg_3(AsO_4)_2$)—is an arsenical (arsenate) insecticide of minor value (40).

TRIPHENYL TIN ACETATE: (Brestan)—belongs to the organotin compounds and is similar in composition to copper compounds (201).

TRIPHENYL TIN CHLORIDE: belongs to organotin compound and is used in the antifouling of paints (201).

TRIPHENYL TIN DISULPHIDE: $(C_6H_5)_3SnS$—S Sn $(C_8H_5)_3$ (201).

TRIPHENYL TIN HYDROXIDE: (Duter, TPTH, $(C_4H_8)_3SnOH$)—belongs to the organotin compounds. It does not show any phytotoxic effect (200).

TRIPHENYL TIN SULPHIDE: ($(C_6H_5)_3$ SnS Sn $(C_6H_5)_3$)—belongs to the organo tin compounds of fungicides having less limited spectrum of activity than brestan (201).

TRITHION: (Carbophenothion, 0,0-diethyl-S-(4-chlorophenylthiomethyl) dithiophosphate)—is a powerful contact organophosphorus acaricide and insecticide (136).

TROLENE: *see* Ronnel (119).

USPULAN: is the first organomercurial fungicide belonging to aromatic series suggested for *see*d treatment (191).

UV LIGHT: Radiation in the region of spectrum of lower wave length (higher frequency) than the visible violet light extending from the end of visible light about 3900-4000 A to down 200 A light.

VAIGFRU TAMRAGHOL: *see* Copper oxychloride (185).

VAMIDOTHION: (0,0-dimethyl-S-2-(l-N-methyl carbomyl ethyl mer- capto)-ethyl thiophosphate)—is an organophosphorus insecticide and approaches the methyl demeton in insecticidal activity (128).

VANCIDE 89: *see* Captan (197).

VAPAM: (SDMC; VPM; CHEMRAPE; Sodium-N-methyl-dithiocar-bamate)—possesses fungicidal, nematicidal and herbicidal properties (177).

VISCOSITY: is the internal friction or resistance to flow shown by a liquid (62).

VITAVAX: It is an oxathiin derivative developed by von Schmelung and Kulka. It is specific against Basidiomycetes (207).

VPM: *see* Vapam (17).

WARFARIN: is a rodenticide manufactured in India (4).

WETTABILITY: The extent to which a solid is wetted by a liquid, measured by the force of adhesion between the solid and liquid phases (19).

WETTABLE POWDER: (water dispersible powder concentrates)—similar to dusts in appearance but contain certain wetting agents which make them go into dispersion or suspension when they are mixed with water and applied as spray (19).

WOFATOX: *see* Methyl parathion (115).

WO TYPE: Out of the two types of emulsions, in the WO type (water in oil type) oil is the internal and water is the external phase (23).

ZECTRAN: 4(N,N-dimethylamino) 3,5-dimethylphenyl-N-methyl carbamate)—is a wide spectrum carbamate insecticide having only poor contact insecticidal activity (153).

ZERLATE: *see* Ziram (173).

ZINC DIMETHYL DITHIOCARBAMATE: *see* Ziram (173).

ZINC ETHYLENE BIS DITHIOCARBAMATF: *see* Zineb (175).

ZINC FLUORIDE: (ZnF_2)—is a minor fluorine insecticide (44).

ZINC OXIDE: is used as a corrective in the formulation of arsenical insecticides (41, 176).

ZINC PHOSPHIDE: is a rodenticide now manufactured in India (4).

ZINEB: (Dithane Z. 78; Parazate; DuPont fungicide A. Lonocol; Zinc ethylene bis dithiocarbamate)—is the most important dithiocarbamate fungicide and is of great use in zinc deficient soils also (175).

ZINOPHOS: (Potsane, Thionazin, 0,0-diethyl-0-pyrozinylthiophos- phate)—is a systemic insecticide, nematicide and acaricide (123).

ZIRAM: (Zerlate, Methasan, Fuklasin, Karbam white, Milbam, Niagara-G; Zirberk, Cuman, Zinc dimethyldithiocarbamate)—is the most stable dithiocarbamate fungicide (173).

ZIRBERK: *see* Ziram (173).

IMPORTANT REFERENCE BOOKS AND JOURNALS

1. Advances in Pest Control Research. Ed. Metcalf, R.L., Interscience Publishers Inc. New York, U.S.A.
2. Analytical Abstracts. Published by W. Heffer and Sons Ltd., 4. Petty Curry, Cambrideg England.
3. Annual Review of Entomology. Ed. Edward, A., Steinhaus and Ray F. Smith, Annual Reviews Inc. Stanford, (California) U.S.A.
4. Annual Review of Phytopathology.
5. Beroza M., Chemicals Controlling Insect Behaviour, Academic Press, New York and London.
6. British Insecticides and Fungicides for Crop, Protection. Association of British Insecticides Manufacturers, England.
7. Chichester, C.O., Research in Pesticides. Academic Press, New York. U.S.A
8. De Ong, E.R., Chemistry and Uses of Pesticides. Reinhold Publishing Co. New York, U.S.A.
9. Edwards, C.A., Environmental Pollution by Pesticides. Plenum Press, London and New York.
10. Frear, D.E.H., Agricultural Chemistry, Von Nostrand Co. New York, U.S.A.
11. Frear, D.E.H., Chemistry of Pesticides, Von Nostrand Co. New York, U.S.A.
12. Fungicides—An Advance Treatise. Ed. Ferguson, D.C., Academic Press, New York and London.
13. Gunther, F.A. Pesticides Reviews, Springer Verlag, New York, U.S.A.
14. Gunter Zweig. Analytical Methods for Pesticides, Plant Growth Regulators and Food Additives, (Volumes I to V) Academic Press, New York, U.S.A.
15. Hartley. G.S. and T.F. West. Chemicals for Pest Control. Pergamon Press, New York, U.S.A.
16. Heath, D., Organophosphorus Poisons, Pergamon Press, New York, U.S.A.
17. Indian Journal of Entomology, Entomological Society of India, New Delhi.
18. Indian Journal of Plant Protection, Plant Protection Association of India, Hyderabad.
19. India Pesticide Industry—Ed. B. Vasanthraj David—Vishvas Publications.
20. Indian Phytopathology, Pathological Society of India, New Delhi.
21. Indian Standards Institute Specifications, Indian Standards Institute, New Delhi.

22. Jacobson and D.G. Crosby. Naturally Occurring Insecticides, Marcel Dekker Inc. New York, U.S.A.
23. Journal of Agriculture and Food Chemistry. American Chemical Society, 1155, Sixteenth Street, Washington D.C., U.S.A.
24. Journal of Association of Official Agricultural Chemists, Box 540, Benjamin Franklin Station, Washington, U.S.A.
25. Journal of Economic Entomology, Entomological Society of American Executive Section 1530, Washington D.C., U.S.A.
26. Journal of Food and Agricultural Chemistry.
27. Kilgore, W.W. and R.L. Doutt. Pest control, Biological, Physical and Selective Chemical Methods. Academic Press, New York U.S.A.
28. Lukens, R.L, Chemistry of Fungicidal Action, Springer Verlag, New York, U.S.A.
29. Mazumdar, S.K., Pesticides, Academy of Pest Control Sciences, Pesticides Biddings, Mysore-2A, India.
30. Marsh, R.W., Systemic Fungicides, John Wiley and Sons, New York, U.S.A.
31. Martin, H., Guide to Chemicals Used in Crop Production, Canadian Department of Agriculture, Canada.
32. Martin, H., The Scientific Principles in Crop Protection, St. Martins Press, New York.
33. Melnikov, N.N., Chemistry of Pesticides. Springer-Verlag, New York, U.S.A.
34. Metcalf, R.L. Organic Insecticides, Interscience, New York, U.S.A.
35. Negheriborn, W.O., Handbook of Toxicology III, Insecticides. W.B. Saunders and Co. Philadelphia, U.S.A.
36. Nene, Y.L., Fungicides in Plant Disease Control, Oxford and IBH Publishing Co., New Delhi, India.
37. Niagara, Formulation Information for Thiodan, FMC Corp. Niagara Chemical Division. Middleport, New York.
38. O' Brien, R.D., Toxic Phosphorus Esters, Academic Press, New York and London.
39. O' Brien, R.D., Insecticides, Action and Metabolism, Academic Press, New York and London.
40. O' Brien, R.D., and Izuru Yamamato, Biochemical Toxicology of Insecticides, Academic Press, New York, U.S.A.
41. O'Brien, R.D. and L.S. Wolfe. Radiation, Radioactivity and Insects, Academic Press, New York and London.
42. PANS—Centre for Overseas Pest Research, 56 Grey's Inn R.I. London, WCIX-8LV. England.
43. Pesticides—Colour Publications Pvt. Ltd. 126, A. Dhuruwadi Off. Dr. Mariman Road, Bombay, 25, DC. India.
44. Phytopathology—Phytopathological Society of America. U.S.A.
45. Pyenson, L.L., Elements of Plant Protection, John Wiley and Son, New York, U.S.A.

46. Residue Reviews, Springer-Verlag, New York, U.S.A.
47. Rose, G.J. Crop Protection, Leonard Hill (Books) Ltd. London.
48. Rudd, R.L. Pesticides and the Living Landscape, Univ. of Wisconsin Press, Madison, U.S.A.
49. Shell. Handbook of Phosdrin Insecticide Formulation. Sc. 57-28; Shell Chemical Corp., New York, U.S.A.
50. Shell. Handbook of Aldrin, Dieldrin and Endrin formulations. Shell Chemical Corp., New York, U.S.A.
51. Shepherd, H.H., Chemistry of Insecticides. McGraw-Hill Book Co., New York, U.S.A.
52. Shorvelle, K.U., Nature and Uses of Modern Fungicides. Burgess Minneapolis, Minnesota, U.S.A.
53. Street, J.C., Pesticide Selectivity, Marcel Dekker Inc., New York and Basel.
54. Smith, R.F. and R. Van den Bosch. Pest Control, Biological Physical and Selected Chemical Methods, Academic Press, New York.
55. Summary of tolerances and Exceptions for residus of pesticides chemicals, United States Department of Health, Education and Welfare, Food and Drug Administration. Washington D.C., U.S.A.
56. Tahori, A.S., Pesticides Chemistry Vol. I., Gordon and Breach, London, England.
57. The Analyst, Society of Analytical Chemists. 9.10. Savile Rev. London.
58. Thorn, G.D. and R.A. Ludwig. The Dithiocarbamates and Related Compounds, Elsevier.
59. Torgeson, D.C. Fungicides I and II. Academic Press, New York, U.S.A.
60. Union Carbide. Sevin Carbaryl Insecticide Manual for formulations, F/41-54. Union Carbide Corporation, New York, U.S.A
61. United States Department of Agriculture. A Summary of Certain Pesticides Chemicals and Uses, USDA Pesticides Regulatory Section, Washington D.C., U.S.A.
62. Vasanthraj David, B. and T. Kumaraswami. Elements of Economic Entomology, Popular Book House, Madras, India.
63. Velsicol. Formulation Manual no. 514-1, Parathion (Technical) and Methyl Parathion, Velsicol Chemical Corporation, Chicago, Illinois, U.S.A.
64. Wade Van Valkenburg. Pesticide formulation. Marcel Dekker Inc., New York, U.S.A.
65. Weidhaas, D.E. and J.L. Brann, Jr. Handbook of Insecticide, Dust, Diluents and Carriers, Dorland Caldwell, N.J., U.S.A.
66. West, T.F. and G.A. Campell. DDT and Newer Persistent Insecticides, Chemical Publishing Co. Inc., New York, U.S.A.
67. White Stevens. Pesticides in the Environment, Volumes I (Parts I and II), Mercer Dekker Inc. New York, U.S.A.
68. World Health Organisation. Specifications for Pesticides, World Health Organisation, Geneva, Switzerland.

INDEX